METHODS IN MOLECULAR BIOLOGY

For further volumes:
http://www.springer.com/series/7651

NMDA Receptors

Methods and Protocols

Edited by

Nail Burnashev and Pierre Szepetowski

Mediterranean Institute of Neurobiology (INMED), Aix-Marseille University
INSERM U901, Marseille, France

Humana Press

Editors
Nail Burnashev
Mediterranean Institute
of Neurobiology (INMED)
Aix-Marseille University
INSERM U901
Marseille, France

Pierre Szepetowski
Mediterranean Institute
of Neurobiology (INMED)
Aix-Marseille University
INSERM U901
Marseille, France

ISSN 1064-3745 ISSN 1940-6029 (electronic)
Methods in Molecular Biology
ISBN 978-1-4939-7320-0 ISBN 978-1-4939-7321-7 (eBook)
DOI 10.1007/978-1-4939-7321-7

Library of Congress Control Number: 2017949681

Printed on acid-free paper

This Humana Press imprint is published by Springer Nature
The registered company is Springer Science+Business Media LLC
The registered company address is: 233 Spring Street, New York, NY 10013, U.S.A.

Preface

This book is on *N*-methyl-D-aspartate receptors (NMDARs), and more precisely on the methods that can be used to study NMDARs at various levels. NMDARs are cation channels that are gated by the major excitatory neurotransmitter glutamate. NMDAR-mediated signaling is involved in normal development, plasticity, learning, memory, and high cognitive functions. NMDARs play an important role in temporal integration of neuronal network activity and long-term alterations in synaptic structure and function. There is long known association of NMDAR dysfunctions with various neurodevelopmental disorders and excitotoxicity. Recent human genetic studies revealed the existence of multiple alterations in NMDARs subunits genes in numerous brain diseases, such as intellectual disability, autism spectrum disorders, or epilepsy, hence nominating NMDARs as promising targets for pharmacological treatment.

This book puts together the methods and techniques that are in use in leading laboratories to ensure a better understanding of the NMDAR structure-function relations, principles, and rules by which NMDARs operate in brain processing under normal and pathological conditions. Written by world-renowned experts, developers, and experienced users, this book provides thorough and detailed coverage of readily reproducible laboratory protocols that will be of interest to scientists, clinicians, or industry professionals working in this field.

Indeed, the book covers many aspects of NMDARs biology and the different ways this can be addressed. In a very comprehensive and up-to-date introductory Chapter 1, Hansen, Yi, Perszyk, Menniti, and Traynelis summarize everything you always wanted to know about NMDARs—and were not afraid to ask. They review the relationship between NMDARs structure and function, the diversity and significance of NMDAR subtypes in the CNS, as well as principles and rules by which NMDARs operate in the CNS under normal and pathological conditions. NMDARs subunits have different temporal and spatial expression patterns in the brain and NMDAR subtypes also vary according to the cell types and subcellular localization; accordingly, in Chapter 2 Pallesi-Pocachard describes the way transcriptional expression of the different genes encoding those various subunits can be quantified using reverse-transcription PCR. Whereas defects in several NMDARs subunits have been consistently reported in different types of brain disorders, notably GluN2A in epileptic encephalopathies of the epilepsy-aphasia spectrum, unexpected findings point to the high rate of somatic mutations of this GluN2A subunit in melanoma. The way such somatic mutations can be found in melanoma samples is described here by Prickett, Gartner, and Samuels in Chapter 3. NMDARs functioning can also be disturbed by nongenetic factors. One such factor is the presence of autoantibodies directed against NMDARs. How the existence of NMDARs autoantibodies can be searched and studied in the context of encephalitis is reported here by Gastaldi, Waters, and Vincent in Chapter 4. NMDARs structure-function relations and genetic variations can be studied by reconstituting them in cells transfected with the appropriate constructs for the expression of the corresponding subunits. This can be done in non-neuronal cells not expressing any native NMDARs as described by Bruneau and Szepetowski in Chapter 5, notably using magnetofection, allowing to study in isolation subunit-specific properties of NMDARs and their mutants in a physiological context. On the other hand transfection of NMDARs in primary cultured neuronal cells expressing NMDARs endogenously, as shown in Chapter 6 by Marwick and Hardingham, allows to experimentally vary the NMDAR composition in the neurons, in order to investigate NMDAR trafficking and regulation in a physiological context, as well as synapse formation, synaptic activity, and cell survival. In Chapter 7 Yi, Traynelis, and Hansen describe the method to selectively express in heterologous expression systems recombinant triheteromeric GluN1/2A/2B receptors without interfering co-expression of diheteromeric GluN1/2A and GluN1/2B receptors; this enables quantitative evaluation of functional and pharmacological properties of triheteromeric GluN1/2A/2B receptors, which are presumably the most abundant native NMDARs in the adult cortex and hippocampus. This chapter is followed by logically linked chapters describing the methods for studying NMDARs function using electrophysiological and imaging techniques. In Chapter 8 Rozov and Jappy describe a method for electrophysiological functional analysis of recombinant channels expressed in host cells using a fast agonist application system. This method allows adequate assessment of ion permeation, kinetics, drug sensitivity, and other electrophysiological properties of the NMDAR-mediated currents using a wide range of the receptor activation protocols. Perrio, Nicole, and Buisson in Chapter 9 describe the methods for preparation of the fluorescent GluN2B specific probe, evaluation of GluN2B binding and inhibition potency of the probe by calcium imaging, and the labeling and visualization of GluN2B by confocal GluN2B imaging in living cortical neurons. This approach provides important insights into receptor dynamics and trafficking as well as into biophysical microenvironment of the binding site. In Chapter 10 Tian and Ye provide a protocol for the incorporation of two UV-sensitive crosslinking unnatural amino acids into NMDARs using the genetic code expansion technique to engineer light-sensitive NMDARs for imaging studies. Reduced recombinant systems provide a unique opportunity to study the biophysical properties of NMDAR channels with known subunit compositions. For a detailed molecular, morphological, and cellular analysis of NMDARs and the consequences of their dysfunction, they have to be studied in their natural environment in a particular cell type and/or in synapses and neuronal networks in animal models. Hence in Chapter 11 Sprengel, Eltokhi, and Single describe in detail the different methodological steps for successful gene targeting and generation of conditional NMDAR mutant mouse lines in which the hypomorphic Grin allele can be activated at specific time points and in specific cell types allowing functional analysis of the mutated NMDARs in living animals *in vivo* and in brain slices *in vitro*. In Chapter 12 Pons-Bennaceur and Lozovaya describe the methods to study electrophysiological properties of NMDAR-mediated component of spontaneous or evoked excitatory postsynaptic currents by extracellular stimulation or by stimulating synaptically connected neurons in brain slices. This approach allows pinpointing the basic functional properties of NMDARs that are specific to identified brain regions, neurons, and synapses in wild type or genetically manipulated mice. NMDARs are not only present in neuronal cells but also found in a wide variety of non-neuronal cells, within the brain or even in peripheral tissues. In Chapter 13 Kirchhoff describes in detail the methodology on how to study the physiology of NMDARs in astrocytes. This includes electrophysiological characterization of NMDAR-mediated currents in brain slices, analysis of the NMDAR kinetics in acutely isolated, single astrocytes, and visualization of NMDARs localization by Ca^{2+} imaging using confocal laser-scanning microscopy. In Chapter 14 Peineau, Degos, Verney, and Gressens describe the complementary steps to demonstrate the presence and functions of endogenous NMDARs in microglia using a wide range of biological tools with a dedicated strategy. Makhro, Kaestner, and Bogdanova in Chapter 15 present the methodological approaches for detection of the erythroid NMDARs activity in red blood cell fractions of low, medium, and high density, or in single cells. In Chapter 16 Khazipov describes a procedure using NMDARs single-channel activity recorded in neurons in cell-attached mode as a tool for noninvasive measurements of the neuronal membrane potential in living cells. Finally, in Chapter 17 Keller, Bouteiller, and Berger introduce a computational approach as a complementary tool to explore NMDARs function based on the knowledge provided by various experimental results. They describe a general computational method aiming at developing kinetic Markov chain-based models of NMDARs subtypes that can be used to make predictions on NMDARs properties and on their role in synaptic function under various physiological and pharmacological conditions.

With this book, we expect that the readers will find different and complementary ways NMDARs can be studied nowadays. The different chapters do not only reflect the diversity of the corresponding approaches, the importance of NMDARs, and their complexity, but also the progresses that have been and are still being made at various and intertwined levels.

Marseille, France

Nail Burnashev
Pierre Szepetowski

Contents

Contributors

THEODORE W. BERGER • *Department of Biomedical Engineering, Center for Neural Engineering, University of Southern California, Los Angeles, CA, USA*
ANNA BOGDANOVA • *Red Blood Cell Research Group, Institute of Veterinary Physiology, University of Zurich, Zurich, Switzerland; The Zurich Center for Integrative Human Physiology (ZIHP), University of Zurich, Zurich, Switzerland*
JEAN-MARIE C. BOUTEILLER • *Department of Biomedical Engineering, Center for Neural Engineering, University of Southern California, Los Angeles, CA, USA*
NADINE BRUNEAU • *INSERM U901, Mediterranean Institute of Neurobiology (INMED), Aix-Marseille University, Marseille, France*
ALAIN BUISSON • *Grenoble Institut des Neurosciences, Centre Inserm U1216, Equipe Neuropathologies et Dysfonctions Synaptiques, Université Grenoble Alpes, Grenoble, France*
VINCENT DEGOS • *PROTECT, INSERM, Université Paris Diderot, Paris, France*
AHMED ELTOKHI • *Max Planck Research Group, Institute for Anatomy and Cell Biology, Heidelberg University, Heidelberg, Germany; Institute of Human Genetics, Heidelberg University, Heidelberg, Germany*
JARED J. GARTNER • *National Cancer Institute, Surgery Branch, US National Institutes of Health, Bethesda, MD, USA*
MATTEO GASTALDI • *Nuffield Department of Clinical Neurosciences (NDCN), University of Oxford, Oxford, UK; IRCCS National Neurological Institute C. Mondino, Pavia, Italy; University of Pavia, Pavia, Italy*
PIERRE GRESSENS • *MRC Centre for Synaptic Plasticity, School of Physiology, Pharmacology, Neurosciences, University of Bristol, Bristol, UK; Centre for the Developping Brain, King's College, St Thomas' Campus, London, UK*
KASPER B. HANSEN • *Department of Biomedical and Pharmaceutical Sciences, University of Montana, Missoula, MT, USA; Center for Biomolecular Structure and Dynamics, University of Montana, Missoula, MT, USA*
GILES E. HARDINGHAM • *UK Dementia Research Institute, Edinburgh Medical School, University of Edinburgh, Edinburgh, UK*
DAVID JAPPY • *OpenLab of Neurobiology, Kazan Federal University, Kazan, Russia*
LARS KAESTNER • *Theoretical Medicine and Biosciences, Saarland University, Homburg, Germany; Dynamics of Fluids, Experimental Physics, Saarland University, Saarbrücken, Germany*
A. FLORENCE KELLER • *Mulhouse, France*
ROUSTEM KHAZIPOV • *Mediterranean Institute of Neurobiology (INMED), INSERM U-901, Marseille Cedex, France; Aix-Marseille University, Marseille, France; Laboratory of Neurobiology, Kazan Federal University, Kazan, Russia*
FRANK KIRCHHOFF • *Molecular Physiology, Center for Integrative Physiology and Molecular Medicine (CIPMM), University of Saarland, Homburg, Germany*
NATALIA LOZOVAYA • *Neurochlore c/o Mediterranean Institute of Neurobiology (INMED), INSERM U901, Marseille Cedex, France*

ASYA MAKHRO • *Red Blood Cell Research Group, Institute of Veterinary Physiology, University of Zurich, Zurich, Switzerland*
KATIE F.M. MARWICK • *Centre for Discovery Brain Sciences, University of Edinburgh, Edinburgh, UK*
FRANK S. MENNITI • *MindImmune Therapeutics, Inc., George & Anne Ryan Institute for Neuroscience, Kingston, RI, USA*
OLIVIER NICOLE • *Université de Bordeaux, CNRS UMR 5293, Institut des Maladies Neurodégénératives, Bordeaux, France*
EMILIE PALLESI-POCACHARD • *Molecular and Cellular Biology Platform, INSERM U901, Mediterranean Institute of Neurobiology (INMED), Aix-Marseille University, Marseille, France*
STÉPHANE PEINEAU • *MRC Centre for Synaptic Plasticity, School of Physiology, Pharmacology, Neurosciences, University of Bristol, Bristol, UK; PROTECT, INSERM, Université Paris Diderot, Paris, France; INSERM ERI 24 GRAP, Centre Universitaire de Recherche en Santé, Université de Picardie Jules Verne, Amiens, France*
CÉCILE PERRIO • *Normandie University, UNICAEN, CEA, CNRS, UMR6301-ISTCT, LDM-TEP, Cyceron, Caen, France*
RILEY E. PERSZYK • *Department of Pharmacology, Emory University School of Medicine, Atlanta, GA, USA*
ALEXANDRE PONS-BENNACEUR • *INSERM U901, Mediterranean Institute of Neurobiology (INMED), Aix-Marseille University, Marseille, France*
TODD D. PRICKETT • *National Cancer Institute, Surgery Branch, US National Institutes of Health, Bethesda, MD, USA*
ANDREI ROZOV • *OpenLab of Neurobiology, Kazan Federal University, Kazan, Russia; Department of Physiology and Pathophysiology, University of Heidelberg, Heidelberg, Germany*
YARDENA SAMUELS • *Department of Molecular Cell Biology, Weizmann Institute of Science, Rehovot, Israel*
FRANK N. SINGLE • *Department of Molecular Neurobiology, Max Planck Institute for Medical Research, Heidelberg, Germany; Miltenyi Biotec GmbH, Bergisch Gladbach, Germany*
ROLF SPRENGEL • *Department of Molecular Neurobiology, Max Planck Institute for Medical Research, Heidelberg, Germany; Max Planck Research Group, Institute for Anatomy and Cell Biology, Heidelberg University, Heidelberg, Germany*
PIERRE SZEPETOWSKI • *Mediterranean Institute of Neurobiology (INMED), Aix-Marseille University, INSERM U901, Marseille, France*
MEILIN TIAN • *Shanghai Key Laboratory of Brain Functional Genomics, East China Normal University, Shanghai, China; INSERM U1024, CNRS UMR 8197, Ecole Normale Supérieure, Institut de Biologie de l'Ecole Normale Supérieure (IBENS), Paris, France; Institut National de la Santé et de la Recherche Médicale, Paris, France; Centre National de la Recherche Scientifique, Paris, France*
STEPHEN F. TRAYNELIS • *Department of Pharmacology, Emory University School of Medicine, Atlanta, GA, USA*
CATHERINE VERNEY • *PROTECT, INSERM, Université Paris Diderot, Paris, France*
ANGELA VINCENT • *Nuffield Department of Clinical Neurosciences (NDCN), University of Oxford, Oxford, UK*

PATRICK WATERS • *Nuffield Department of Clinical Neurosciences (NDCN), University of Oxford, Oxford, UK*
SHIXIN YE • *INSERM U1024, CNRS UMR 8197, Ecole Normale Supérieure, Institut de Biologie de l'Ecole Normale Supérieure (IBENS), Paris, France; Institut National de la Santé et de la Recherche Médicale, Paris, France; Centre National de la Recherche Scientifique, Paris, France; Laboratory of Computational and Quantitative Biology (LCQB), Institute of Biology, Paris-Seine, Pierre et Marie Currie University, Paris, France*
FENG YI • *Department of Biomedical and Pharmaceutical Sciences, University of Montana, Missoula, MT, USA; Center for Biomolecular Structure and Dynamics, University of Montana, Missoula, MT, USA*

Chapter 1

NMDA Receptors in the Central Nervous System

Kasper B. Hansen, Feng Yi, Riley E. Perszyk, Frank S. Menniti, and Stephen F. Traynelis

Abstract

NMDA-type glutamate receptors are ligand-gated ion channels that mediate a major component of excitatory neurotransmission in the central nervous system (CNS). They are widely distributed at all stages of development and are critically involved in normal brain functions, including neuronal development and synaptic plasticity. NMDA receptors are also implicated in the pathophysiology of numerous neurological and psychiatric disorders, such as ischemic stroke, traumatic brain injury, Alzheimer's disease, epilepsy, mood disorders, and schizophrenia. For these reasons, NMDA receptors have been intensively studied in the past several decades to elucidate their physiological roles and to advance them as therapeutic targets. Seven NMDA receptor subunits exist that assemble into a diverse array of tetrameric receptor complexes, which are differently regulated, have distinct regional and developmental expression, and possess a wide range of functional and pharmacological properties. The diversity in subunit composition creates NMDA receptor subtypes with distinct physiological roles across neuronal cell types and brain regions, and enables precise tuning of synaptic transmission. Here, we will review the relationship between NMDA receptor structure and function, the diversity and significance of NMDA receptor subtypes in the CNS, as well as principles and rules by which NMDA receptors operate in the CNS under normal and pathological conditions.

Key words Ionotropic glutamate receptor, Neurotransmitter, NMDA, Ion channel, Regulation, Structure-function, Disease, Synaptic transmission

1 Introduction

Glutamatergic neurotransmission in the CNS is mediated by metabotropic glutamate receptors (mGluRs) and ionotropic glutamate receptors (iGluRs). The iGluRs are ligand-gated ion channels permeable to cations (Na^+, K^+, and Ca^{2+}) that can be divided into three functional classes, namely the α-amino-3-hydroxy-5-methyl-4-isoxasolepropionic acid (AMPA) receptors, kainate receptors, and *N*-methyl-D-aspartate (NMDA) receptors [1, 2] (Fig. 1a, b). These functional classes were historically named on the basis of their pharmacological properties (i.e., the activating agonist), but the division was firmly established by subsequent cloning that

Nail Burnashev and Pierre Szepetowski (eds.), *NMDA Receptors: Methods and Protocols*, Methods in Molecular Biology, vol. 1677, DOI 10.1007/978-1-4939-7321-7_1,

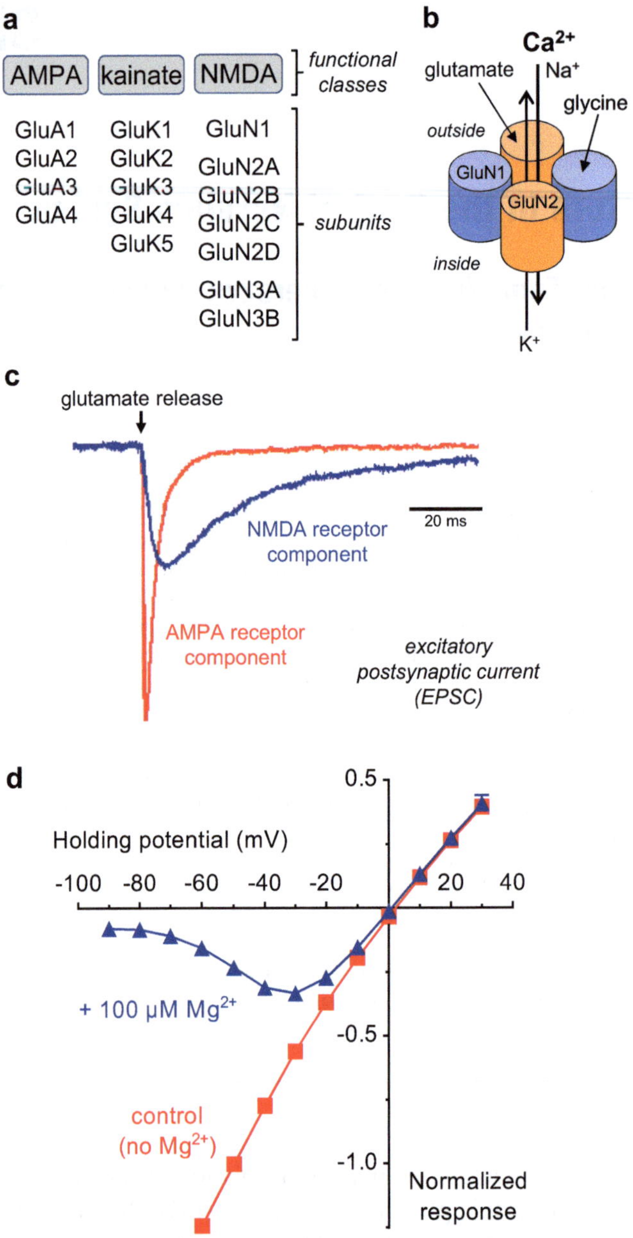

Fig. 1 Functional classes of ionotropic glutamate receptors. (**a**) Ionotropic glutamate receptors are divided into three functional classes, namely AMPA, kainate, and NMDA receptors. Multiple subunits have been cloned in each of these classes. (**b**) The majority of NMDA receptors in the CNS are composed of two glycine binding GluN1 and two glutamate binding GluN2 subunits, which form a central cation-permeable channel pore. (**c**) AMPA and NMDA receptor-mediated components of the EPSC at a central synapse. The slow NMDA receptor-mediated component is isolated in the absence of Mg^{2+} using the AMPA receptor antagonist CNQX, whereas the fast AMPA receptor-mediated component is isolated using the NMDA receptor antagonist AP5. The figure shows unpublished data from Lonnie P. Wollmuth and is adapted with permission from Traynelis et al. [1]. (**d**) Relationship between NMDA receptor current response and membrane potential (i.e., I/V-relationship) in the presence and absence of 100 μM extracellular Mg^{2+}. Voltage-dependent Mg^{2+}-block is relieved with depolarization of the membrane potential (i.e., as the membrane potential approaches 0 mV). Unpublished data from Feng Yi and Kasper B. Hansen

demonstrated strong correlation between the sequence identity and the pharmacological properties of subunits in these functional classes. The δ (delta) receptors are also considered iGluRs, primarily based on sequence identity, but their function is not fully understood [3–5]. The δ (delta) receptors appear to play important roles in synapse organization and some forms of synaptic plasticity [6–8], but it is uncertain whether they are capable of forming functional ion channels [9–11]. NMDA receptors exhibit voltage-dependent Mg^{2+}-block, high permeability to Ca^{2+}, and require simultaneous binding of the co-agonists glycine and glutamate for activation. These hallmark features distinguish NMDA receptors from AMPA/kainate receptors (i.e., non-NMDA receptors) and have profound impact on their physiological roles in the CNS.

In most central synapses, the release of glutamate activates excitatory postsynaptic currents (EPSCs) with a time course that can be described primarily by two exponential components corresponding to the activation of AMPA and NMDA receptors. Activation of AMPA receptors mediates a fast component with rapid rise time and decay, whereas activation of NMDA receptors mediates a slower component with slower rise time and a time course lasting for tens to hundreds of milliseconds [12–15] (Fig. 1c). Activation of postsynaptic kainate receptors typically results in EPSCs with a slower time course than AMPA receptors and a comparable, but generally faster time course than NMDA receptors [16]. At resting membrane potential, the NMDA receptor ion channel is blocked by physiological levels of extracellular Mg^{2+}, but synaptic release of glutamate and the resulting rapid activation of AMPA/kainate receptors can depolarize the membrane potential and thereby relieve voltage-dependent Mg^{2+}-block of NMDA receptors [17, 18] (Fig. 1d). Thus, NMDA receptors serve as coincidence detectors that require simultaneous presynaptic release of glutamate and postsynaptic depolarization to produce the slow Ca^{2+}-permeable component of the EPSC [19, 20].

The NMDA receptors can mediate substantial Ca^{2+}-influx during the EPSC due both to their high Ca^{2+} permeability and to prolonged time course. The resulting increase in intracellular Ca^{2+} can trigger multiple downstream signaling events in the postsynaptic neuron, which are central to the roles of NMDA receptors under both normal and pathophysiological conditions [1, 2, 21–23]. The rise in intracellular Ca^{2+} triggers both short-term and long-term effects, which are accompanied by changes in synaptic efficacy and neuronal morphology (i.e., synaptic plasticity) [24]. Robust NMDA receptor-mediated Ca^{2+}-influx for a short duration can lead to long-term potentiation (LTP) of synaptic efficacy, whereas less pronounced Ca^{2+}-influx for a longer duration can result in long-term depression (LTD) [25, 26]. Thus, the frequency and duration of synaptic NMDA receptor activation can result in either potentiation or depression of synaptic efficacy, which is considered a cellular correlate of memory and learning [27, 28].

Glutamate is sufficient for the activation of AMPA and kainate receptors, whereas the NMDA receptors are unique in that they require simultaneous binding of two distinct agonists, glutamate and glycine/D-serine, for activation [29–35]. In the CNS, NMDA receptors mainly rely on synaptic release of glutamate for activation, since extracellular glycine (or D-serine) is thought to be continuously present. Whether glycine or D-serine serves as the endogenous co-agonist may depend on brain region and subcellular compartment [36–38]. For example, it has recently been suggested that D-serine is the predominant co-agonist in synapses, whereas glycine is more prevalent at extrasynaptic sites [39]; more work is needed to determine whether this is a principle that transcends anatomical regions. Furthermore, glycine and D-serine are not present at concentrations that fully saturate the NMDA receptor co-agonist binding sites, at least in some brain regions [40, 41]. Thus, the co-agonist requirement enables an additional mechanism of NMDA receptor regulation, in which activation is controlled by phasic changes in glutamate concentrations (i.e., synaptic release), but the magnitude of activation can be modulated by changes in the tonic concentration of glycine/D-serine.

Given the central roles of NMDA receptors in normal brain function, it is not surprising that their dysregulation has been linked to a number of pathophysiological conditions [1, 2, 23, 42]. In acute conditions, such as ischemia, seizures, and traumatic brain injury, the increase in extracellular glutamate that follows increased release and impaired uptake can produce profound NMDA receptor-mediated Ca^{2+}-flux into the neuron, which may promote neuronal death [43–46]. Impairment of neuronal health by glutamate-mediated signaling is often referred to as "excitotoxicity" [47]. Under chronic conditions of enhanced neuronal susceptibility, as in Parkinson's and Alzheimer's diseases, the NMDA receptor-mediated excitotoxicity may contribute to impairment of neuronal health over many years (e.g., *see* [48]). NMDA receptor antagonists have been proposed to be beneficial under such conditions involving excitotoxicity (e.g., *see* [49]), but side effects, such as psychosis, memory impairment, anesthesia, and neuronal cell death, can accompany strong and nonselective NMDA receptor blockade, thereby limiting the clinical usefulness of such drugs for chronic conditions [50, 51].

Interestingly, the side effects of high-affinity NMDA receptor channel blockers resemble symptoms exhibited by patients suffering from schizophrenia. The observations of these "psychotomimetic" properties of the channel blockers PCP and ketamine have led to the "NMDA receptor hypofunction model of psychosis," which proposes that multiple symptoms associated with schizophrenia may be caused by lower than normal NMDA-receptor-mediated glutamatergic activity in key brain circuits [51–53]. In theory, enhancing NMDA receptor function, perhaps selectively in key brain circuits, could be beneficial for treating cognitive disorders and schizophrenia. However, NMDA receptor agonists have not been fully studied in this context due to the risk that excessive stimulation may cause excitotoxicity, and indirect methods to enhance NMDA receptor function through block of glycine uptake have been inconclusive. Moreover, only very recently have subunit-selective NMDA receptor positive allosteric modulators been identified that allow this idea to be further investigated (see below). In this regard, subunit-selective modulators of NMDA receptors may be therapeutically beneficial in some CNS disorders, since these modulators would target NMDA receptor subtypes in specific neuronal populations or brain regions associated with the disease without affecting NMDA receptors in other regions [54–56].

1.1 NMDA Receptor Subunit Diversity

The arrival of the action potential at the presynaptic terminal triggers the release of glutamate into the synaptic cleft. Termination of glutamatergic neurotransmission is mediated by diffusion and rapid removal of glutamate from synaptic and extrasynaptic sites via reuptake by excitatory amino acid transporters (EAATs; i.e., glutamate transporters) [57]. Synaptically released glutamate reaches a very high peak concentration (~1 mM) for a brief duration (~1 ms) [58]. In this short period of time, glutamate will bind iGluRs and initiate receptor conformational changes that lead to opening of the ion channel (i.e., ion channel gating). However, the NMDA receptor-mediated component of the EPSC continues for tens to hundreds of millisecond after synaptic glutamate is removed, during which time, NMDA receptors transition between glutamate-bound open and closed conformational states until glutamate eventually unbinds and the EPSC is terminated. Thus, the time course of the EPSC is governed by glutamate binding affinity, the connectivity, and lifetime of the receptor in pre-gating conformations that must be traversed before unbinding, and the rates into and out of the desensitized states following agonist binding [59–61]. For NMDA receptors, these functional properties are controlled by the subunit composition [62–64] (Fig. 2). Subunit diversity among NMDA receptors and assembly of different receptor subtypes with distinct functional properties enable precise tuning of the synaptic response and allow variation in the physiological roles of NMDA receptors at synaptic versus extrasynaptic sites, in different neuronal cell types and brain regions, and during neuronal development.

Seven genes that encode NMDA receptor subunits have been identified, which include GluN1, four different GluN2 (GluN2A-D), and two GluN3 subunits (GluN3A-B) [1, 2] (Fig. 1a). All the NMDA receptors are obligatory heteromeric assemblies of four subunits that form a central ion channel pore, and the majority of NMDA receptors in the CNS are composed of two glycine binding GluN1 and two glutamate binding GluN2 subunits

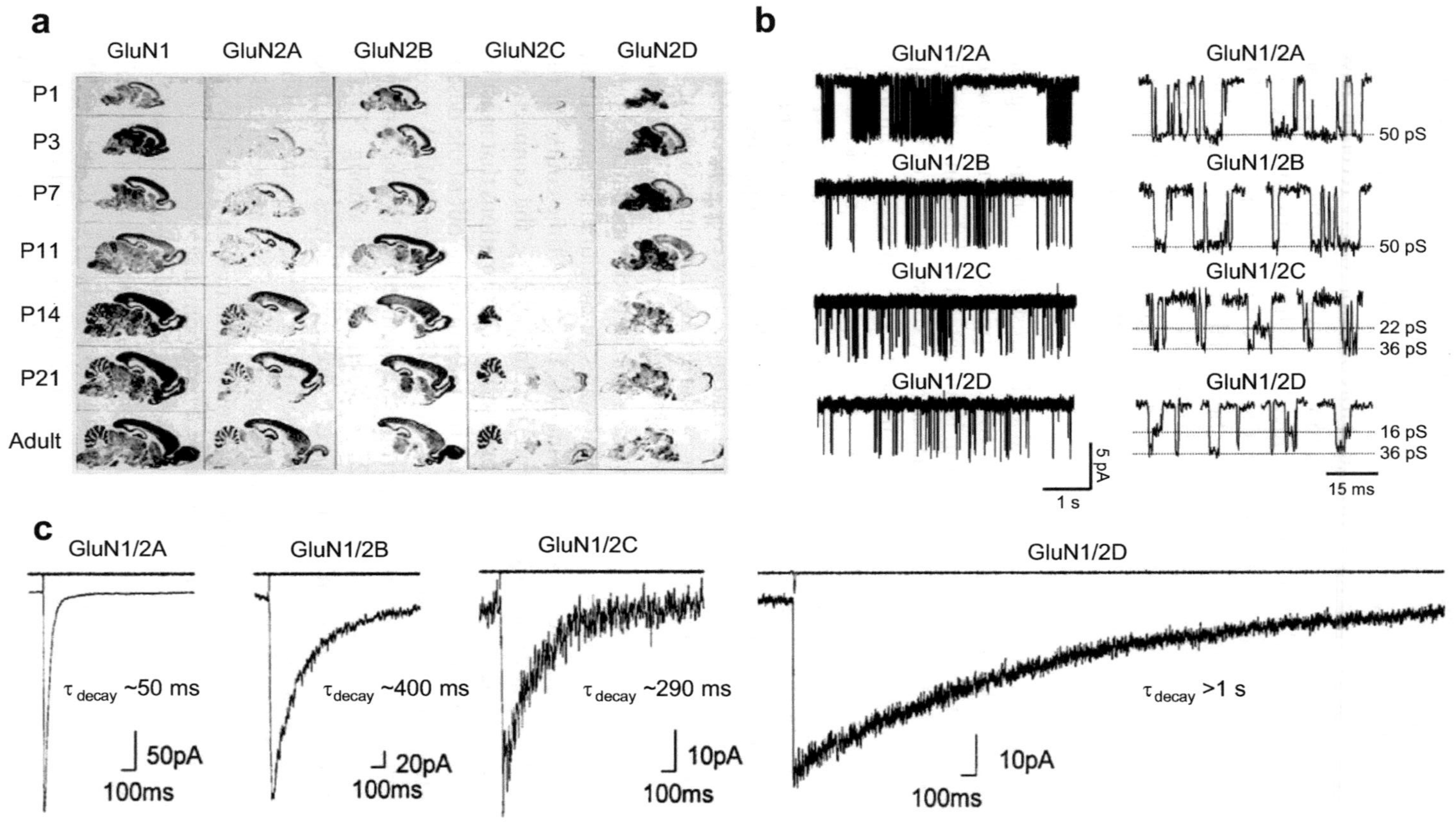

Fig. 2 GluN2 subunit-specific expression and functional properties of recombinant NMDA receptor subtypes. (**a**) Regional and developmental expression of GluN2 subunits in rat brain revealed in autoradiograms using in situ hybridizations of oligonucleotide probes for the relevant mRNAs to parasagittal sections. Modified with permission from Akazawa et al. [92]. (**b**) Single-channel recordings of currents from diheteromeric NMDA receptor subtypes expressed in HEK293 cells (outside-out membrane patches). Open probability is ~0.5 for GluN1/2A, ~0.1 for GluN1/2B, and $<$0.05 for GluN1/2C and GluN1/2D. Highlights of individual openings are shown on the *left*. GluN1/2A and GluN1/2B have higher channel conductance (~50 pS) compared to GluN1/2C (~22 and ~36 pS) and GluN1/2D (~16 and ~36 pS). Adapted with permission from Yuan et al. [523]. (**c**) Whole-cell patch-clamp recordings of responses from brief application of glutamate (1 ms of 1 mM glutamate) to recombinant diheteromeric NMDA receptor subtypes expressed in HEK293 cells. The open tip current indicating the duration of the drug application is shown in the *upper trace*. Adapted with permission from Vicini et al. [62]

(i.e., GluN1/2 receptors) [65–67] (Fig. 1b). However, the glycine binding GluN3 subunits can also assemble with GluN1 and GluN2 subunits to form GluN1/2/3 receptors or with GluN1 alone to form GluN1/3 receptors [68–72].

1.2 The GluN1 Subunit

The glycine/D-serine binding GluN1 subunit is ubiquitously distributed in the brain and is an obligatory subunit in all NMDA receptor subtypes. GluN1 has eight different isoforms that arise from alternative splicing of three exons within a single gene product [73–76] (Fig. 3a, b). Exon 5 encodes 21 highly charged amino acids in the GluN1 amino-terminal domain (ATD) named the N1 cassette, exon 21 encodes 37 amino acids in the carboxyl-terminal domain (CTD) named the C1 cassette, and exon 22 encodes 38 amino acids in the CTD named the C2 cassette. Deletion of exon 22 eliminates a stop codon and causes a reading frame shift, which results in the inclusion of 22 alternative amino acids named the C2′ cassette. Different GluN1 splice variants have distinct regional and developmental expression patterns [77–79] and display differences in NMDA receptor function and pharmacology (see below; Fig. 3b, c).

NMDA receptors with GluN1 subunits that contain the N1 cassette (i.e., exon 5) have reduced agonist potency (i.e., increased EC_{50}) and are less sensitive to inhibition by protons and extracellular zinc [80, 81]. Consistent with the effect on agonist potency, the presence of the N1 cassette accelerates deactivation of glutamate-activated NMDA receptor responses, which shortens the time course of EPSCs [82–84] (Fig. 3c). Furthermore, the N1 cassette has a negative impact on modulation of NMDA receptor function by GluN2B-selective antagonists, such as ifenprodil [85, 86], and potentiation by extracellular polyamines, such as spermine [74, 81, 82, 87, 88] (Fig. 3d, e). Alternative splicing of exons 21 and 22 dramatically alters the length and the amino acid sequence of the intracellular GluN1 CTD, which mediates interactions with several intracellular proteins, including PSD-95, calmodulin, and the neurofilament subunit NF-L. [1] Many of these proteins are involved in surface trafficking and anchoring of receptors at synaptic sites, and alternative splicing of exons 21 and 22 can therefore mediate changes in the subcellular distribution of NMDA receptors [89–91]. Contrasting with the changes observed upon inclusion of exon 5, there has not been convincing demonstration that alternative splicing of exons 21 and 22 has strong effects on the functional and pharmacological properties of NMDA receptors.

1.3 The GluN2 Subunits

The four different glutamate binding GluN2 subunits (GluN2A-D) have pronounced differences in both developmental and regional expression levels and endow NMDA receptors with strikingly different pharmacological and functional properties [62–64, 79, 92–95] (Fig. 2). Thus, the GluN2 subunits essentially dictate

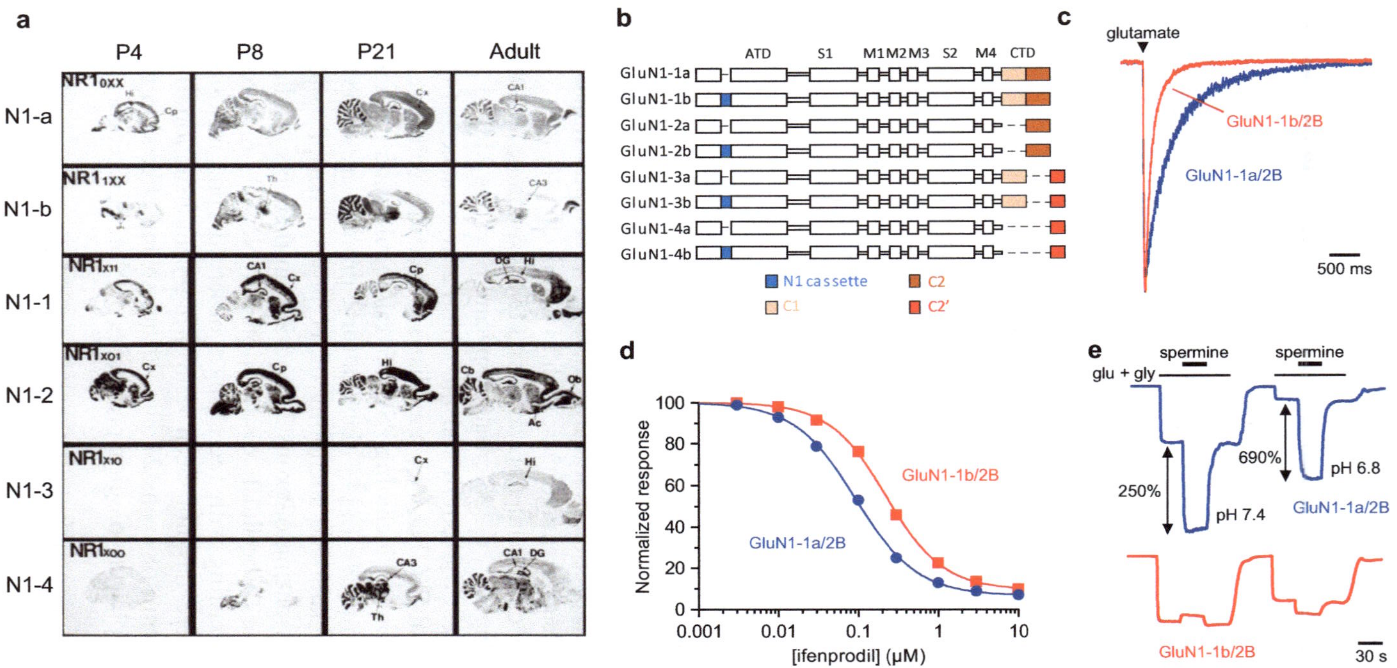

Fig. 3 Expression and functional properties of GluN1 splice variants. (**a**) Regional and developmental expression of GluN1 splice variants in rat brain revealed in autoradiograms using in situ hybridizations of oligonucleotide probes for the relevant mRNAs to parasagittal sections. *Ac* nucleus accumbens, *Cb* cerebellum, *Cp* caudate-putamen, *Cx* cortex, *DG* dentate gyrus, *DP* dorsal pons, *Hi* hippocampus, *Ob* olfactory bulb, *Th* thalamus, *VPn* ventro-posterial thalamic nuclei. Modified with permission from Paupard et al. [78]. (**b**) Linear representation of the GluN1 polypeptide chain for eight alternative splice variants. GluN1 subunits are composed of the amino-terminal domain (ATD), S1 and S2 segments that form the ligand binding domain (LBD), three transmembrane helices (M1, M3, and M4) and a membrane reentrant loop (M2), and the intracellular carboxyl-terminal domain (CTD). The N1 cassette (*blue*) is 21 amino acids in the ATD encoded by exon 5. The C1 cassette (*yellow*) is 37 amino acids in the CTD encoded by exon 21, while the C2 cassette (*orange*) is 38 amino acids in the CTD encoded by exon 22. Deletion of exon 22 creates a shift in the open reading frame, resulting in the alternate exon 22′ that encodes the C2′ cassette (*red*; 22 amino acids). (**c**) Whole-cell patch-clamp recordings of responses from brief application of glutamate (1 ms of 1 mM glutamate) to recombinant GluN1-1a/2B and GluN1-1b/2B receptors expressed in HEK293 cells. NMDA receptors containing exon 5 (e.g., as in GluN1-1b) display faster deactivation time course compared to receptors lacking exon 5 (e.g., as in GluN1-1a). (**d**) Ifenprodil concentration-inhibition relationships for recombinant GluN1-1a/2B and GluN1-1b/2B receptors expressed in *Xenopus* oocytes. Ifenprodil potency is lower for receptors containing exon 5. (**e**) Representative recordings for spermine potentiation of responses from recombinant GluN1-1a/2B and GluN1-1b/2B receptors expressed in *Xenopus* oocytes. Spermine sensitivity is dramatically reduced for receptors containing exon 5. Data in (**c**)–(**e**) are unpublished from Feng Yi and Kasper B. Hansen

the physiological roles of NMDA receptor subtypes in the CNS. Since GluN1 is an obligatory subunit in all NMDA receptors, the significant sites of structural variation among subtypes are located in the GluN2 subunits of the receptor complex. Efforts to pharmacologically manipulate specific NMDA receptor subtypes in the CNS therefore focus on the development of ligands that can distinguish NMDA receptors on the basis of GluN2 subunits [54–56, 96, 97]. In recent years, the therapeutic rationale for the development of GluN2-selective ligands has been reinforced by increasingly precise localization of GluN2 subunits in neuronal populations relevant to CNS diseases.

The GluN2 subunits impart NMDA receptor subtypes with differences in sensitivity to voltage-dependent Mg^{2+} block [63, 64, 98, 99] and inhibition by endogenous modulators, such as protons and extracellular Zn^{2+} [80, 81, 100]. The potency of glutamate and glycine/D-serine, as well as other agonists, is influenced by the GluN2 subunits [101–103]. For example, the EC_{50} for glutamate activating NMDA receptors containing two GluN1 and two GluN2D subunits (i.e., diheteromeric GluN1/2D receptors) is six to tenfold lower (i.e., glutamate is more potent) compared to that for GluN1/2A, whereas GluN1/2B and GluN1/2C receptors show intermediate EC_{50} values [1, 2, 104–106]. GluN1/2A and GluN1/2B have higher single channel conductances compared to GluN1/2C and GluN1/2D receptors [1, 2, 104–106]. In addition, the probability that the channel will be open when all the agonist binding sites are fully occupied by agonists (i.e., the open probability) is ~0.5 for recombinant GluN1/2A, ~0.1 for GluN1/2B, and <0.05 for GluN1/2C and GluN1/2D [1, 2, 104–106] (Fig. 2). Importantly, the time constants of deactivation (τ_{decay}) are also highly dependent on the GluN2 subunits; τ_{decay} is ~50 ms for GluN1-1a/2A, ~400 ms for GluN1-1a/2B, ~290 ms for GluN1-1a/2C, and >1 s for GluN1-1a/2D [62, 64, 94] (Fig. 2). Thus, the GluN2 subunits control functional NMDA receptor properties relevant to synaptic transmission. Furthermore, the amino acid sequence of the intracellular CTD displays pronounced variation among GluN2 subunits and harbors distinct interaction sites for phosphatases, kinases, and proteins responsible for surface trafficking and anchoring at synaptic sites [1]. The GluN2 subunits therefore also affect subcellular localization, cell-surface expression, and recycling/degradation of NMDA receptor subtypes.

1.4 The GluN3 Subunits

GluN3A and GluN3B were cloned based on similarity to GluN1 and GluN2 subunits and were the last NMDA receptor subunits to be discovered (reviewed in [68–72, 107]). The GluN3 subunits bind glycine and D-serine [108–110], but the functional properties and physiological roles of GluN3-containing NMDA receptors remain elusive.

Triheteromeric NMDA receptors that are assembled from a combination of GluN1, GluN2, and GluN3 subunits have been consistently reported in both the CNS and heterologous expression systems on the basis of biochemical and functional experiments [111–122]. It is therefore surprising that the subunit stoichiometry of triheteromeric GluN3-containing NMDA receptors has not been resolved; it is unknown whether GluN3 replaces GluN1, GluN2, or both GluN1 and GluN2 in these receptors. Despite this gap in our understanding, numerous studies suggest that the inclusion of GluN3 in the NMDA receptor complex reduces Mg^{2+}-block and Ca^{2+}-permeability as well as response amplitudes (reviewed in [68–72, 107]). Thus, the GluN3 subunits appear to have dominant-negative effects on NMDA receptor function.

The GluN3 subunits can also assemble with GluN1 in heterologous expression systems to form functional diheteromeric NMDA receptors that contain two GluN1 and two GluN3 subunits [65, 123, 124], but their existence in the CNS has not been firmly established. These GluN1/3 receptors have been termed "excitatory glycine receptors," since they can be activated by glycine alone without the requirement of glutamate binding [123]. In recombinant systems, the GluN1/3 receptors are characterized by low Ca^{2+}-permeability and relative insensitivity to extracellular Mg^{2+} [123, 125–127]. Interestingly, agonist binding to the GluN1 subunit of GluN1/3 receptors triggers strong desensitization, whereas agonist binding to the GluN3 subunits is sufficient for activation [127–130]. Thus, in contrast to the more conventional GluN1/2 receptors [30, 35], simultaneous binding of agonist to all subunits does not appear to be a requirement for activation of GluN1/3 receptors. While many aspects of the physiological roles of GluN3-containing NMDA receptors remain elusive, it is clear that GluN3 subunits endow NMDA receptors with strikingly unique functional properties.

1.5 Diheteromeric and Triheteromeric NMDA Receptors

The seven NMDA receptor subunits can assemble to produce receptor subtypes with distinct physiological roles across neuronal cell types and brain regions, thereby mediating changes in synaptic transmission and subcellular localization during neuronal development. At least two different GluN2 subunits are expressed in most, if not all NMDA receptor-expressing cells, and a large proportion of native NMDA receptors in the adult CNS are triheteromers that contain GluN1 and two different GluN2 subunits [84, 131–142]. Examples of triheteromeric NMDA receptors identified in an increasing number of studies are the GluN1/2A/2B, GluN1/2A/2C, GluN1/2B/2D subtypes, but the existence of NMDA receptors with other compositions of GluN2 subunits has not been ruled out [131–140, 142–149]. These subtypes, which are expressed in distinct neuronal populations, have been detected using co-immunoprecipitation and by intriguing functional and pharmacological observations that are not mediated by diheteromeric NMDA receptors.

Despite their prevalence in the CNS, there is a gap in our knowledge of triheteromeric NMDA receptors due to our inability to study a homogenous population of these receptors in heterologous expression systems [62, 140, 148, 150, 151]. The nature of the problem is that co-expression of GluN1 with two different GluN2 subunits (e.g., GluN2A and GluN2B) generates three populations of functional NMDA receptors, which are composed of two different diheteromeric receptors (e.g., GluN1/2A and GluN1/2B) as well as triheteromeric receptors (e.g., GluN1/2A/2B) [62, 150, 151]. The majority of our knowledge regarding function, pharmacology, and regulation of recombinant NMDA receptors is therefore almost exclusively derived from studies on diheteromeric receptors that are assembled from GluN1 and only one type of GluN2 (e.g., GluN1/2A). The properties of diheteromeric NMDA receptors are well described, but little is known about how co-assembly of two different GluN2 subunits affects properties, such as deactivation time course, Mg^{2+}-block, and the activity of subunit-selective allosteric modulators. Similarly, phosphorylation sites and trafficking properties of the intracellular GluN2 CTDs have been extensively studied, but the regulation of triheteromeric NMDA receptors that possess two distinct GluN2 CTDs is largely unknown (*see* [152]).

Recently, significant insight into functional and pharmacological properties of triheteromeric NMDA receptors has been gained using a method to tightly control cell surface expression of NMDA receptors with defined subunit composition [151, 153]. This method has provided evidence of surprising pharmacological and functional properties of triheteromeric NMDA receptors that are distinct from the properties of the respective diheteromeric receptors [151, 153–157] (Fig. 4). Furthermore, the method is enabling exciting, new opportunities to develop therapeutic agents that target disease-relevant triheteromeric NMDA receptors [155–159].

2 NMDA Receptor Structure and Function

AMPA, kainate, and NMDA receptor subunits share a common membrane topology; each subunit consists of a large extracellular ATD, a bi-lobed ligand binding domain (LBD), a transmembrane domain (TMD), and an intracellular CTD (Fig. 5a). The TMD is formed by three transmembrane helices (M1, M2, and M4) and a membrane re-entrant loop (M2). The ion channel pore of iGluRs is mainly lined with residues in the membrane re-entrant loop from all four subunits. Among NMDA receptor subtypes, the residues in the pore region, which determines basic ion permeation properties,

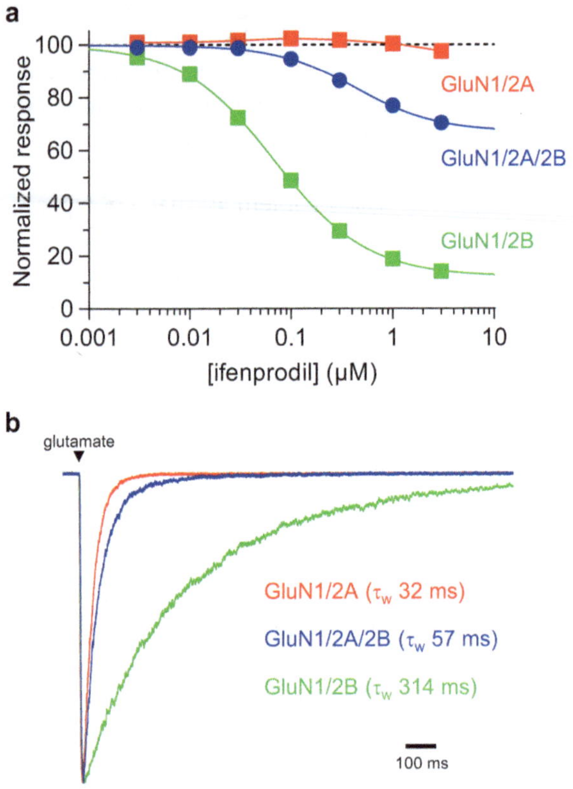

Fig. 4 Functional properties of triheteromeric GluN1/2A/2B receptors. (**a**) Ifenprodil concentration-inhibition relationships for recombinant diheteromeric GluN1/2A and GluN1/2B receptors and triheteromeric GluN1/2A/2B receptors expressed in *Xenopus* oocytes using a method to control subunit composition of NMDA receptors [151]. Ifenprodil efficacy and potency are reduced for triheteromeric GluN1/2A/2B receptors that only contain one binding site for ifenprodil. (**b**) Whole-cell patch-clamp recordings of responses from brief application of glutamate (1 ms of 1 mM glutamate) to recombinant diheteromeric GluN1/2A and GluN1/2B receptors and triheteromeric GluN1/2A/2B receptors expressed in HEK293 cells. The deactivation time course of triheteromeric GluN1/2A/2B receptors is similar to diheteromeric GluN1/2A and strikingly different from diheteromeric GluN1/2B. Data are adapted with permission from Hansen et al. [151]

are highly conserved. One key determinant of ion permeation, which partially defines Ca^{2+} permeability and Mg^{2+}-block, resides at the apex of the membrane re-entrant loop M2 and is sometimes referred to as the Q/R/N site on the basis of amino acid residues found at this position in AMPA, kainate, and NMDA receptors. The ATD adopts a clamshell-like structure formed by the first ~350 amino acids of the subunit and plays an important role in subunit assembly and as a modulatory NMDA receptor domain. In NMDA receptors, the ATD harbors binding sites for several allosteric

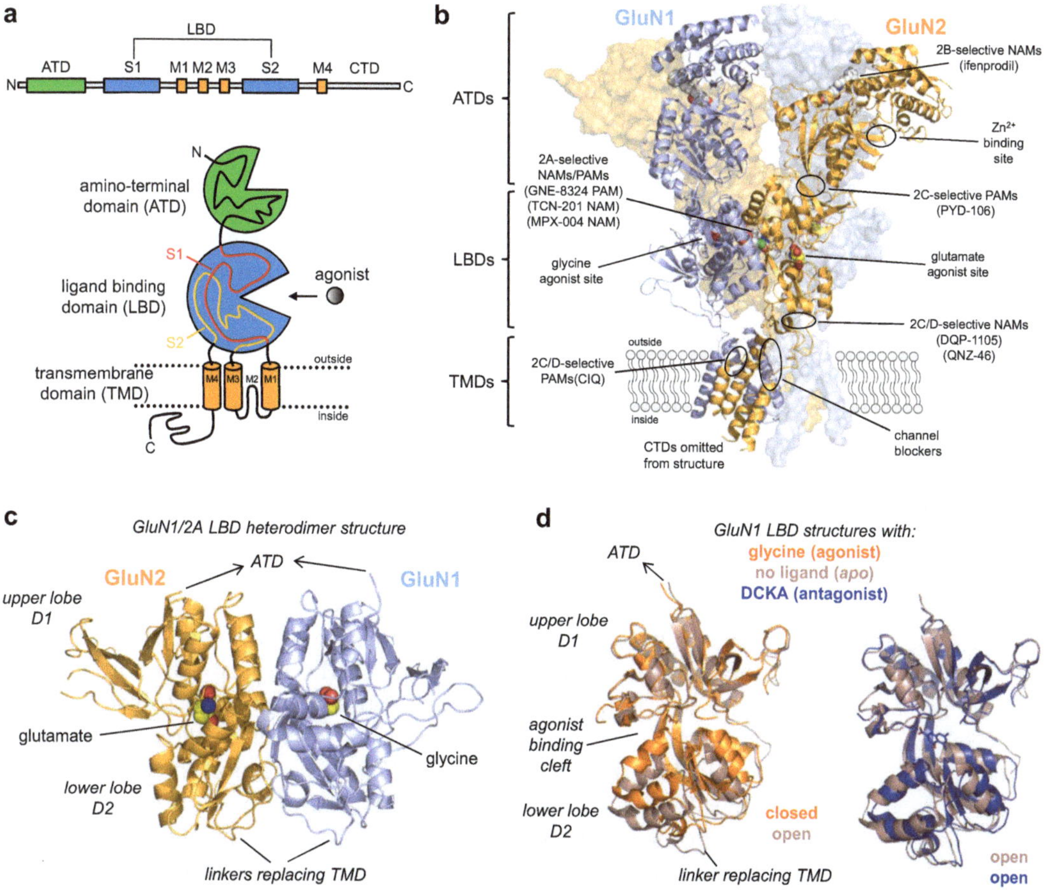

Fig. 5 NMDA receptor structure and ligand binding sites. (**a**) Linear representation and cartoon illustration of the polypeptide chain in iGluR subunits. Each subunit consists of a large extracellular amino-terminal domain (ATD), a bi-lobed ligand binding domain (LBD), a transmembrane domain (TMD), and an intracellular CTD. The TMD is formed by three transmembrane helices (M1, M2, and M4) and a membrane re-entrant loop (M2). The LBD is formed by two segments of the polypeptide chain (S1 and S2), which fold into a kidney-shaped structure composed of an upper lobe (D1) and lower lobe (D2) relative to the cell membrane, and the agonist binding site is located in the cleft between the two lobes. (**b**) Crystal structure of the GluN1/2B NMDA receptor (PDB ID 4PE5 [67]), illustrating the subunit arrangement and the layered domain organization composed of the TMD layer and two extracellular layers formed by LBDs and ATDs. Agonist binding sites as well as known and predicted binding sites for positive and negative allosteric modulators (PAMs and NAMs) are *highlighted*. (**c**) Crystal structure of the soluble GluN1/2A LBD heterodimer (PDB ID 5I57 [155]), showing the subunit interface and back-to-back dimer arrangement of the LBDs. Soluble LBD proteins composed of the S1 and S2 segments of the polypeptide chain are produced by deleting the ATD and replacing the TMD with a di-peptide linker. (**d**) Overlay of crystal structures of the soluble GluN1 LBD in the *apo*-form (PDB 4KCC [167];) or in complex with the agonist glycine (PDB ID 5I57 [155];) or competitive antagonist DCKA (PDB ID 4NF4 [165];). The upper D1 lobes are aligned to illustrate the similar conformations of antagonist-bound and *apo*-form structures. Agonist binding induces considerable closure of the LBD compared to the antagonist-bound and *apo*-form structures, and agonist-induced closure of the LBD is required for activation of NMDA receptors. Competitive antagonists bind the LBD without inducing domain closure, thereby preventing agonist binding and receptor activation.

modulators, including extracellular zinc and polyamines, as well as GluN2B-selective antagonists (e.g., ifenprodil) (Fig. 5b). The LBD is formed by two segments of the polypeptide chain (S1 and S2), which fold into a kidney-shaped structure composed of an upper lobe (D1) and lower lobe (D2) relative to the cell membrane, and the agonist binding site is located in the cleft between the two lobes (Fig. 5c, d). The relationships between domain structures, their intra- and inter-subunit interactions, and receptor function and pharmacology have been extensively studied for more than two decades. Recently, crystallographic and cryo-EM studies have provided the first glimpses at the domain organization of NMDA receptors and mechanisms by which these domains and allosteric modulators influence receptor function [66, 67, 160, 161].

2.1 Structure and Function of GluN1 and GluN2 Ligand Binding Domains

Recombinant proteins that comprise the NMDA receptor LBDs have been generated by combining S1 and S2 segments with a short artificial di-peptide linker [109, 162, 163] (Fig. 5c, d). These water-soluble LBD proteins retain ligand binding activities comparable to those in full-length NMDA receptors, indicating structural identity between the binding pockets of isolated LBDs and the corresponding intact receptor. LBD crystal structures from GluN1, GluN2, and GluN3 subunits have been solved both in complexes with agonists, antagonists, as well as allosteric modulators [108, 109, 155, 157, 162–170]. In addition to NMDA receptor subunits, numerous crystal structures for AMPA and kainate receptor subunits have been determined (reviewed in [171–173]). These studies have provided insight into the molecular determinants of full agonists and partial agonists, as well as the mechanism of action for competitive antagonists. Furthermore, the LBD structures have afforded new opportunities to consider the molecular determinants of subunit selectivity.

The first structure of the GluN2A LBD was solved in complex with the GluN1 LBD, and provided the first view of the glutamate binding site of the NMDA receptor and the first structural information about a GluN1-GluN2 protein-protein interface within the NMDA receptor complex [163]. Glycine and glutamate bind in the cleft between the two clearly defined lobes (D1 and D2) of the GluN1 and GluN2A LBDs, respectively (Fig. 5c, d). Residues from the loops within the upper lobe (D1) form most of the upper half of the binding pocket, and residues from the lower lobe (D2) form most of the lower half of the pocket. Despite being tucked away between the lobes, multiple water molecules are found in close vicinity of the agonists, some of which form hydrogen bonds with the ligand. The glycine binding pockets in GluN1 and GluN3 are smaller and more hydrophobic compared to the GluN2 glutamate binding pocket [108, 162–164, 167]. Residues lining the glutamate binding pocket are fully conserved among the GluN2 subunits, and consequently, agonists or competitive antagonists with strong selectivity among the different GluN1/2 receptor subtypes have not been identified. In order to develop subunit-selective ligands, it is presumably necessary to engage other regions of the NMDA receptor with structural variation, such as inter- or intra-subunit interfaces that are non-conserved among GluN2 subunits.

Interestingly, the heterodimer interface between GluN1 and GluN2 modulates receptor function. Three separate areas of contact between GluN1 and GluN2A are identified in the LBD heterodimer crystal structures (sites I, II, and III) [163]. The two outer regions (sites I and III) consist of hydrophobic residues from GluN1 and GluN2, and nonpolar interactions between these residues mediate heterodimerization of the soluble GluN1 and GluN2A LBDs [163]. Mutations of Y535 in GluN1 demonstrate that modification of site II, which is located at the center of the LBD heterodimer interface, results in increasing or decreasing rates of NMDA receptor deactivation [163, 174]. Crystallographic studies recently showed that site II of the GluN1/2A LBD heterodimer harbor binding sites for both positive and negative allosteric modulators with strong selectivity for GluN2A-containing NMDA receptors [155, 157, 169]. These results suggest that the stability and dynamics of the LBD interface can control NMDA receptor function, similar to what has been found for AMPA and kainate receptors.

NMDA receptors are sensitive to the redox potential and reducing conditions can cause marked enhancement of NMDA receptor function [175–178]. This sensitivity is mainly mediated by a pair of cysteine residues within the GluN1 subunit (C744 and C798) that are conserved among all iGluR subunits [179, 180]. These two residues form a disulfide bond (i.e., are oxidized) in the GluN1/2A LBD heterodimer structure, and relief (by reduction) of the conformational constraints imparted by this disulfide bond in GluN1, but not GluN2, enhances receptor function [179, 181]. Several other disulfide bonds exist in LBD crystal structures of GluN1 and GluN2 subunits, but functional effects of their reduction or oxidation have not yet been described.

Multiple ligands have each been crystallized in complex with GluN1 and GluN2 LBDs, providing a structural basis for the effects of partial agonists, agonists, and competitive antagonists [162–168]. Binding of glycine and glutamate to their respective binding sites is associated with a rapid LBD rearrangement, involving closure of the angle between the two lobes D1 and D2, akin to a clamshell-like closure (Fig. 5c, d). This agonist-mediated LBD closure mediates formation of interactions between residues from the upper and lower lobes that further stabilizes the agonist-bound LBD structure [182, 183]. The energy associated with agonist binding and LBD closure causes the receptor to undergo a series of conformational changes that can lead to opening of the ion channel pore (i.e., channel gating). Thus, LBD closure as a result of agonist binding is the initial conformational change that triggers ion channel gating. Binding of competitive antagonists, such as the glycine site antagonist DCKA and the glutamate site antagonist D-AP5, stabilizes a more open cleft conformation of the bi-lobed LBD that is incapable of triggering ion channel gating and presumably resembles the LBD conformation in the absence of bound ligand (i.e., *apo*-state) [162, 165, 167, 170]. The stabilization of the LBD in a closed conformation by agonists and an open conformation by competitive antagonists in NMDA receptors is similar to what has been found for soluble LBDs from AMPA and kainate receptor subunits (reviewed in [171–173]). However, despite this similarity, the domain closures in structures of GluN1 and GluN2 LBDs bound by partial agonists are not similar to those observed for AMPA and kainate receptor subunits. While most structures of AMPA receptor LBDs show partial domain closure that correlates with the efficacy of the partial agonist (reviewed in [171]), no such relationship is observed in GluN1 and GluN2 LBD structures. Multiple structures show that partial agonists, such as D-cycloserine, ACPC, and ACBC in GluN1 as well as NMDA and Pr-NHP5G in GluN2, bind with virtually identical degrees of domain closure in GluN1 and GluN2 LBDs compared to the complexes with full agonists glycine and glutamate, respectively [164, 166, 168]. However, while crystal structures capture only one low-energy conformation of the LBDs, recent single-molecule FRET and molecular dynamics studies have provided insights into the dynamic behavior of the NMDA receptor LBDs [167, 184–187]. These studies suggest that the LBDs fluctuate between open and closed conformations in the absence of ligand (i.e., in the *apo*-state). The probability that the LBD adopts a fully closed conformation increases by binding of full agonist, whereas binding of partial agonists mainly enables the LBD to adopt conformations with intermediate domain closure. That is, a conformational selection mechanism can presumably account for partial agonism in NMDA receptors, since LBD conformations capable of triggering ion channel gating are selected with greater propensity by full agonists compared to partial agonists.

2.2 Ligand Binding to GluN3 Subunits

Glycine-activated diheteromeric NMDA receptors assembled from two GluN3 and two GluN1 subunits have been widely studied in heterologous expression systems (reviewed in [69–72]). However, structural and functional properties of triheteromeric GluN1/2/3 receptors are largely unresolved. For example, it is unknown how agonist or antagonist binding to the GluN3 LBD affects the function of GluN1/2/3 receptors, in terms of their activation, deactivation, and desensitization properties. However, LBD crystal structures have established that structural variation exists between the agonist binding sites of GluN1 and GluN3 subunits, even though they are both glycine binding subunits [108, 162, 164,

167]. Functional studies on recombinant GluN1/3 receptors suggest that these structural differences can be exploited for the development of GluN3-selective ligands targeting the agonist binding site [130, 170].

As mentioned above, agonist binding to the GluN3 subunits is sufficient for the activation of GluN1/3 receptors. This feature has enabled pharmacological studies on the GluN3 agonist binding site in isolation by abolishing ligand binding to GluN1 using mutagenesis [130]. This approach identified agonists and antagonists with moderate preferences between the agonist binding sites of GluN1 and GluN3 by comparing ligand activities between wild-type GluN1/2 receptors and GluN1/3 receptors with mutations that render GluN1 incapable of ligand binding (hereafter denoted as GluN1*/3) [130]. In addition, these studies brought to light interesting discrepancies between ligand binding to the isolated, soluble LBD proteins, and full-length subunits in intact receptors. The isolated GluN1 LBD protein binds glycine with lower affinity (26 μM) compared to the isolated GluN3A LBD (0.04 μM) [109]. By contrast, the potency of glycine is higher for full-length GluN1 (e.g., in GluN1/2A the glycine EC_{50} is 1.2 μM) compared to for full-length GluN3A (e.g., GluN1*/3A EC_{50} is 57 μM) [130]. The competitive glycine site antagonist DCKA bind with higher affinity to the isolated GluN1 LBD (0.54 μM) compared to the isolated GluN3A LBD (647 μM), and the binding affinities are estimated to be 0.07 μM for GluN1 in GluN1/2A and 35 μM for GluN3A in GluN1*/3A [109, 170]. Thus, in case of GluN1 and GluN3 subunits, there are marked differences in the pharmacology of isolated, soluble LBDs, and full-length subunits in intact receptors. The underlying basis of these differences is poorly understood and raises caveats to evaluation of pharmacology in isolated, soluble LBDs that are excised from the intact NMDA receptor.

2.3 Structures of Intact Tetrameric NMDA Receptors

Crystal structures of GluN1/2A LBD heterodimers and GluN1/2B ATD heterodimers have provided important insight into the overall receptor structure (reviewed in [188]). However, it is only recently that the first structures of intact GluN1/2B receptors firmly established the subunit arrangement and domain organization [66, 67]. These structures show that subunits in GluN1/2B receptors are arranged in an alternating pattern (i.e., 1-2-1-2) (Fig. 6). In addition, the NMDA receptor structure shares many of the characteristics of AMPA and kainate receptors. First, the receptor seemingly adopts a layered structure composed of the TMD layer and two extracellular layers formed by LBDs and ATDs. Second, there is a symmetry mismatch between the TMDs and the extracellular portion of the receptor; the TMDs have a quasi-four-fold symmetry, whereas the extracellular portion has a twofold symmetry (Fig. 6). The TMDs are arranged symmetrically around the ion channel pore, but the extracellular portion of the

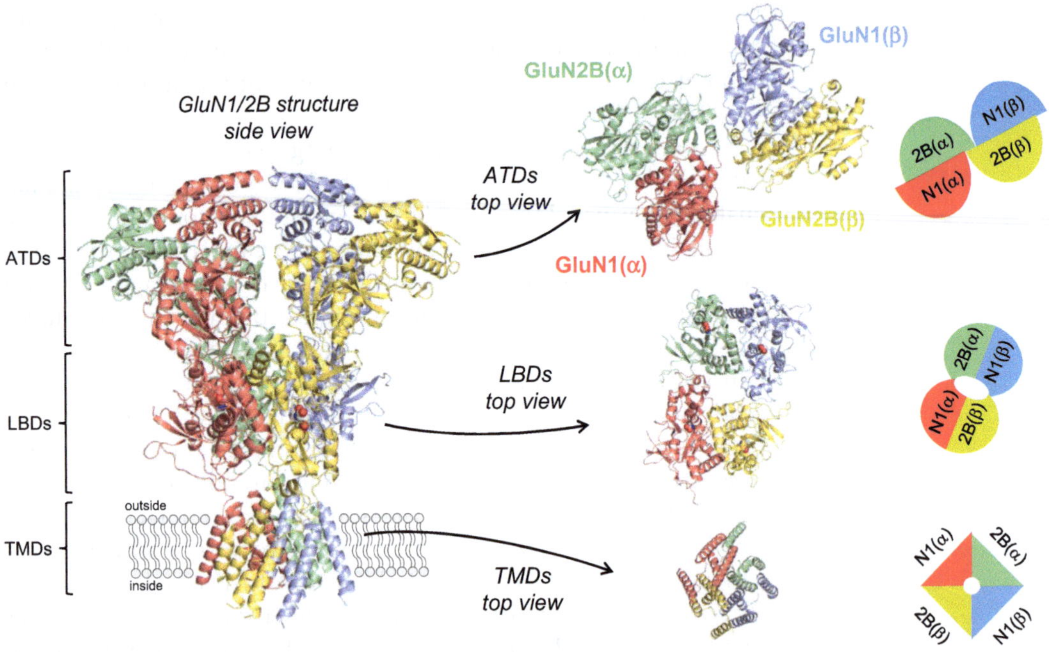

Fig. 6 Subunit crossover and symmetry mismatch in the NMDA receptor structure. Side view of the GluN1/2B NMDA receptor structure (PDB ID 4PE5 [67];) and top views of the ATD, LBD, and TMD layers. The subunits in GluN1/2 receptors are arranged in an alternating pattern (i.e., 1-2-1-2) and there is a symmetry mismatch between the TMDs and the extracellular LBDs and ATDs of the receptor. The TMDs are arranged symmetrically around the ion channel pore with a quasi-four-fold symmetry, whereas the extracellular portion adopts a dimer-of-dimer arrangement (i.e., two GluN1/2 heterodimers) with a twofold symmetry. There is a subunit crossover between the LBD layer and the ATD layer in that the GluN1(α) ATD forms a local dimer with the GluN2B(α) ATD, whereas the GluN1(α) LBD forms a local dimer with the GluN2B(β) LBD.

receptor adopts a dimer-of-dimer arrangement (i.e., two GluN1/2 heterodimers). Third, there is a subunit crossover between the LBD layer and the ATD layer (Fig. 6). Furthermore, the NMDA receptor structure has several unique features compared to the structures of AMPA and kainate receptors [66, 67]. First, there are extensive contacts between the two GluN1/2 LBD heterodimers in the intact NMDA receptor, which are not observed in AMPA receptor structures. Second, the NMDA receptor ATDs are arranged differently and have unique subunit interfaces compared to AMPA and kainate receptors. Third, the ATDs form extensive contacts with the upper lobe of the LBD, giving the NMDA receptor a more compact appearance compared to AMPA and kainate receptors and creating a protein-protein interface at which modulators can bind [159]. Thus, the crystal structures of the intact NMDA receptor reveal multiple unique intra- and interdomain contacts that can provide explanations to allosteric interactions between subunits and allosteric modulation by small-molecule ligands.

Although the crystal structures of intact NMDA receptors provided major advances in our understanding of the structure-function relationship, they are limited by capturing only one low energy conformational state among many in the NMDA receptor activation cycle. In the crystal structures, agonists were bound to GluN1 and GluN2B subunits and GluN2B-selective negative allosteric modulators (NAMs) were bound at interface between GluN1 and GluN2B ATDs [66, 67]. The structures therefore represented the agonist-bound inhibited state of the receptor with the ion channel in the closed conformation. However, recent breakthroughs in the cryo-EM methodology have enabled determination of structures at resolutions sufficient to assign the relative positioning of domains, thereby affording insights into agonist-bound active and inactive states as well as NMDA receptors in states inhibited by competitive antagonists or GluN2B-selective NAMs [160, 161]. These cryo-EM structures provide the first dynamic pictures of conformational changes in intact NMDA receptors and provide insight into the structural mechanism of ion channel gating (i.e., receptor activation) and allosteric modulation.

2.4 Channel Gating in NMDA Receptors

The portion of the receptor that controls whether the ion channel pore is open or closed with respect to ion permeation is often referred to as the activation gate. All three transmembrane helices (M1, M3, and M4) and the membrane-reentrant pore forming loop (M2) are involved in the process that transitions the NMDA receptor pore to a permeant configuration, a process often referred to as gating [189–196]. Among these regions, the transmembrane helix M3 forms a helical bundle crossing that occludes the pore, and thus the M3 helices must move in order to allow permeation of ions through the channel pore [197–199] (Fig. 7). M3 contains a highly conserved nine amino acids motif (SYTANLAAF), and structural and functional studies have demonstrated that the activation gate is located within this motif [197]. Dilation of the helix bundle formed by M3 from each of the four NMDA receptor subunits is presumably the conformational change that opens the ion channel and allows ion permeation [197, 200, 201]. However, the structural mechanisms that control opening and closing of the NMDA receptor ion channel are not fully understood. Agonist binding induces closure of the bi-lobed NMDA receptor LBD, and this LBD closure initiates a sequence of conformational changes that result in multiple short-lived, intermediate conformations during the transition from the closed to the open state of the ion channel [60, 202–206], akin to a wave of conformational changes connecting agonist binding to ion channel gating. However, there is as yet poor understanding of which elements and conformations represent rate limiting steps *en route* to gating. Moreover, the lifetimes of some of these intermediate conformations are too brief (i.e., they are unstable) to be observed in crystal structures or functional experiments,

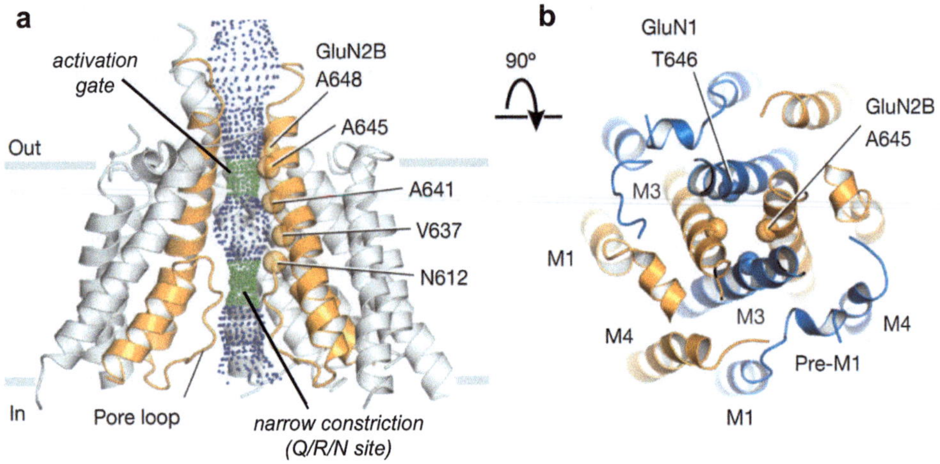

Fig. 7 Structural determinants in the NMDA receptor ion channel pore. (**a**) View parallel to the membrane of the TMDs in the GluN1/2B NMDA receptor structure (PDB ID 4TLM [66];). The solvent accessible surface is carved along the pore axis using the computer program HOLE and shows the M3 bundle crossing near the extracellular side of the membrane, which presumably forms the activation gate, and the narrow constriction in the pore (Q/R/N site). *Green dots* indicate a pore radius of 1.15–2.3 Å and *blue dots* indicate a pore radius >2.3 Å. (**b**) View of the TMDs from the extracellular side of the membrane along the pore axis. GluN1 and GluN2B subunits are blue and orange, respectively. The α-carbon of residues T646 and A645, which appear to define the activation gate, is highlighted as *spheres*. Adapted with permission from Lee et al. [66]

which has confounded attempts to link the sequence of protein conformational changes to kinetically distinct functional pre-gating steps. However, the field is poised for major advances that should occur as new, more detailed structural information emerges and efforts to conceptualize functional models take stock of structural principles. Nonetheless, the presence of intermediate states can be detected using Hidden Markov modeling of single-channel recordings, and the lifetimes of these states differ significantly among NMDA receptor subtypes in a GluN2 subunit-dependent manner [60, 83, 206–208].

Agonist binding steps and the sequence of protein conformational changes that lead to gating can be described as reaction schemes representing agonist binding as well as the transition between different conformational states of the receptor. The first widely accepted kinetic model for the NMDA receptor gating cycle was proposed by Lester and Jahr [61]. This model was designed solely to account for the macroscopic current response waveform, and consisted of two independent, but identical glutamate binding sites, one open state, one closed state, and one desensitized state. This simple formulation adequately described key features of the macroscopic time course for NMDA responses, but was not designed to capture the complexity observed in single-channel recordings. The utility of the model was further limited given the lack of glycine binding steps, which are required for receptor activation. Benveniste et al. [209] developed models that took into account glutamate and glycine binding steps, as well as allosteric interaction between the glutamate and glycine binding domains. These models captured additional features of NMDA receptor pharmacology and response time course, including an apparent glycine-dependent desensitization (see below).

Newer and more complex models, which incorporate both glutamate and glycine binding steps, have been proposed that provide a better description of single-channel data by the incorporation of multiple steps between binding and gating [60, 202–205]. In some studies, single-channel and macroscopic responses to full and partial agonists suggest that agonist binding to either GluN1 or GluN2 controls different steps in the receptor gating scheme [60, 202–205]. These models can also account for some of the actions of allosteric modulators by explicitly representing the modulator bound and unbound receptors as independent states [210–212]. Additional models that exclusively enable modulators to bind to the open state have also been described for channel blockers and other use-dependent modulators [213–218].

The modular nature of the glutamate receptor structure, coupled with the established ability of AMPA receptors subunits to operate independently [219–221], raises the possibility that subunit-independent conformational changes may progress within the sequence of steps leading to channel opening. Some studies suggest that subunit-specific structural changes are required in all four subunits for channel opening, and that these conformational changes occur in any order to arrive at an intermediate state, which can then transition to the open state of the ion channel [60, 202, 204, 205, 222]. However, other models can account for single-channel and macroscopic properties by incorporating just a few sequential gating steps in a linear reaction scheme with an implicit order for fast and slow gating steps [203, 206]. Importantly, all kinetic models for NMDA receptor gating, which faithfully represent both single channel data and macroscopic responses, require multiple pre-gating steps as well as multiple open states. Thus, opening of the NMDA receptor ion channel is not directly coupled to agonist-induced closure of the LBD, but rather the receptor proceeds through a sequence of protein conformational changes that connects agonist binding to ion channel gating.

2.5 Structural Determinants of Ion Permeation and Channel Block

In the open conformation, the NMDA receptor ion channel pore can be divided into the extracellular vestibule and the intracellular vestibule, separated by a narrow constriction. The ion permeation pathway is formed by pore-lining residues that determine ion selectivity and channel conductance. The narrow restriction, also referred to as the selectivity filter, resides at the apex of the membrane re-entrant loop M2 (i.e., the Q/R/N site), approximately halfway across the transmembrane electric field, and is a key determinant of single-channel conductance, Ca^{2+}-permeability, and channel block by Mg^{2+} and organic cations (reviewed in [1, 223]) (Fig. 7). In both the GluN1 and GluN2 subunits, the residue at the position of the Q/R/N site is an asparagine (N), whereas this residue is glycine (G) in GluN3 subunits. Interestingly, the contribution of residues at the apex of M2 to ion permeation is asymmetric between the GluN1 and GluN2 subunits [224–227]. The narrow constriction is mainly formed by the Q/R/N site asparagine in GluN1, whereas in GluN2, it is formed by the asparagine residue adjacent to the Q/R/N site (i.e., Q/R/N + 1 site). Thus, the narrow constriction is formed by non-homologous residues in GluN1 and GluN2 subunits. For example, mutations at the Q/R/N site in GluN2 dramatically reduce Mg^{2+}-block and have weak effects on Ca^{2+}-permeability, but the same mutations at the Q/R/N site in GluN1 have the opposite effects [226, 227]. In addition, substitutions of the asparagine residue at the Q/R/N + 1 site in GluN2 strongly reduce Mg^{2+}-block [227]. Thus, functional data suggest a structural asymmetry, in which the apexes of M2 in GluN1 and GluN2 are slightly staggered [224].

In terms of physiologically relevant ions, the NMDA receptor ion channel is permeable to Ca^{2+}, Na^{+}, and K^{+} ions. GluN1/2 receptors have similar permeability to K^{+} and Na^{+} ions ($P_K/P_{Na} = 1.14$), but are ~2–5 times more permeable to Ca^{2+} relative to monovalent ions ($P_{Ca}/P_X = 1.8$–4.5), depending on the GluN2 subunit [228–232]. Interestingly, despite being highly permeable to Ca^{2+}, NMDA receptors also exhibit voltage-dependent block by external Ca^{2+}, which is readily observed in single-channel recordings as a reduction in channel conductance [207, 233–235]. The concurrent high Ca^{2+}-permeability and Ca^{2+}-block of NMDA receptors are not necessarily incompatible properties, but could be expected if multiple Ca^{2+} binding sites exist within the ion channel pore [228, 233]. One Ca^{2+} binding site is presumably located at the Q/R/N site in the pore, and a cluster of charged GluN1 residues, the DRPEER motif, have been suggested to form another, more external Ca^{2+} binding site [67, 236]. The external Ca^{2+} binding site is located C-terminal to the transmembrane helix M3 in GluN1 at the external entrance to the ion channel. Removal of the net negative charge in DRPEER using mutagenesis (i.e., ARPAAR) reduces the fractional Ca^{2+} currents in NMDA receptors, consistent with an important role of this motif in mediating high Ca^{2+}-permeability [236].

It has been suggested that diheteromeric GluN1/3 receptors form a unique narrow constriction in the extracellular vestibule of the ion channel pore [237]. This narrow constriction, which is presumably not found in GluN1/2 receptors, is proposed to be a main structural determinant for the dramatically reduced Ca^{2+}-permeability and minimal Mg^{2+}-block of GluN1/3 receptors [237]. Co-expression of GluN3 subunits with GluN1 and GluN2 subunits also produces NMDA receptors with reduced single-channel conductance, decreased Ca^{2+}-permeability, and diminished Mg^{2+}-block (reviewed in [68–72]). However, it is unknown whether the GluN3-specific narrow constriction is formed in the extracellular vestibule of these NMDA receptors, which are presumably triheteromeric GluN1/2/3. Furthermore, the extent and mechanisms by which GluN3 subunits influence permeation properties of triheteromeric GluN1/2/3 receptors have not been quantitatively evaluated and remain poorly understood.

NMDA receptor ion channels are blocked by divalent cations Zn^{2+} and Mg^{2+} in a membrane potential-dependent manner (i.e., voltage-dependent) (Fig. 1d). GluN1 and GluN2A mutations in the re-entrant pore loop M2 that reduce channel block by extracellular Mg^{2+} have similar effects on Zn^{2+}-block, suggesting shared molecular determinants [100, 238]. While Mg^{2+}-block of NMDA receptors is centrally implicated in neuronal function, the channel block by Zn^{2+} is low affinity, rapidly reversing, and has far less physiological implications [239, 240]. GluN1/2A and GluN1/2B are more strongly blocked by external Mg^{2+} compared to GluN1/2C and GluN1/2D [63, 98, 99, 241, 242]. At a holding potential of −100 mV, the IC_{50} values for block by Mg^{2+} are 2.4 μM, 2.1 μM, 14.2 μM, and 10.2 μM for GluN1/2A, GluN1/2B, GluN1/2C, and GluN1/2D, respectively [99]. The GluN2 subunit-specific effects on Mg^{2+}-block are likely influenced by multiple structural elements, but a main determinant is a single residue, which is a serine in GluN2A/B and a leucine in GluN2C/D (i.e., the S/L-site) [242]. The S/L-site does not face the ion channel pore, but is located on the internal side of the M3 transmembrane helix, and mutagenesis data suggest that this residue interacts with tryptophan residues in the GluN1 membrane re-entrant loop M2 [242]. In addition to channel block by Mg^{2+}, the subunit-subunit interaction between GluN1 and the GluN2 S/L site is also a main determinant of GluN2 subunit-specific variation in Ca^{2+}-permeability and channel conductance [242]. The structural mechanism by which the GluN2 subunits control block by external Mg^{2+} is unknown, but it is possible that the GluN2 S/L site and other structural elements influence the binding sites for permeant ions in the channel pore, since these binding sites are different between GluN2 subunits and have been shown to profoundly modulate Mg^{2+}-block [243–247].

Numerous organic cations with diverse chemical structures are capable of binding and blocking the NMDA receptor ion channel pore in a voltage-dependent manner [218, 248, 249]. Most, if not all, of these compounds are positively charged at physiological pH, and almost exclusively block activated NMDA receptors with open channels. This mechanism of channel block has been termed "use-dependent" or "uncompetitive." The open channel blockers are further classified into three categories based on their interaction with the channel: (1) "Sequential" or "foot-in-the-door" blockers, such as aminoacridine derivatives, bind to the channel only when it is open and prevent channel closure [250–253]. (2) Partial trapping blockers, such as memantine and amantadine, impede channel closure without completely preventing it [216, 217, 254–257]. (3) Trapping blockers, such as MK-801, phencyclidine (PCP), and ketamine, are trapped inside the pore as the channel returns to the closed state and agonists unbind [258]. Some channel blockers can also interact with the gate to facilitate channel closure [217, 254].

Open channel blockers are generally considered nonselective among NMDA receptor subtypes [215]. However, some channel blockers, at least ketamine and memantine, may display some selectivity under physiological conditions, since five to tenfold selectivity for GluN2C/D-containing receptors over GluN2A/B-containing receptors has been reported in the presence of 1 mM extracellular Mg^{2+} [259]. This observation may be clinically significant, since NMDA receptor channel blockers have been shown to have neuroprotective effects in animal models of CNS disorders that involve excessive stimulation of NMDA receptors, such as traumatic brain injury, epilepsy, and stroke. Unfortunately, human clinical trials have been disappointing due to patient heterogeneity, dose-limiting side effects, and a narrow temporal window for intervention, which may have confounded interpretation. High-affinity NMDA receptor channel blockers, such as phencyclidine (PCP) and ketamine, are dissociative anesthetics, but their clinical use is limited by strong psychomimetic side effects (see below). By contrast, low-affinity channel blockers, which show fast blocking/unblocking kinetics [260], appear to have a greater therapeutic index with respect to psychomimetic effects, which may be due to less channel block under conditions of normal synaptic transmission [261]. One such low-affinity blocker, memantine, has been approved for clinical use in the treatment of moderate to severe Alzheimer's disease. However, the mechanism by which NMDA receptor channel block by memantine may contribute to a symptomatic benefit for advanced Alzheimer's disease patients is not well understood.

2.6 Modulation by the Amino-Terminal Domain

Similar to the LBD, the ATD also adopts a bilobed kidney-shaped structure with upper and lower lobes termed R1 and R2, respectively [262, 263]. Crystal and cryo-EM structures of intact iGluRs revealed a unique dimer-of-dimer arrangement of the NMDA receptor ATDs compared to those in AMPA and kainate receptors [66, 67, 160, 161, 264, 265]. This arrangement, which is also present in crystal structures of heterodimers formed by soluble GluN1 and GluN2B ATDs, is characterized by a subunit interface formed by extensive contacts between the upper R2 lobes from GluN1 and GluN2, whereas the lower R1 lobes, which connect to the LBDs, are almost completely separated. The ATDs are resting immediately above the LBDs and strong interactions are formed between the LBD and ATD layers. By contrast, AMPA and kainate receptor ATDs associate through interactions between both the upper R1 and lower R2 lobes in a back-to-back fashion and there is very little contact between the LBD and ATD layers. Numerous studies have revealed important roles of the NMDA receptor ATD as a modulatory domain that affects function and harbors modulatory binding sites for ions and small-molecule ligands (reviewed in [55, 56, 188, 266]). Modulatory roles or ligand binding sites have not been identified for AMPA and kainate receptor ATDs, even though molecular dynamics simulations predict they should be capable of similar conformational changes as NMDA receptor ATDs [267, 268]. Consistent with these differences, mutant subunits with the deletion of the ATD have dramatic impact on the functional properties of NMDA receptors [94], whereas little to no changes are observed in AMPA and kainate receptors [269].

Many of the GluN2-specific differences between NMDA receptor subtypes are in large part due to variation in the weakly conserved GluN2 ATDs [94, 270]. Studies with NMDA receptors containing chimeric GluN2 subunits have revealed that swapping of the ATD between GluN2A and GluN2D, which have widely different properties, shifts the open probability, deactivation time course, agonist potency in the direction of the subunit contributing the ATD [94, 270]. Little is known about how the ATD controls NMDA receptor function, but the mechanism presumably involves a combination of intra- and inter-subunit allosteric interactions between the ATDs and LBDs that can affect the dynamic behavior and stability of the GluN1/GluN2 LBD heterodimer [271, 272]. Functional and structural studies suggest that the ATDs adopt distinct conformations, depending on the GluN2 subunit, which may underlie some GluN2-specific functional and pharmacological NMDA receptor properties [151, 273].

Extracellular Zn^{2+} is an endogenous modulator that inhibits NMDA receptors in a voltage-independent manner through a binding site in the GluN2A and GluN2B ATDs [80, 262, 274–279]. The affinity of Zn^{2+} to the GluN2A ATD is in the low nanomolar range, whereas the affinity to the GluN2B ATD is in the low micromolar range. Crystal structures and functional data have identified the binding site for Zn^{2+}, which is located at the mouth of the cleft formed by the two lobes R1 and R2 [262]. Several experimental observations support a mechanism of Zn^{2+}-modulation that involves opening and closing motions of the angle between the two lobes R1 and R2 as well as twisting motions around the hinge region of the ATD clamshell [262, 272, 279]. Binding of Zn^{2+} stabilizes a conformation of the GluN2 ATD, which presumably is accompanied by structural changes at the GluN1/2 LBD subunit interface [272].

Crystal structures of both isolated ATDs and intact NMDA receptors established that GluN2B-selective NAMs, such as ifenprodil and Ro 25-6981, bind the subunit interface between GluN1 and GluN2B ATDs [66, 67, 263, 280]. Interestingly, only one residue in the ifenprodil binding pocket is different between GluN2A and GluN2B, but sensitivity to ifenprodil is not introduced by converting this or other residues in GluN2A to that in GluN2B [263, 281]. This observation further supports that the ATD arrangement in GluN2A- and GluN2B-containing receptors is likely fundamentally different and highlights that the mechanism of subunit-selectivity for ifenprodil and its analogs remains unresolved. Recent cryo-EM structures of intact NMDA receptors, supported by functional studies and computational analyses, suggest that the mechanism of ifenprodil inhibition involves closure of the GluN2B ATD clamshell and changes in the arrangement of the GluN1/2B ATD heterodimers [160, 281] (see below).

Polyamines, such as spermine and spermidine, enhance NMDA receptor function in a GluN2B-selective manner through a binding site, suggested to be located in the vicinity of clusters of negatively charged residues in the lower R2 lobes of GluN1 and GluN2B ATDs [282]. Although the precise location of this binding site for positive allosteric modulation remains to be identified, it has been shown using FRET that spermine binding opens the GluN2B ATD clamshell [283]. Furthermore, a model has been proposed where the positively charged spermine shields the negatively charged residues in GluN1 and GluN2B ATDs, thereby potentially eliminating electrostatic repulsion between the two lower R2 lobes [282]. Consistent with this model, other cations can also potentiate GluN2B-containing NMDA receptors in a manner similar to spermine; for example, millimolar concentrations of extracellular Mg^{2+} enhance GluN1/2B responses under conditions with no channel block [284].

Functional and structural investigations appear to converge on a structural model for NMDA receptor modulation by Zn^{2+}, ifenprodil, and spermine, in which modulator binding regulates receptor function through GluN2 ATD clamshell opening and closing motions and rearrangement of the ATD layer. It is not fully understood how these conformational changes affect other structural elements of the receptor, but several studies propose that downstream changes occur at the subunit interface of GluN1/2 LBDs. Interestingly, the activity of all three allosteric modulators (Zn^{2+}, ifenprodil, and spermine) is reduced for NMDA receptors containing GluN1 with exon 5 (e.g., the GluN1-1b splice variant) [81, 85]. Recent structures of intact NMDA receptors show that the 21 amino acids, which are encoded by exon 5, are located just above the GluN1/2 LBD heterodimer interface between the ATD and LBD layers, well positioned to influence allosteric coupling between GluN2 ATD clamshell motions and GluN1/2 LBDs

[66, 67]. In addition, mutational analyses identified GluN2C residues from both the ATD and LBD that influenced the activity of PYD-106, which is a recently developed GluN2C-selective positive allosteric modulator (PAM), and molecular modeling proposed a binding site located in a pocket residing at the intra-subunit ATD/LBD interface of GluN2C [159]. Thus, the ATD is the major structural determinant of GluN2-specific variation in functional and pharmacological properties of NMDA receptors. The mechanism of allosteric modulation by the NMDA receptor ATD remains an important focus in structure-function studies, and drug discovery efforts are poised to identify novel ATD ligands with therapeutic potential. In particular, it is unknown how structure and ATD arrangement differs among the various NMDA receptor subtypes.

2.7 Control of Assembly by the Amino-Terminal Domain

The mechanism and progression of subunit assembly of two GluN1 and two GluN2 subunits in an alternating 1-2-1-2 arrangement around the central ion channel pore is not well understood. Three main models of the steps required for NMDA receptor assembly have been proposed: (1) It has been suggested that GluN1-GluN1 and GluN2-GluN2 homodimers initially form and then associate to form the tetrameric receptor [285–288]. (2) Alternatively, two initial GluN1-GluN2 heterodimers are formed that subsequently associate to generate the tetrameric arrangement [289]. (3) Lastly, it has been suggested that a GluN1-GluN1 homodimer is initially formed to which GluN2 subunits are sequentially added to form the tetrameric NMDA receptor [290, 291]. While there is some supporting experimental data for each model, this data is as yet insufficient to make a clear distinction between these models. Regardless of sequence, it appears that the NMDA receptor ATD is the main determinant of the initial subunit dimer formation [285, 287, 291]. This feature of the ATD in NMDA receptors appears to be shared in AMPA and kainate receptors, where the role of the ATD in the subunit assembly has been extensively studied [292, 293]. Interestingly, the NMDA receptor ATD may also influence receptor trafficking, since the GluN2A ATD has been shown to contain a retention signal that prevents exit from the endoplasmic reticulum unless it is masked by assembly with the GluN1 ATD [294].

3 Mechanisms of NMDA Receptor Regulation

Many functional and membrane trafficking properties of NMDA receptors are regulated by extracellular ions, phosphorylation, and intracellular binding proteins. Here, we will describe regulation of NMDA receptor function by extracellular ions and molecules, and highlight key phosphorylation sites and their implications for protein-protein interactions important for neuronal functions.

3.1 Desensitization of NMDA Receptors

The definition of desensitization is a decrease in the receptor response in the continued presence of a stimulus. NMDA receptors exhibit several different types of desensitization with distinct mechanisms, including glycine-dependent desensitization, Zn^{2+}-dependent desensitization, Ca^{2+}-dependent desensitization, and glycine/Ca^{2+}/Zn^{2+}-independent desensitization.

Glycine-dependent NMDA receptor desensitization can be observed in the presence of sub-saturating glycine concentrations, and is abolished in a saturating concentration of extracellular glycine [295]. This type of desensitization occurs due to a negative allosteric interaction between the GluN1 and GluN2 subunits such that the binding of glutamate decreases the affinity for glycine [209, 296]. Thus, when glutamate binds GluN2, the affinity for the glycine binding site in GluN1 becomes lower, and in the absence of high concentrations of glycine, the current diminishes and relaxes to a new equilibrium as glycine unbinds from the receptor. The time course for the desensitization therefore reflects glycine unbinding, which is within the range of the synaptic NMDA receptor time course, suggesting glycine-dependent desensitization could impact synaptic signaling. Recent crystal and cryo-EM structures of intact NMDA receptors offer plausible structural models for the negative allosteric coupling between the glutamate and glycine binding sites [66, 67, 160, 161], but the structural mechanism of glycine-dependent desensitization is still not fully understood.

Extracellular Zn^{2+} mediates a rapid component of desensitization that occurs by a mechanism similar to glycine-dependent desensitization [297]. It has been proposed that a positive allosteric interaction exists between the glutamate binding site in the GluN2 LBD and the Zn^{2+} binding site in the GluN2A ATD, which enables binding of glutamate to enhance Zn^{2+} binding [298, 299]. Thus, glutamate binding will, in the presence of subsaturating concentrations of Zn^{2+}, cause a relaxation of the receptor response to a new equilibrium as more Zn^{2+} ions bind and inhibit the receptor in a concentration-dependent fashion. The time course of Zn^{2+}-dependent desensitization therefore reflects the time course for Zn^{2+} binding following a glutamate-dependent shift into a Zn^{2+} binding site with higher affinity.

NMDA receptors also undergo Ca^{2+}-dependent inhibition, which is often referred to as Ca^{2+}-dependent desensitization or inactivation, since this type of desensitization requires intracellular Ca^{2+} and develops slowly over seconds [300–303]. The magnitude of Ca^{2+}-dependent desensitization varies among GluN2 subunits, and is more prominent for GluN2A-containing than for GluN2D-containing receptors and appears to be absent for GluN2B- and GluN2C-containing NMDA receptors [304, 305]. It has been hypothesized that a local increase in the intracellular Ca^{2+} concentration occurs in the immediate vicinity of the NMDA receptor, which results in inhibition by stimulating uncoupling of the receptor from filamentous actin in a manner sensitive to second messenger systems [306]. Furthermore, calmodulin binding to the GluN1 CTD has been suggested to play an important role in this form of desensitization. Thus, Ca^{2+}-dependent desensitization is abolished in NMDA receptors containing GluN1 splice variants in which calmodulin binding sites are absent [307, 308], and mutations within calmodulin binding sites in the GluN1 CTD similarly disrupt Ca^{2+}-dependent desensitization [309, 310].

Most ligand-gated channels undergo a form of desensitization that reflects a conformational change to a relatively stable and sometimes long-lived agonist-bound receptor state with a closed ion channel. NMDA receptors can also desensitize in the continued presence of agonist by a mechanism that is independent of glycine, Zn^{2+}, and Ca^{2+} (i.e., the types of desensitization discussed above). This desensitization develops with time, is sensitive to intracellular dialysis, and is thus more prominent in excised outside-out membrane patches compared to whole-cell patches [311, 312]. However, desensitization can also be influenced by mutations in the conserved SYTANLAAF motif, the preM1 region, and other positions deep within the ion channel pore, the LBD, and the TMD-LBD interface [194, 313, 314], suggesting that this desensitization reflects a conformational change in the agonist-bound receptor.

3.2 Regulation of NMDA Receptor Function by Protons

Extracellular protons potently (IC_{50} = ~50 nM) and completely inhibit NMDA receptor function [315–318]. Thus, neuronal NMDA receptors are tonically inhibited by protons at physiological pH 7.4, which corresponds to approximately the proton IC_{50}. NMDA receptors can therefore respond to small changes in extracellular pH under physiological conditions. Moreover, extracellular pH is dynamic and changes with neuronal activity, given that synaptic vesicles are acidic and various transporters can generate proton gradients [319]. Furthermore, pathological conditions, such as seizure or ischemia, reduce extracellular pH (i.e., increase proton concentration) to levels that are sufficient to strongly inhibit NMDA receptor function [319].

As with many other NMDA receptor properties, the inhibition by extracellular protons depends on the GluN2 subunit [81]. GluN2A-, GluN2B-, and GluN2D-containing NMDA receptors are inhibited with proton IC_{50} values near physiological pH (7.2–7.4), whereas GluN2C-containing receptors are much less sensitive to protons (IC_{50} value at pH 6.2) [81, 320]. In addition, proton inhibition is reduced for NMDA receptors with the GluN1-1b isoform, which has an additional 21 amino acids inserted in the ATD [81]. Proton inhibition is voltage-independent and is also independent of actions at the agonist binding site. The location of the structural determinant for proton inhibition (i.e., the proton sensor) is unknown and it is possible that multiple sites within the

NMDA receptor work in concert to mediate the actions of protons. However, residues within the ion channel gate, near the linkers that couple the TMD to the LBD, and in the GluN1-GluN2 LBD dimer interface have been shown by mutagenesis to influence pH sensitivity [272, 320], suggesting that NMDA receptor gating elements are tightly coupled to the proton sensor. This idea is supported by evidence that channel blockers are sensitive to the protonation state of the receptor while entering the pore [215].

Several studies suggest that actions of ATD modulators may reflect a subtle change in the pKa of the proton sensor that either enhances or reduces tonic proton inhibition at physiological pH (see below). In this regard, both extracellular Zn^{2+} and ifenprodil appear to enhance proton sensitivity, which will increase tonic inhibition at physiological pH, whereas binding of extracellular polyamines, such as spermine, reduces proton sensitivity, which results in potentiation. For example, spermine potentiation of GluN1/2B strongly correlates with the degree of proton inhibition and is most robust at pH values that produce strong tonic inhibition (i.e., pH < 8). This is consistent with a mechanism in which polyamines enhance receptor function by relieving proton inhibition [81, 321, 322]. Similar functional evidence supports a mechanism for inhibition by extracellular Zn^{2+} and ifenprodil in which receptor function is reduced by enhancing proton inhibition [80, 85, 86, 211, 274, 323].

3.3 Regulation of NMDA Receptor Function by Extracellular Zn^{2+}

GluN2A-containing NMDA receptors are highly sensitive to extracellular Zn^{2+}, and numerous studies have reported variable IC_{50} values in the low nanomolar range (e.g., [80, 100, 277, 297]). A key provision in these studies was the need for a buffer system to accurately control Zn^{2+} concentration and unambiguously determine the IC_{50} value for Zn^{2+} inhibition, since hundreds of nanomolar Zn^{2+} contaminate physiological saline solutions under most experimental conditions [80, 100]. Thus, in order to remove effects of extracellular Zn^{2+} in functional experiments, many studies include Zn^{2+}-chelators, such as tricine or EDTA, in the extracellular recording solution. The high affinity of these chelators for Zn^{2+} means that ten of micromolar of chelator will bind virtually all of the nanomolar contaminating Zn^{2+} ions, but minimally alter millimolar concentrations of Ca^{2+} or Mg^{2+} (e.g., *see* [324]). The concentration-inhibition relationship for Zn^{2+} at GluN1/2A receptors is biphasic, since Zn^{2+} binding to the high affinity site in the ATD causes incomplete inhibition, whereas higher micromolar concentrations of Zn^{2+} result in voltage-dependent channel block [277]. The incomplete inhibition by high affinity Zn^{2+} binding is related to enhancement of proton sensitivity, since Zn^{2+} binding causes a leftward shift of the proton inhibition curve such that inhibition is more complete at acidic pH compared to at alkaline pH [80, 211, 274, 276]. For example, maximal inhibition of

GluN1/2A by extracellular Zn^{2+} is ~62% at physiological pH 7.3 compared to ~76% at pH 6.8 [276]. Interestingly, high affinity Zn^{2+} inhibition is maintained in triheteromeric GluN1/2A/2B receptors, albeit with less maximal inhibition and a somewhat different relationship to the extracellular pH [150, 151, 153].

3.4 NMDA Receptor Phosphorylation and Membrane Trafficking

The intracellular CTDs of NMDA receptor subunits contain numerous sites for posttranslational modifications (e.g., phosphorylation, nitrosylation, and palmitoylation) and for protein-protein interactions, which have important implications for receptor localization, trafficking, and signaling (reviewed in [1, 325, 326]). The intracellular CTDs display very little conservation among subunits, and subcellular localization and trafficking of each subunit therefore appear to be uniquely regulated. Furthermore, the intracellular CTD of GluN1 is modified by alternative RNA splicing, which removes or inserts regulatory sites with important effects on receptor trafficking.

Experimental evidence suggests that the precise subcellular localization of NMDA receptor subtypes is determined by protein-protein interactions between the extreme C-terminus of GluN2 subunits and PDZ domain-containing proteins, including the MAGUK proteins PSD-93, PSD-95, SAP97, and SAP102. Members in the MAGUK protein family have widely different subcellular localization and exhibit GluN2-specific variation in their preferential association with NMDA receptor subtypes. For example, the scaffolding protein PSD-95 is primarily expressed at the postsynaptic density (PSD), whereas SAP102 is more evenly distributed between the synaptic and extrasynaptic sites. Furthermore, PSD-95 and SAP102 have been proposed to preferentially bind GluN2A and GluN2B subunits, respectively [327]. The differential interaction of GluN2A and GluN2B subunits with MAGUKs is not firmly established and has been questioned in several studies (e.g., [131]), but has been suggested to underlie differences in the subcellular localization of these subunits. For example, GluN2B-containing receptors appear to move more freely in and out of synaptic sites compared to GluN2A-containing receptors [328]. It has therefore been suggested that GluN2B can be found at both the extrasynaptic and synaptic sites, whereas GluN2A is enriched at synaptic sites [328–330]. In addition to MAGUK proteins, numerous other binding partners have been implicated in the subcellular localization and membrane trafficking properties of NMDA receptors (reviewed in [1]).

The cytoplasmic CTDs of NMDA receptor subunits are differentially regulated by posttranslational modifications including phosphorylation, palmitoylation, and nitrosylation (reviewed in [1]). These modifications can affect the ability to bind intracellular proteins involved in membrane trafficking, and can therefore mediate changes in subcellular localization and surface expression. An example of the type of regulation, which has important implications on synaptic plasticity, is CaMKII phosphorylation of GluN2B subunits on Ser1303, which is located in the CaMKII binding site [331, 332]. Transient NMDA receptor-mediated influx of Ca^{2+} induces autophosphorylation of CaMKII, which enhances its enzymatic activity and results in persistent activation of CaMKII long after cytoplasmic Ca^{2+} levels return to baseline [332, 333]. Once activated, CaMKII rapidly and reversibly undergoes a translocation to the spine, where it binds the CTD of the GluN2B subunit [334–337]. Multiple lines of evidence show that disruption of autophosphorylation and activation of CaMKII, as well as its binding to the GluN2B subunit, significantly impairs NMDA receptor-dependent LTP and affects memory in mice, consistent with a key role of CaMKII as a key mediator of some types of synaptic plasticity (reviewed in [338]). The mechanism by which the interaction between the NMDA receptor and CaMKII contributes to synaptic plasticity is largely unresolved and therefore continues to be a primary focus in studies that aim at advancing our understanding of NMDA receptor-dependent synaptic plasticity. Numerous other kinases (e.g., PKA, PKB, and PKC), many protein tyrosine kinases (e.g., Fyn and Src), and phosphatases (e.g., STEP) have been implicated in the regulation of NMDA receptors, and the consequences of modification by these proteins on neuronal function continue to be extensively studied (reviewed in [339]). In addition to CaMKII, other calcium-sensing proteins can interact with NMDA receptor subtypes to mediate downstream signaling and regulate synaptic plasticity. RAS-GRF1 and RAS-GRF2 are two such calcium sensors that selectively bind the GluN2B CTD and thereby initiate ERK- and CREB-mediated signaling pathways in response to NMDA receptor-mediated Ca^{2+}-influx [340–342]. Thus, the implications of NMDA receptor phosphorylation and membrane trafficking on neuronal function are incredibly complex and highly dependent on NMDA receptor subunit composition.

There is a growing body of evidence for metabotropic signaling through the NMDA receptor (i.e., not mediated by ion flux), resulting from direct changes in the interaction of the receptor with other signaling complexes [343, 344]. Conformational changes induced by agonist binding are required to gate the NMDA receptor ion channel, but emerging evidence suggests that the conformational changes induced by these various ligands may also have effects that are independent of ion channel flux. For example, glycine binding, but not glutamate binding, have been shown to prime NMDA receptors for internalization [345]. In this case, glycine binding to GluN1 promotes association of the NMDA receptor with clathrin-mediated endocytic machinery that is independent of glutamate binding and receptor activation. Receptor endocytosis is then triggered upon binding of both glutamate and glycine and receptor activation. More recently, evidence has emerged that metabotropic NMDA receptor signaling may play a role in synaptic depression [345]. In this case, the metabotropic effects appear dependent on glutamate binding to GluN2 [346], which results in rearrangement of the associations of protein phosphatase 1 (PP1) and CaMKII with the C-terminal tails of the NMDA receptor to modulate kinase activity [347]. It has been proposed that the Aβ peptide, a putative pathogen in Alzheimer's disease, may cause synaptic depression and dysfunction via this mechanism [343]. At the physiological level, it may be speculated that these subunit specific metabotropic signaling mechanisms interact with ionotropic signaling mechanisms. Thus, metabotropic signaling provides another layer of signal integration by these important NMDA receptor complexes.

4 Pharmacological Manipulation of NMDA Receptor Subtypes

Small-molecule modulators with selectivity for the different NMDA receptor subtypes (i.e., the GluN2 subunits) are powerful pharmacological tools that can be used to dissect the roles of different NMDA receptors in neurophysiology, behavior, development, and diseases. In this regard, past studies aimed at developing glutamate-site agonists, competitive antagonists, and channel blockers have not identified such pharmacological tools with sufficient GluN2 subunit-selectivity; in part due to the fact that these sites are highly conserved among GluN2 subunits. However, extensive pharmacology has been developed around ifenprodil that was shown to be a GluN2B-selective NAM in 1993 [348], and until recently, ifenprodil and analogs were the only available and widely used pharmacological tool compounds with strong GluN2 subunit-selectivity. Since approximately 2010, there has been an acceleration in the discovery of novel NMDA receptor allosteric modulators with GluN2 subunit-selectivity and multiple new binding sites of positive and negative allosteric modulators have been identified [54–56, 349] (see below).

4.1 GluN2A-Selective Allosteric Modulators

NVP-AAM077 (also known as PEAQX) is a competitive antagonist that interacts with the glutamate binding site. Although it was initially reported as having a high degree of selectivity for GluN2A over GluN2B [350], subsequent evaluation of the binding affinity of NVP-AAM077 at GluN1/2A and GluN1/2B receptors found more modest selectivity (K_B values were 15 nM for GluN1/2A and 78 nM for GluN1/2B) [351]. This, and other studies, suggested that the level of selectivity (fivefold) of NVP-AAM077 is insufficient for dissection of synaptic responses mediated by GluN2A- and GluN2B-containing receptors [351, 352]. Many studies evaluating native receptors and excitatory synaptic transmission have been performed using NVP-AAM077; however, the results of this body of work should be carefully interpreted with regards to the experimental design, level of selectivity assumed, and conclusions drawn.

TCN-201 and TCN-213 were the first non-competitive GluN2A-selective inhibitors (or GluN2A-selective NAMs) that were identified [353]. TCN-201 has a binding affinity of 27–70 nM at GluN2A-containing receptors, with >1000-fold selectivity over other GluN2 subunits [155, 354, 355]. Inhibition by TCN-201 is surmounted by glycine binding, which is paradoxical since the subunit selectivity depends on the glutamate binding GluN2 subunit [353–355]. However, the TCN-201 binding site is located in the LBD heterodimer interface between GluN1 and GluN2A subunits, with key interacting residues around 16 Å from the glycine binding site in GluN1 [155, 157, 355]. Quantitative analyses show that the functionally observed interaction between TCN-201 and glycine was best described by an allosteric model of antagonism rather than a direct competition model [354, 355]. Recent analyses of crystal structures of receptor states that are activated and inhibited by the GluN2A-selective NAMs demonstrated a mechanism in which NAM binding to the modulatory site stabilizes the open conformation of the GluN1 LBD, thereby facilitating glycine unbinding and receptor inactivation [155]. Furthermore, these structures revealed that two residues in the interface between GluN1 and GluN2A LBDs play principal roles in the allosteric mechanism of GluN2A-selective NAMs by forming a molecular switch that controls the difference between low- and high-affinity NAM binding; this difference is the primary driving force for allosteric inhibition. MPX-004 and MPX-007 are newer GluN2A-selective NAMs closely related to TCN-201 that have improved potency (79 nM and 27 nM, respectively, determined in 3 μM glycine) compared to TCN-201 (340 nM in 3 μM glycine) [356]. The MPX compounds provide nearly complete block of GluN2A-containing NMDA receptors and have improved solubility compared to TCN-201 [155, 356]. In recent years, the GluN2A-selective NAMs have been used to probe the GluN2B to GluN2A developmental switch, the expression of GluN2A in subcortical and subthalamic nuclei, as well as the role of GluN2A in nicotine reinstatement, cortical spreading depression, and hippocampal plasticity [84, 354, 356–360].

A high-throughput screen performed by Genentech to identify GluN2A-selective positive allosteric modulators (PAMs) identified several structurally related compounds, here referred to as GNE compounds. These compounds are GluN2A-selective PAMs with at least tenfold selectivity over other GluN2 subunits that bind the LBD heterodimer interface between a GluN1 and GluN2A subunits, similar to the GluN2A-selective NAMs [157, 169]. Interestingly, the GNE compounds interact with the same residue (GluN2A V783) in the GluN2A subunit that mediates the selectivity and inhibition by TCN-201, MPX-004, and MPX-007 [155, 157]. This valine is non-conserved across the GluN2 subunits, and introduction of this residue into GluN2B via site-directed mutagenesis is sufficient to confer both inhibition and potentiation to GluN2B-containing NMDA receptors [157, 355]. GluN1/2A LBD heterodimer crystal structures in complex with GNE compounds and the GluN2A-selective NAMs show that the binding modes of both positive and negative allosteric modulators are distinct within this pocket, a finding reinforced by the results of mutagenesis studies [155, 157, 169]. Interestingly, some GluN2A-selective PAMs (i.e., GNE compound analogs) also affect the function of AMPA receptors with similar potencies as for NMDA receptors [157]. The different GNE compounds display variation in the efficacy of GluN1/2A receptor potentiation (up to sixfold potentiation of receptors activated by EC_{30} of glutamate) and potency (EC_{50} values between 0.02 and 60 μM) [169]. Furthermore, the series of modulators show reduced efficacy when receptors are activated by saturating concentration of agonist. For example, GNE-0723 shows approximately fivefold potentiation of an EC_{30} response compared to approximately twofold potentiation of an EC_{100} response, which is presumably due to an increase in agonist potency mediated by the modulator [157, 169]. A complex relationship appears to exist between the GNE compound structure, efficacy, and the degree of prolongation of glutamate deactivation rate, which could reflect increased glutamate affinity and potency [157, 169]. Two GNE compounds (GNE-6901 and GNE-8324) were evaluated on NMDA receptor-mediated responses in hippocampal neurons [157]. These GNE analogs differed in their ability to prolong the deactivation rate of NMDA receptors, and also showed differences in their ability to alter short- and long-term synaptic plasticity. More studies could help fully elucidate the mechanism of action of this interesting series of GluN2A-selective PAMs and demonstrate their usefulness in studies on the physiological roles of GluN2A-containing NMDA receptors. The pharmacology of GluN2A-selective modulators is summarized in Table 1.

4.2 GluN2B-Selective Allosteric Modulators

Ifenprodil and its mechanistically similar analogs have been tremendously useful tool compounds since the discovery in 1993 that they are non-competitive GluN2B-selective inhibitors (i.e., GluN2B-selective NAMs) [348]. The IC_{50} for ifenprodil is in the nanomolar range with 200–400-fold selectivity for the GluN1/2B receptor over GluN1/2A [348]. The observed inhibition of GluN1/2A at high concentrations is caused by low-affinity nonselective channel block [348]. The high-affinity ifenprodil binding site is located in the interface between the GluN1 and GluN2B ATD heterodimer [263, 361]. Ifenprodil inhibition is dependent on agonist concentrations; at saturating glutamate and glycine concentrations,

Table 1
Summary of GluN2A-selective modulators

Compound			Activity at GluN1/2× (in μM)				
			2A	2B	2C	2D	
NVP-AAM077	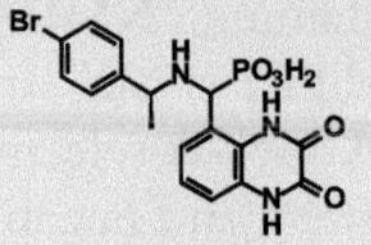	K_B[a]	0.015	0.078	–	–	[351]
TCN-201	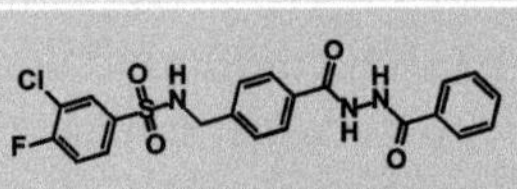	K_B[a,b]	0.045 0.070 0.027	NE	NE	NE	[355] [354] [155]
MPX-004	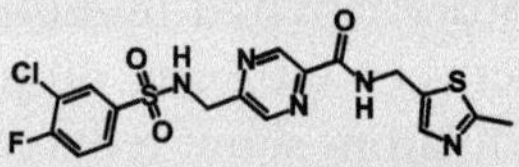	IC_{50}[c,d]	0.079[c] 0.198[e]	NE[c] NE[e]	–[c] NE[e]	NE[c] NE[e]	[356]
MPX-007	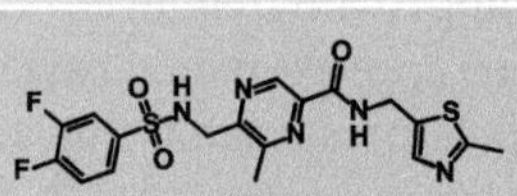	IC_{50}[c,d]	0.027[c] 0.143[e]	NE[c] ND[e]	–[c] NE[e]	NE[c] NE[e]	[356]
GNE-3419	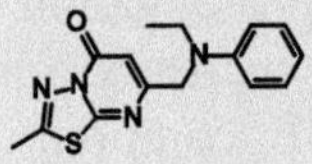	EC_{50}[c]	2.03	NR	NR	NR	[157]
GNE-6901	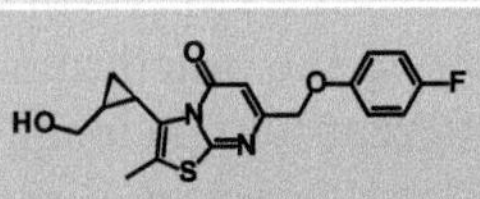	EC_{50}[c]	0.33	NR	NR	NR	[157]
GNE-0723	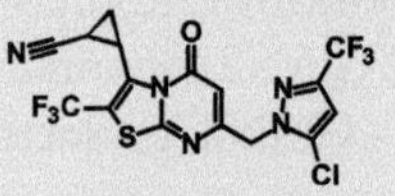	EC_{50}[c]	0.021	ND	7.4	6.2	[169]
GNE-8324	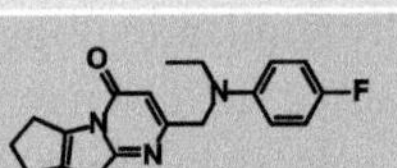	EC_{50}[c]	2.43	NR	NR	NR	[157]

– Denotes not determined, NE denotes no effect at the highest concentrations evaluated, and ND indicates that the compound displayed some activity, but the affinity or potency could not be determined. NR denotes some activity, but that the numerical affinity value was not reported. Unless otherwise stated, the values were determined using two-electrode voltage-clamp experiments with *Xenopus* oocytes

[a]Denotes when Schild analysis was used for affinity determination

[b]Three independent studies are published reporting the K_B of TCN-201 at GluN1/2A

[c]Denotes that potency (i.e., half maximally effective concentration) was determined using a Ca^{2+} imaging assay

[d]Experiments using MPX compounds were performed in 3 μM glycine

maximally effective concentrations of ifenprodil produce incomplete inhibition, with 10–20% residual response [85, 348, 361, 362], whereas the glycine concentration is inversely correlated with the extent of observed inhibitory effect [348]. The actions of ifenprodil at different glutamate concentrations are also complex; ifenprodil causes an increase in glutamate-site agonist affinity, which produces less inhibition with lower agonist concentrations [362]. This positive allosteric interaction between ifenprodil and glutamate binding is similar to that observed for Zn^{2+} acting at the GluN2A ATD and can lead to apparent potentiation at low agonist concentrations [100, 299, 362]. Many newer GluN2B-selective NAMs acting at the ifenprodil site have been synthesized with improved potency and selectivity (e.g., Ro 25-6981 and CP-101,606) [363, 364], and additional mechanistic features such as pH-sensitivity (e.g., *see* [365]). Recently, crystal structures suggest that GluN2B-selective NAMs can be divided into two classes with distinct binding modes at the GluN1-GluN2B ATD heterodimer interface; one class containing ifenprodil, CP-101,606 and Ro 25-6981 and a second class typified by EVT-101 [280]. EVT-101 is an orally active compound with potent inhibition at low nanomolar concentrations [280]; however, a thorough study of EVT-101 properties and mechanism of action has not been published. GluN2B-selective NAMs have been intensely studied by academic research groups and pharmaceutical companies in an effort to identify new series with therapeutic benefits as well as to expand our understanding of the role of GluN2B in normal physiology and disease, a topic thoroughly summarized in a number of excellent reviews [366–368]. GluN2B-selective inhibitors have also been evaluated in clinical trials, with mixed and complex results [368–371]. The pharmacology of GluN2B-selective modulators is summarized in Table 2.

4.3 GluN2C/D-Selective Allosteric Modulators

Spurred by the description of NVP-AAM077 and other glutamate-site competitive antagonists, studies of related compounds were performed to find similar antagonists with variation in the selectivity at NMDA receptor subtypes [372, 373]. This effort leads to the discovery of PPDA, which is similar to the earlier identified competitive antagonist PBPD [374], that had differential selectivity and showed high potency [373]. Several structural modifications were pursued in subsequent studies, yielding compounds that consistently displayed a preference for GluN2C- and GluN2D-containing NMDA receptors over GluN2A- and GluN2B-containing receptors [375]. UBP-141 was observed to have five to sevenfold selectivity for GluN1/2D over GluN1/2A and GluN1/2B, but was less potent than PPDA. Several studies have used UBP-141 to probe the role of GluN2D in certain physiological processes [143, 376–379]. These studies provided important insight into the physiological roles of GluN2D in central neurons, but they also should

Table 2
Summary of GluN2B-selective modulators

			Activity at GluN1/2× (in μM)				
Compound			**2A**	**2B**	**2C**	**2D**	
Ifenprodil	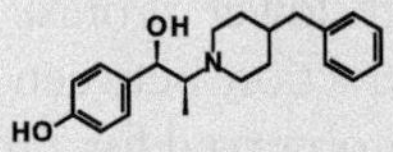	IC_{50}	39.5	0.114	29.1	75.9	[524]
CP-101,606	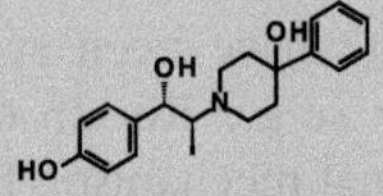	IC_{50}	NE	0.039	NE	NE	[85]
Ro 25-6981	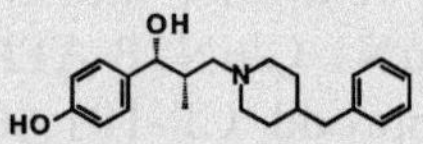	IC_{50}	52	0.009	–	–	[364]
EVT-101	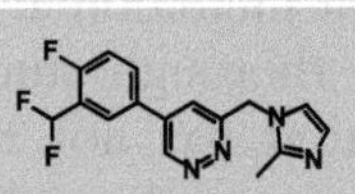	IC_{50}	–	0.012	–	–	[280]

– Denotes not determined and NE denotes no effect at the highest concentrations evaluated. The potency (i.e., half maximally inhibiting concentration) was determined using two-electrode voltage-clamp experiments with *Xenopus* oocytes

be interpreted with the caveat of modest subunit-selectivity, which is apparently inherent to glutamate-site competitive antagonist.

Further expansion and exploration of the chemical space around the compounds related to UBP-141 led to an investigation of related scaffolds, and the subsequent discovery of several mixed-action modulators, including UBP-710 and UBP-551 [380]. UBP-710 shows divergent action at concentrations of 100 μM and higher, resulting in potentiation of GluN1/2A and GluN1/2B, but inhibition of GluN1/2C and GluN1/2D [380]. UBP-551 appears to be uniquely selective for GluN2D-containing NMDA receptors and potentiates current responses with a biphasic concentration-effect relationship, with maximal potentiation of GluN1/2D observed at 30 μM, a concentration at which other NMDAR diheteromeric receptors are inhibited [380]. The mixed-action UBP compounds possess remarkably unique actions, but their utility is hampered by poor physicochemical properties of the parent scaffold and a lack of high affinity actions or high subunit-selectivity [380, 381]. It will be interesting to learn more about mechanism and site of action of this class as molecules with higher potency, selectivity, and improved physicochemical properties are developed.

A series of quinazolin-4-ones (QNZ) are negative allosteric modulators of NMDA receptors that show ~50-fold selectivity for GluN2C- or GluN2D-containing NMDA receptors [382, 383]. The prototypical compound QNZ-46 has an IC_{50} of 7.1 μM and 3.9 μM at GluN1/2C and GluN1/2D, respectively, and has minimal effects on AMPA and kainate receptors. QNZ-46 does not compete with glutamate or glycine binding and inhibition is voltage-independent. Interestingly, the inhibition by QNZ-46 is dependent on glutamate binding, but not glycine binding, and the potency of QNZ-46 is increased when glutamate is bound [383]. Glutamate deactivation is prolonged in the presence of QNZ-46, consistent with a mechanism in which QNZ-46 must unbind before glutamate can unbind [383]. Structural determinants of action appear to reside in the lower lobe of the GluN2D LBD; however, the precise binding site for this series remains to be determined [383].

A series of dihydroquinolone-pyrazoline (DQP) analogues are, like QNZ-46, GluN2C- and GluN2D-selective NAMs [384]. The prototypical analogue in this series, DQP-1105, is ~50-fold selective for GluN2C/D-containing receptors with IC_{50} values of 7.0 μM and 2.7 μM at recombinant GluN1/2C and GluN1/2D, respectively [384]. Inhibition by DQP-1105 is voltage-independent and is not surmounted by increased concentrations of glutamate or glycine, consistent with a non-competitive mechanism of action [384]. Inhibition by DQP-1105 is dependent on glutamate binding [384], a property it shares with the QNZ class of inhibitors. Similarly, the structural determinants of DQP-1105 action resided in the lower lobe of the GluN2D LBD and largely overlapped with those of the QNZ class of inhibitors [384]. The finding that the QNZ and DQP series share similar structural determinants on GluN2C/D-containing receptors raises the possibility their binding sites may overlap and that the binding pocket could be exploited by a wide array of ligands with distinct binding modes. Further exploration of the DQP structure-activity relationship led to the synthesis of chiral compounds with nanomolar activity at GluN2C/D-containing receptors, making the DQP series more potent and selective than the QNZ series [385]. DQP-1105 has been used in several recent studies illustrating roles for GluN2C and GluN2D in normal physiology as well as pathophysiology in various nuclei of the brain [84, 378, 386–388].

A series of tetrahydroisoquinoline PAMs are highly selective for GluN2C/D-containing NMDA receptors [389]. Further exploration of the structure-activity relationship for this class of PAMs resulted in a large family of stereo-selective analogues with strong selectivity for GluN2C/D-containing receptors, some of which have nanomolar EC_{50} values [390, 391]. Separation of CIQ, an early prototype in this class of GluN2C/D-selective PAMs, into its two stereoisomers showed that (+)-CIQ contains all the activity observed for the racemic mixture [390, 391] and has reduced off-target actions [392]. CIQ has similar potency and efficacy at GluN2C- and GluN2D-containing receptors, as do virtually all related analogues studied to date [390, 391]. Importantly, it has been demonstrated that CIQ also potentiaties responses from tri-heteromeric GluN1/2A/2C and GluN1/2B/2D receptors, albeit with some reduction in efficacy [389]. CIQ has no effect on the deactivation time course for GluN1/2D, but prolongs glutamate deactivation for GluN1/2C [389]. Chimeric and mutational studies suggest that the potentiation by CIQ is dependent on residues in the M1 transmembrane helix and a short pre-M1 helix in the GluN2 subunit [190, 389]. However, whether these structural determinants correspond to the CIQ binding site remains to be determined. Racemic CIQ and (+)-CIQ have been used in several studies probing the expression and role of GluN2D in synaptic transmission in various nuclei across the brain and spinal cord [84, 386, 392–396].

One series of GluN2C-selective PAMs has been described (i.e., PYD compounds) [159, 397]. The structure-activity relationship for this series revealed stereo-selective actions and additional analogues with enhanced potency [397]. To date, the PYD series is the only highly selective positive modulator series that discriminates between GluN2C- and GluN2D-containing NMDA receptors. The prototypical analogue, PYD-106, has an EC_{50} of 16 μM at GluN1/2C and maximally potentiates receptor responses to 200% of control [159]. PYD-111, a closely related analogue, is slightly more potent with an EC_{50} of 4 μM [397]. Interestingly, PYD-106 is highly selective for the diheteromeric GluN1/2C receptors, but has no effect on triheteromeric GluN1/2A/2C receptors [159]. PYD-106 has a weak allosteric effect on glutamate potency and modestly prolongs the glutamate deactivation time-course (in the sustained presence of glycine) [159]. Chimeric and mutational studies identified structural determinants of PYD-106 actions at a unique site residing at the interface of the GluN2C ATD and the upper lobe of the GluN2C LBD [159]. Modeling of the GluN1/2C structure on the basis of the GluN1/2B crystal structure [67] revealed that residues that affect PYD-106 actions line a large pocket, suggesting a novel modulatory site on the NMDA receptor [159]. The pharmacology of GluN2C/D-selective modulators is summarized in Table 3.

5 NMDA Receptor Subtypes in the CNS

As described above, the different GluN2 subunits endow the NMDA receptor subtypes with distinct functional properties, unique pharmacology, and markedly different mechanisms of regulation. This feature is a major determinant of the variation observed

Table 3
Summary of GluN2C/D-selective ligands

Compound			Activity at GluN1/2× (in μM)				
			2A	2B	2C	2D	
PBPD	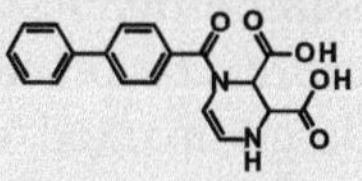	K_i[a]	15.8	5.0	9.0	4.3	[374]
PPDA	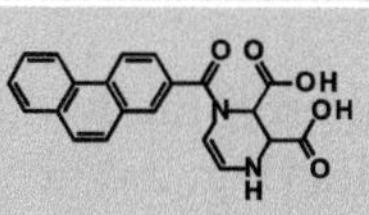	K_i[a]	0.55	0.31	0.096	0.125	[373]
UBP141	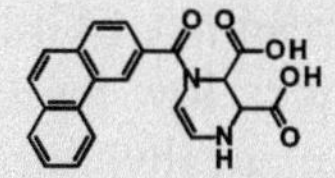	K_i[a]	14.2	19.3	4.2	2.8	[375]
QNZ-46	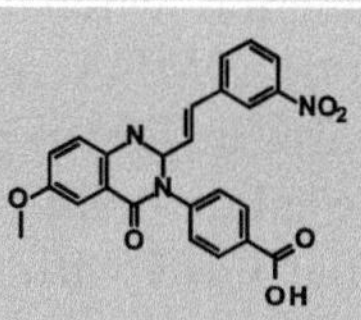	IC_{50}	229 182	ND 193	6 7.1	3 3.9	[382] [383]
DQP-1105	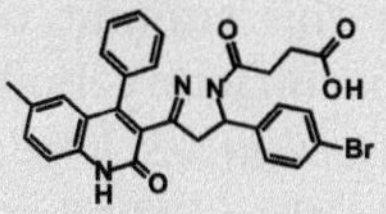	IC_{50}	ND	113	7.0	2.7	[384]
CIQ (+)-CIQ[b]	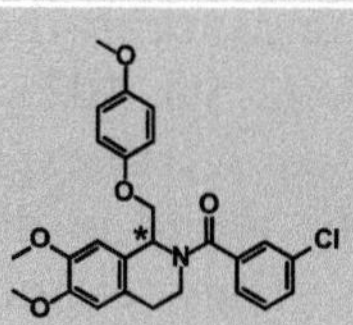	EC_{50}	NE	NE	2.7 9.0[c]	2.8 8.0[c]	[389] [390, 391]
PYD-106	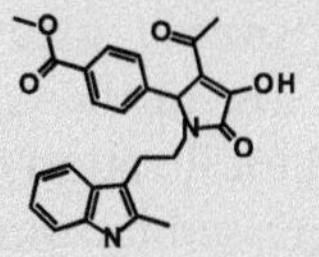	EC_{50}	NE	NE	16	NE	[159]

NE denotes no effect at the highest concentrations evaluated, and ND indicates that the compound displayed some activity, but the affinity or potency could not be determined. The values were determined using two-electrode voltage-clamp experiments with *Xenopus* oocytes

[a] K_i values were estimated using the Cheng-Prusoff correction of the measured IC_{50} values

[b] The chiral carbon of (+)-CIQ, the active enantiomer, is denoted by the ***asterisk*** in the chemical structure

[c] The apparent lower potency for (+)-CIQ compared to the racemic mixture is likely due to better estimation of maximum potentiation, since the active enantiomer has increased abundance in solution at concentrations close to the solubility limit (i.e., the pure enantiomers can be evaluated at higher concentrations compared with racemic CIQ)

between distinct neuronal cell types with respect to the time course of the synaptic NMDA receptor response as well as their changes in response to neuronal activity or other stimuli. Thus, the neuronal cell types in the different brain regions or nuclei can precisely tune their functional properties by expressing different complements of GluN2 subunits. Furthermore, the expression profiles of the different NMDA receptor subtypes undergo marked changes during development to enable modifications of neuronal functions during critical neurodevelopmental periods and maturation of the CNS.

5.1 Distinct Expression Profiles of NMDA Receptor Subunits

The different GluN2 subunits have profoundly different regional and developmental expression profiles (Fig. 2a). The GluN2B subunit is widely expressed in the embryonic brain, but becomes restricted to the forebrain in the adult rodent brain [63, 92, 93, 95, 398, 399]. By contrast, the expression of GluN2A subunit is ubiquitous in the CNS, initially at very low levels at birth, after which the expression increases dramatically during the second postnatal week (P7-P14). Thus, in some regions, such as the cortex and hippocampus, there is a developmental switch in the expression of GluN2B to GluN2A, and synaptic NMDA receptors change from mainly containing GluN2B early in life to also containing GluN2A (see below) [135]. In the adult brain, the GluN2A is present in virtually all regions of the CNS with particular high abundance in the cortex, hippocampus, and cerebellum [63, 92, 93, 95, 398, 399]. Expression of GluN2C is undetectable at birth, but in the second postnatal week this subunit becomes highly enriched in the cerebellum and the olfactory bulb [63, 92, 93, 399–401]. Similar to GluN2B, the GluN2D subunit is widely expressed early in development, but then expression fades in the second postnatal week. The GluN2D subunit remains expressed into adulthood with the highest abundance in the diencephalon, mesencephalon, and spinal cord [63, 92, 93, 402].

In addition to the aforementioned overall expression profiles, the different GluN2 subunits can be found in distinct neuronal populations in some brain regions. Thus, although the overall expression of a GluN2 subunit may appear low in a specific region, the expression can still be high in a small subpopulation of neurons in that region. For example, the overall expression levels of GluN2C and GluN2D appear to be low in the cortex and hippocampus, but more precise anatomical localization of these subunits suggests that they are specifically expressed in some populations of glial cells and interneurons in these regions [63, 392, 400, 403, 404]. Similarly, GluN2B and GluN2D are highly expressed in cerebellar Golgi cells, although they are considered to have less overall abundance in the cerebellum [137, 405]. In recent years, increasingly precise identification of GluN2 subunit expression and subcellular localization in distinct neuronal populations have been reported as more refined methods of detection and pharmacological tool compounds become available.

Weak expression of GluN3A can be detected in several regions of the embryonic brain, and expression increases throughout the brain during the early postnatal development [116, 406–410]. GluN3A expression peaks around postnatal day 8 (P8) in rodents, but then diminishes with time. By adulthood, GluN3A is weakly, but widely, expressed in the CNS. By contrast, expression of GluN3B slowly increases during the late stages of postnatal development and becomes widely expressed in the adult CNS [119, 411–413]. Thus, there is an apparent developmental switch from expression of GluN3A to GluN3B in the rodent brain during the first two postnatal weeks. In addition, GluN3B is also highly expressed in motoneurons in the rodent spinal cord, but here expression starts at embryonic day 16 (E16) [412]. Recent studies suggest that the GluN3 subunits also have distinct subcellular distributions with GluN3B found primarily in the postsynaptic membrane and GluN3A found mostly at extrasynaptic and presynaptic sites [414]. The contrasting expression profiles of the GluN3 subunits suggest they serve distinct physiological roles in the CNS. It should be noted, however, that the expression profiles of GluN3 subunits appear to vary markedly between brain regions [116, 119, 406–413], and also appear to be different in rodents compared to primates and humans [111, 415–417].

Functional properties and trafficking of NMDA receptor subtypes are influenced by alternative splicing of the GluN1 mRNA. Differences between the regional and developmental distributions of GluN1 isoforms in the CNS have been described (Fig. 3a) [63, 77, 78, 92, 93]. However, the functional significance of these differences remains unclear and not as well characterized as those of GluN2 subunits. Consistent with its inclusion in all NMDA receptor subtypes, the GluN1 subunit is ubiquitously expressed in the CNS throughout development [63, 77, 78, 92, 93]. The GluN1-2 isoforms are widely distributed in the rodent brain, whereas low expression of GluN1-3 isoforms appears to be restricted to the sensorimotor cortex, the neocortex, hippocampus, and selected thalamic nuclei at later developmental stages. There is an apparent complementary distribution of GluN1-1 and GluN1-4 isoforms with GluN1-1 primarily expressed in more rostral regions, such as the cortex and hippocampus, and GluN1-4 in more caudal regions, such as the basal ganglia and cerebellum. The expressions of GluN1-a and GluN1-b isoforms largely overlap, but marked variation in the relative abundance is observed between regions and even between neuronal cell types in the same region. For example, GluN1-a and GluN1-b isoforms have strikingly distinct developmental expression profiles in the hippocampus, and in the adult rodent brain, GluN1-b appears to be the major isoform in the CA3, while GluN1-a is the major isoform in the CA1 and dentate gyrus [78]. These differences in expression profiles are likely to have functional significance, since the deactivation time course of NMDA receptors containing GluN1-b (e.g., GluN1-1b) is accelerated compared to receptors containing GluN1-a (e.g., GluN1-1a) (Fig. 3c) [82, 83], and GluN1-1b-containing receptors are less sensitive to endogenous negative allosteric regulators [80, 81].

5.2 The GluN2B to GluN2A Developmental Switch

The increase in GluN2A expression during the second postnatal week in the rodent cortex and hippocampus results in a switch in the subunit composition of synaptic NMDA receptors from primarily being GluN2B-containing to also being GluN2A-containing. This switch is accompanied by the appearance of triheteromeric GluN1/2A/2B receptors, which contain two GluN1, one GluN2A, and one GluN2B subunit [132, 135]. At early developmental stages, the time course of the EPSC (i.e., deactivation time constant) and the sensitivity to GluN2B-selective NAMs, such as ifenprodil, suggest that diheteromeric GluN1/2B is the primary NMDA receptor subtype in the central synapses of the cortex and hippocampus [132, 149, 418–420]. However, the marked acceleration of the EPSC time course and reduced ifenprodil sensitivity observed during the second postnatal week are consistent with a switch in the synaptic content from GluN2B-containing to GluN2A-containing NMDA receptors. That is because the deactivation time constants of both diheteromeric GluN1/2A and triheteromeric GluN1/2A/2B are markedly faster than diheteromeric GluN1/2B (Fig. 4) [133, 151]. Furthermore, inhibition by GluN2B-selective NAMs is retained, but maximal inhibitory efficacy is significantly reduced for triheteromeric GluN1/2A/2B compared to diheteromeric GluN1/2B [150, 151].

The "GluN2B to GluN2A developmental switch" is evolutionarily conserved and occurs in many brain areas of frogs, birds, and mammals, including cortex, hippocampus, amygdala, and cerebellum. Numerous studies have reported that the timing of the switch, which varies between brain regions, is coincident with changes in specific learning abilities. The prevalent hypothesis is therefore that the GluN2B to GluN2A developmental switch is a major factor in the synaptic maturation, which is important for the refinement and fine tuning of neuronal circuits. The developmental switch in NMDA receptor subunit composition closes a critical period for the refinement of connections in the key brain regions, resulting in changes in synaptic plasticity [421, 422]. However, in some brain circuits, the changes in synaptic plasticity during critical developmental periods are not corresponding to the switch from GluN2B to GluN2A expression (e.g., [418]), suggesting that other NMDA receptor subunits (e.g., GluN2C/D or GluN3 subunits) may have important roles in the refinement and fine tuning of these neuronal circuits.

The switch in the GluN2 subunit composition of synaptic NMDA receptors is experience-dependent and can occur acutely following synaptic activity or sensory input. For example, the change from synaptic GluN2B- to GluN2A-containing NMDA receptors is not observed in the visual cortex of dark-reared rats until they are exposed to light [423]. Thus, the EPSCs in the visual cortex of dark-reared rats have slower time course and higher sensitivity to ifenprodil compared to light-reared rats. Remarkably, returning the animals to the dark can restore the synaptic content of GluN2B to levels observed in animals that have not been exposed to light [424]. Thus, the experience-dependent GluN2 subunit switch appears to be bi-directional, at least in some brain regions [419].

The mechanisms that mediate the exchange of synaptic GluN2B-containing NMDA receptors with GluN2A-containing receptors are not fully understood and this remains an area of intense investigation. Similarly, detailed insights into the consequences of changes in GluN2 subunit composition on the refinement of synaptic plasticity and neuronal circuits are still lacking. However, many excellent reviews discuss our accumulated understanding of these processes and highlight important studies in these areas of NMDA receptor research [23, 325, 425].

6 NMDA Receptors in Disease

NMDA receptors have been considered in the context of numerous neurological conditions, either as a potential causative feature, exacerbating component, or therapeutic target [1, 2, 23, 42, 426]. However, the interest in NMDA receptor modulators as therapeutics has grown significantly in recent years. Contributing to this interest has been the growing clinical evidence that the NMDA receptor channel blocker ketamine could act as a radically new treatment for depression. Indeed, discovery of the antidepressant activity of ketamine has been characterized as "the most important psychiatric discovery in half a century" [427, 428]. Here, we will highlight two emerging mechanistic themes in this area of drug discovery. These are the growing awareness of the significance of metaplasticity in the therapeutic response to NMDA receptor modulation and the progress in linking NMDA receptor subtypes to CNS disorders.

6.1 Depression

Short intravenous infusions of the pan-NMDA receptor channel blocker ketamine (0.5 mg/kg over 40 min) have now been repeatedly demonstrated to yield a robust antidepressant response that (1) develops within hours, (2) may last for days to weeks, and (3) is effective in up to 70% of patients [429–431]. This ketamine regimen also is reported to reduce suicidal ideation [432, 433] and has benefit in patients suffering bipolar depression [434], obsessive-compulsive disorder [435], and posttraumatic stress disorder [436]. The antidepressant response appears to be sustainable with repeated doses [437–439] and clinical studies are beginning to define the optimal dose and treatment chronicity [440]. Side effects include those expected for an NMDA receptor channel blocker, including cognitive disruption and neuropsychiatric symptoms; however, these appear to be mild and manageable at effective exposures [441]. Indeed, intravenous infusion may be the most significant limitation to ketamine use and clinical studies are exploring other routes of administration [442, 443]. Furthermore, another NMDA receptor channel blocker, lanicemine [444, 445], and GluN2B-selective negative allosteric modulators (GluN2B NAMs) [446, 447] also are reported to have clinical antidepressant activity. Several detailed reviews of the rapid progress in this area have been recently published [428, 448].

A remarkable aspect of the antidepressant activity of ketamine is that the clinical response develops and is sustained after the drug has been cleared from the body. In contrast, the psychotomimetic effects track closely with drug residence time and typically resolve shortly after cessation of drug infusion [429]. The antidepressant effects of other NMDA antagonists also persist beyond drug clearance from the body. In fact, it has been shown that the brief ketamine exposure is sufficient to induce a long-lasting change in human brain physiology [430]. These results may indicate that the antidepressant effects of these drugs arise from a metaplastic change in synaptic activity. Metaplasticity is "the plasticity of synaptic plasticity" [449]; that is, the effect that an acute change in synaptic function has on the ability of subsequent stimuli to effect further change [450, 451]. The antidepressant effects of NMDA receptor inhibitors may be interpreted as a variation on this theme. The antidepressant response is not the direct result of acute NMDA receptor inhibition, but rather a long-lasting change in synaptic function triggered by the brief inhibition. There is considerable interest in determining the nature of these long-lasting synaptic changes at the molecular level, as these findings might reveal insight into the neurobiology of depression and be applied prospectively to develop new antidepressants. There is speculation that the antidepressant effect of ketamine may be due to an effect of a metabolite that does not inhibit NMDA receptors [452]; however, this has not yet been reconciled with the clinical antidepressant effect of the other chemically and mechanistically diverse NMDA receptor modulators. Indeed, a fruitful avenue of research is through comparative analysis of these different agents to pinpoint common mechanisms that may account for the antidepressant effects.

Both ketamine and GluN2B NAMs induce persistent increases in synaptic strength after drug washout. This is evidenced by mTOR-driven increases in synaptic protein levels in rodents [453–455], an increase in sensitivity to the induction of LTP in rodents [456], and an increase in sensory stimulus-evoked potentials in rodents [457] and in humans [430]. A working hypothesis is that such an upregulation of synaptic strength underlies the antidepressant activity. One hypothesis for the mechanism by which ketamine induces synaptic upregulation derives from its use-dependence for channel block, which confers selectivity for highly active NMDA receptors on PV-positive, fast spiking GABAergic interneurons [458]. It is hypothesized that inhibiting fast-spiking interneurons disinhibit cortical microcircuits, inducing gamma-band cortical activity that drives an LTP-like upregulation of synaptic strength. While attractive, this hypothesis accounts poorly for the putative antidepressant activity of lanicemine [444] and particularly the GluN2B NAMs. These latter compounds do not induce gamma-band activity in rodents [457, 459, 460] or primates [461] even at high levels of receptor occupancy. This functional difference between ketamine and the GluN2B NAMs may be accounted for by differences in brain micro-circuitry modulated by these agents. GluN2B is weakly expressed in interneurons arising from medial ganglionic eminence [462] that include the ketamine-sensitive fast spiking PV- and SST-family interneurons that synapse directly with pyramidal neurons [463]. Lack of predominant GluN2B expression on these interneurons may account for the fact that GluN2B NAMs do not induce gamma-band activity. Instead, GluN2B is expressed by CCK-family interneurons that arise from the caudal ganglionic eminence [462]. These interneurons synapse with the fast-spiking interneuron classes to regulate their activity in response to long-range pyramidal neuron inputs [464]. Thus, GluN2B NAMs may be speculated to increase activity of fast-spiking interneurons by decreasing excitatory drive on CCK-family interneurons, the opposite of the putative effect of ketamine. The effects of ketamine and the GluN2B NAMs are also likely to be divergent on pyramidal neurons. Deployment of GluN2B varies across different pyramidal neuron populations [465] and in different synaptic compartments (reviewed in [466]). The deployment of GluN2B subunits is also activity-dependent and is increased at relatively inactive synapses (reviewed in [425, 467]). Thus, the pan-NMDA receptor antagonist ketamine and the GluN2B NAMs likely inhibit different receptor pools on pyramidal neurons based on the subunit-selectivity of the NAMs, as well as the activity dependence of ketamine. At present there is no obvious point of convergence between these two compound classes that may account for their striking similarity in terms of functional endpoints in preclinical models and clinical antidepressant efficacy (and side effect profile, see below). However, the fact that points of convergence are apparently so few increases the power of comparative analyses to pinpoint the molecular mechanisms of their antidepressant response. This seems a promising area for continued research.

6.2 Neuro-developmental Disorders

NMDA receptor signaling plays a central role in circuit development of the central nervous system. As noted above, during development, high expression of GluN2B and GluN2D NMDA receptor subunits is superseded by the expression of GluN2A [63, 92, 93, 135, 468, 469]. This choreography mediates the transition from high levels of synaptic plasticity as circuits are formed and refined to the circuit stability of the adult brain. Consistent with a fundamental role in this developmental progression, variation in genetic loci encompassing *GRIN2A* and *GRIN2B* (i.e., genes encoding GluN2A and GluN2B, respectively) is identified in genome-wide association studies (GWAS) as contributing to the risk of developing the two major neurodevelopmental disorders, autism and schizophrenia [470–473]. The symptoms of autism manifest early in life, whereas those of schizophrenia do not fully manifest until late adolescence or early adulthood. Thus, these two disorders arise from derangements at different epochs of the brain's developmental program. The association of *GRIN2A* and *GRIN2B* genetic variation in the risks for both disorders highlights a role for NMDA receptor signaling in unfolding the entire developmental program. However, there are different scenarios by which variation in NMDA receptor signaling may contribute to these disorders that are important in considering NMDA receptor modulation as a therapeutic strategy. Defective NMDA receptor signaling could impact a specific segment of the developmental program, in which case therapeutic intervention would need to occur during that developmental epoch. Alternatively, defective NMDA receptor signaling could impact a developmental trajectory and so therapeutic intervention would need to occur at some time before the symptoms begin to manifest. Finally, aberrant NMDA receptor signaling may be a factor in the expression of symptoms, in which case NMDA receptor modulation may be effective as a "symptomatic" therapeutic at any time after symptoms manifest. Of these three scenarios, the most extensively studied therapeutic use of NMDA receptor modulators is as a "symptomatic" approach to schizophrenia.

6.3 Schizophrenia

The association of NMDA receptor dysfunction with schizophrenia initially arose from the clinical observation that NMDA receptor inhibition in healthy individuals induces a spectrum of symptoms that are strikingly similar to those exhibited by patients suffering schizophrenia [474–476]. These "schizophrenomimetic" symptoms (e.g., *see* [477]) correspond closely to NMDA receptor occupancy [476, 478]. This infers that symptom expression in schizophrenia patients may result from hypofunction of NMDA receptor signaling [479–481]. The NMDA receptor hypofunction hypothesis for schizophrenia has driven a great deal of research to develop drugs to potentiate NMDA receptor signaling to overcome the symptoms of this disorder.

The largest body of work aiming to overcome NMDA receptor hypofunction encompasses strategies to increase agonist occupancy of the GluN1 glycine co-agonist binding site. This has included clinical testing of the natural ligands glycine and D-serine, glycine analogs such as D-cycloserine (DCS), and inhibition of the GlyT1 transporter to increase peri-synaptic glycine levels [482, 483]. Unfortunately, the effectiveness of the glycinergic approach has so far proved modest, with the most consistent effect being a reduction in negative symptoms, but with little effect on cognitive or positive symptoms [483, 484]. Nonetheless, this clinical research has yielded significant insight that may be critical to further advancement of NMDA receptor potentiator strategies. It has been suggested that treatment with "glycinergics" may trigger metaplastic changes in glutamate signaling that significantly affect the drug response [485, 486]. These effects on drug response include limited efficacy of continuous drug exposure and can cause complex dose responses, such as observed with the GlyT1 inhibitor bitopertin [487]. To exploit the plasticity induced by modulating NMDA receptors, Goff and colleagues have begun to explore intermittent dosing with the glycinergic DCS. In preliminary clinical studies, intermittent DCS treatment also improved negative symptoms. More significantly, intermittent DCS improved memory performance and reduced delusional severity when combined with cognitive behavioral therapy [486, 488, 489]. Thus, an intermittent DCS dosing regimen may be at least as efficacious as continuous treatment with regard to negative symptoms and may deliver efficacy against positive and cognitive symptoms not observed with continuous exposure regimens. This line of clinical research clearly calls for further study and begs investigation into the underlying molecular mechanisms.

Several mechanisms may contribute to enhanced efficacy with intermittent DCS treatment. Increasing glycine-site occupancy to acutely increase NMDA receptor activity also increases NMDA receptor internalization rate, which may offset positive effects [490]. An intermittent dosing regimen may reduce drive on internalization and thereby tip the balance toward potentiation. More intriguing is the possibility that intermittent dosing enhances plasticity beyond a simple "drug-on" potentiation [486]. The pharmacology of DCS is complex; the compound is a partial glycine site agonist and a single administration may therefore potentiate or inhibit NMDA receptors, and possibly both, over the exposure time course of a single dose. Furthermore, DCS is a super-agonist at GluN2C-containing receptors and will activate a larger current compared to glycine/D-serine [491, 492], suggesting that at these receptors, substitution of DCS for glycine could selectively enhance synaptic NMDA receptor responses. An appropriate single dose may trigger a longer lasting metaplastic change in synaptic activity that results in sustained efficacy. Indeed, it is interesting to draw analogy between intermittent dosing with DCS in schizophrenia and intermittent dosing of NMDA antagonists in depression. In both the cases, it is the metaplastic effect of the brief drug exposure, i.e., the "drug-off" effects, that delivers the efficacy.

It is also of interest to understand the underlying mechanism(s) by which NMDA receptor hypofunction may result in the expression of schizophrenia symptoms. The clinical pharmacology may be informative. First, DCS produce a maximal response that is twice as large as glycine at GluN2C-containing NMDA receptors, resulting in increased NMDA receptor signaling whenever concentrations of DCS allow it to displace glycine from its site these receptors [491, 492]. GluN2C is highly expressed in cerebellum and in the thalamic reticular nucleus [63, 400, 493, 494]. It has been speculated that the efficacy of DCS may be derived from agonist activity GluN2C-containing NMDA receptors in these brain regions, prompting an effort to develop other GluN2C-selective PAMs [159, 389, 393, 397]. Another interesting clue to underlying mechanisms is the clinical observation that the GluN2B-selective NAM, CP-101,606, causes cognitive disruption and dissociative effects similar to those caused by ketamine [447, 495, 496]. Consistent with the clinical data, GluN2B NAMs and NMDA receptor channel blockers share discriminative stimulus properties in animal studies [497, 498]. These findings are interesting with respect to the fact that there is little apparent overlap in the neuronal microcircuitry impacted by these two drug classes, as reviewed above. Thus, comparative analyses of the schizophrenomimetic effects of these drugs may also help pinpoint the microcircuit defects in NMDA receptor signaling relevant to the expression of schizophrenia.

6.4 Epilepsy/Aphasia Syndromes

GRIN1, *GRIN2A*, and *GRIN2B* have been associated with epilepsy [499]. For example, a deterministic link has recently been made between genetic variation in *GRIN2A* and childhood epilepsy/aphasia syndromes [500–503]. The spectrum of these syndromes includes relatively benign Rolandic epilepsy, the more severe continuous spike-and-waves during slow-wave sleep syndrome (CSWSS), and Landau-Kleffner syndrome (LKS), and very severe epilepsies with significant developmental delay, intellectual disability, and dysmorphic features. Manifestation arises between ages 3 and 11 during the developmental epoch that is associated with language development [504]. This is also the epoch over which there is significant pruning of cortical excitatory synapses [505], in which NMDA receptor signaling is fundamentally involved. To date, more than 60 mutations in *GRIN2A* have been identified that appear to be causal to these developmental disorders [506]. Significantly, whereas many of these mutations result in receptor truncation or other losses of function, numerous point mutations result in a gain of function. This includes reduced Mg^{2+}-block, enhanced agonist potency, and increased open probability and open time, at least when the receptors are expressed in heterologous expression systems [506, 507]. Critical questions remain around how specific variations in *GRIN2A*, including both gain and loss of function, relate to the spectrum of severities in a common group of epilepsies and language disorders.

The discovery of the association of *GRIN2A* mutations with epilepsy/aphasia syndromes immediately suggested NMDA receptor modulators as potential therapeutics. In a first case, Pierson, Yuan, and colleagues [158, 508] identified a child through the NIH Undiagnosed Diseases Program suffering early-onset epileptic encephalopathy, manifest as profound cognitive and motor development and intractable seizures resistant to standard anticonvulsant therapies, who had a point mutation in *GRIN2A*. Analyses in heterologous expression systems revealed that the mutation resulted in significantly increased activity of GluN2A-containing receptors, suggesting that inhibition of NMDA receptors may have a therapeutic benefit where other conventional therapeutics had failed. The treatment of this patient with the NMDA receptor antagonist memantine (approved for the treatment of Alzheimer's disease) produced a rapid onset and persistent reduction in the number of seizures suffered by the child [508]. This suggests that the altered function of the GluN2A subunit may have contributed to seizure etiology. Unfortunately, memantine did not have an effect on the child's cognitive or motor disability, suggesting that the *GRIN2A* mutation also had effects on the developmental trajectory, which were insensitive to memantine at the time treatment was initiated. It should be noted that this remains only a single case, and considerable work is needed to determine whether viable treatment options can be developed for these patients with specific mutations in NMDA receptor subunits.

6.5 Rett Syndrome

Rett Syndrome (RTT) is another neuro developmental disorder in which NMDA receptor dysfunction is implicated and NMDA receptor modulators are of therapeutic interest. RTT is a severe X-linked neurodevelopmental disorder caused by defects in transcriptional regulation by MeCP2 [509]. Although girls with RTT initially develop on a normal trajectory, developmental stasis and regression begins at 6–18 months that includes a severe reduction in the size and complexity of forebrain pyramidal neuron dendritic arbors, but without apparent reduction in neuron number [510]. Significantly, Bird and colleagues demonstrated in a mouse model that restoration of MeCP2 function in symptomatic animals reverted much of the neurological phenotype [511]. This implies that the fundamental architecture of the brain develops normally prior to the effects of MeCP2 lesion and that restoration of network function is an attainable goal. There are several emerging lines of research that suggest NMDA receptor dysfunction contributes to this network dysfunction and that modulation of these receptors may be an effective therapeutic approach [512]. Blue et al. [513]

reported alteration of NMDA receptor expression in MeCP2 mutant mice. Subsequently, the Fagiolini lab reported an imbalance in GluN2A/GluN2B subunit deployment in both cortical pyramidal neurons and interneurons [514, 515]. Significantly, manipulating the GluN2A/GluN2B balance through hemizygous *GRIN2A* knockout prevented the development of cortical dysfunction and ameliorated some of the RTT-like phenotypes [514]. In another line of research, Katz and colleagues demonstrated that treatment of Mecp2 mutant mice with a low, sub-anesthetic dose of ketamine acutely reversed RTT-like phenotypes, including abnormal patterns of neuronal activation in cortical and subcortical structures as well as sensorimotor dysfunction [516]. Subsequently, Patrizi et al. [517] reported that once daily administration of this same low dose ketamine produced a sustained reduction in RTT-like symptoms and ameliorated structural circuit defects that underlie or contribute to neurological dysfunction. Significantly, in the study by Patrizi et al., neurological testing of mice occurred ~21 h after drug administrations; i.e., after ketamine had been completely eliminated. Thus, it appears that ketamine has beneficial effects in mouse RTT models during both "drug-on" [516] and "drug-off" [517] periods. The latter drug-off effects suggest a potential mechanistic parallel to the effects of ketamine in depression, particularly with respect to the possibility of durability of action beyond the acute period of NMDAR antagonism. Trials have now been initiated to test the safety and efficacy of NMDAR antagonists in RTT patients, including dextromethorphan, a weak NMDAR antagonist, and low-dose ketamine [512].

6.6 NMDA Receptors as Therapeutic Targets

Clearly, the most significant recent advance in the area of therapeutics targeting NMDA receptors has been the emergence of ketamine as a rapidly acting antidepressant. Ketamine is now being used in clinics to treat patients for whom standard of care monoaminergic reuptake inhibitors provide little relief. Ketamine is also groundbreaking from a mechanistic perspective. The therapeutic effect of ketamine is not due to an effect of the drug "on" the brain per se, but to the response of the brain to the drug that manifests after the drug is gone. The concept of synaptic metaplasticity has been evolving for decades [450, 451, 518], and ketamine is the first example where such a metaplastic effect has been harnessed for therapeutic benefits. Indeed, this therapeutic effect may be broad, as clinical data is emerging to suggest that brief ketamine exposure may be beneficial across a range of neuropsychiatric conditions and perhaps in neurodevelopmental disorders. While the discovery of the antidepressant effect of ketamine was serendipitous [429], as were the implications of metaplasticity as its therapeutic mechanism, Goff and colleagues have been developing a parallel theme of inducing plasticity in exploring the utility of intermittent dosing of DCS in schizophrenia and anxiety disorders [486]. These research paths may be the herald of a new era in the development of CNS therapeutics in which we try and work with the brain instead of trying to overpower it.

The above notwithstanding, there is a tremendous amount of work ahead to build on the theme of harnessing the brain's plasticity to therapeutic benefit. From a very practical perspective, the optimal duration of NMDA receptor inhibition and exposure interval remains to be determined to realize therapeutic benefits in depression and other conditions. For example, a single exposure to the GluN2B NAM CP-101,606, a short half-life compound, had a robust and long-lasting antidepressant effect [447], whereas a long half-life GluN2B NAM, CERC 301, dosed for 28 days and likely resulting in continuous NMDA receptor occupancy, was without efficacy. Although it is not possible to draw firm conclusions from single studies, it is tempting to speculate that the difference in efficacy in these studies may be due in part to the difference in exposure duration, with the short duration exposure allowing the metaplastic mechanism to emerge.

It is also important to note the significance of back-translational research in advancing this area. Clinical data on NMDA receptor modulators provides a rich frame for preclinical studies into molecular mechanisms of disease as well as for new therapeutic approaches. As an example mentioned above, the similarity in clinical efficacy and side effect profile between ketamine and the GluN2B NAM CP-101,606, in light of the apparently scant overlap in site of action at the level of neurocircuitry, may be leveraged to gain significant insight into the role of NMDA receptor signaling in both depression and schizophrenia [519].

It also appears that the repertoire of NMDA receptor modulators is expanding rapidly [54, 55, 96, 520]. This includes compounds with unique NMDA receptor subtype selectivity and modes of action that include both augmenting and inhibiting receptor activity. Thus, these compounds will provide new tools to interrogate the physiology of NMDA receptor signaling, which in turn may reveal new therapeutic opportunities. As insightfully pointed out by Köhr [521, 522], it will be important to consider not only the locus of action of such compounds based on NMDA receptor subtype expression pattern, but also the mechanism of pharmacological action, as each may have a significant impact on the functional effects of these new compounds. Based on the new insights gained from the effects of ketamine and DCS, it will also be important to consider the metaplastic effects of these compounds in addition to their more direct effects on signaling.

7 Conclusions

Emerging information from genetic analyses linking NMDA receptors to specific disease conditions and the discovery of antidepressant effects for NMDA receptor antagonists have fortified and reinvigorated the long-standing focus on NMDA receptors as therapeutic targets. Recent years' remarkable acceleration in the discovery of novel allosteric NMDA receptor modulators as pharmacological tools greatly facilitates studies to achieve new levels of understanding of NMDA receptor subtypes in physiology and disease. Many new modulatory binding sites in NMDA receptors have been identified along the way and combined with rapidly improving structural crystallographic and cryo-EM data, we are improving our understanding of how agonist binding is linked to channel gating and how the different subunits contribute to conformational changes during gating and allosteric modulation. These developments in the NMDA receptor field offer new perspectives and exciting opportunities to study unique roles for NMDA receptor subtypes, diheteromeric as well as triheteromeric, in distinct neuronal populations and subcellular locations. Furthermore, the converging advances in NMDA receptor pharmacology and clinical and mechanistic understanding of CNS diseases involving NMDA receptor dysfunction are poised to result in the development of new therapeutic agents.

Acknowledgments

This work was supported by grants from National Institutes of Health to S.F.T. (NS036654 and NS065371) and K.B.H. (GM103546 and NS097536).

References

1. Traynelis SF, Wollmuth LP, McBain CJ, Menniti FS, Vance KM, Ogden KK, Hansen KB, Yuan H, Myers SJ, Dingledine R (2010) Glutamate receptor ion channels: structure, regulation, and function. Pharmacol Rev 62 (3):405–496. doi:10.1124/pr.109.002451
2. Paoletti P, Bellone C, Zhou Q (2013) NMDA receptor subunit diversity: impact on receptor properties, synaptic plasticity and disease. Nat Rev Neurosci 14(6):383–400. doi:10.1038/nrn3504
3. Matsuda K, Yuzaki M (2012) Cbln1 and the delta2 glutamate receptor–an orphan ligand and an orphan receptor find their partners. Cerebellum 11(1):78–84. doi:10.1007/s12311-010-0186-5
4. Yuzaki M (2009) New (but old) molecules regulating synapse integrity and plasticity: Cbln1 and the delta2 glutamate receptor. Neuroscience 162(3):633–643
5. Schmid SM, Hollmann M (2008) To gate or not to gate: are the delta subunits in the glutamate receptor family functional ion channels? Mol Neurobiol 37(2-3):126–141
6. Kakegawa W, Miyazaki T, Kohda K, Matsuda K, Emi K, Motohashi J, Watanabe M, Yuzaki M (2009) The N-terminal domain of GluD2 (GluRdelta2) recruits presynaptic terminals and regulates synaptogenesis in the cerebellum in vivo. J Neurosci 29(18):5738–5748. doi:10.1523/JNEUROSCI.6013-08.2009

7. Matsuda K, Miura E, Miyazaki T, Kakegawa W, Emi K, Narumi S, Fukazawa Y, Ito-Ishida A, Kondo T, Shigemoto R, Watanabe M, Yuzaki M (2010) Cbln1 is a ligand for an orphan glutamate receptor delta2, a bidirectional synapse organizer. Science 328 (5976):363–368
8. Elegheert J, Kakegawa W, Clay JE, Shanks NF, Behiels E, Matsuda K, Kohda K, Miura E, Rossmann M, Mitakidis N, Motohashi J, Chang VT, Siebold C, Greger IH, Nakagawa T, Yuzaki M, Aricescu AR (2016) Structural basis for integration of GluD receptors within synaptic organizer complexes. Science 353 (6296):295–299. doi:10.1126/science.aae0104
9. Kakegawa W, Miyazaki T, Hirai H, Motohashi J, Mishina M, Watanabe M, Yuzaki M (2007) Ca2+ permeability of the channel pore is not essential for the delta 2 glutamate receptor to regulate synaptic plasticity and motor coordination. J Physiol 579(3):729–735. doi:10.1113/jphysiol.2006.127100
10. Schmid SM, Kott S, Sager C, Huelsken T, Hollmann M (2009) The glutamate receptor subunit delta2 is capable of gating its intrinsic ion channel as revealed by ligand binding domain transplantation. Proc Natl Acad Sci U S A 106(25):10320–10325
11. Dadak S, Bouquier N, Goyet E, Fagni L, Levenes C, Perroy J (2017) mGlu1 receptor canonical signaling pathway contributes to the opening of the orphan GluD2 receptor. Neuropharmacology 115:92. doi:10.1016/j.neuropharm.2016.06.001
12. Geiger JRP, Lubke J, Roth A, Frotscher M, Jonas P (1997) Submillisecond AMPA receptor-mediated signaling at a principal neuron-interneuron synapse. Neuron 18 (6):1009–1023. doi:10.1016/S0896-6273(00)80339-6
13. Sah P, Hestrin S, Nicoll RA (1990) Properties of excitatory postsynaptic currents recorded in vitro from rat hippocampal interneurones. J Physiol 430:605–616
14. Hestrin S, Nicoll RA, Perkel DJ, Sah P (1990) Analysis of excitatory synaptic action in pyramidal cells using whole-cell recording from rat hippocampal slices. J Physiol 422:203–225
15. Trussell LO, Zhang S, Raman IM (1993) Desensitization of AMPA receptors upon multiquantal neurotransmitter release. Neuron 10(6):1185–1196
16. Copits BA, Swanson GT (2012) Dancing partners at the synapse: auxiliary subunits that shape kainate receptor function. Nat Rev Neurosci 13(10):675–686. doi:10.1038/nrn3335
17. Nowak L, Bregestovski P, Ascher P, Herbet A, Prochiantz A (1984) Magnesium gates glutamate-activated channels in mouse central neurones. Nature 307(5950):462–465
18. Mayer ML, Westbrook GL, Guthrie PB (1984) Voltage-dependent block by Mg2+ of NMDA responses in spinal cord neurones. Nature 309(5965):261–263
19. Seeburg PH, Burnashev N, Kohr G, Kuner T, Sprengel R, Monyer H (1995) The NMDA receptor channel: molecular design of a coincidence detector. Recent Prog Horm Res 50:19–34
20. Bourne HR, Nicoll R (1993) Molecular machines integrate coincident synaptic signals. Cell 72(Suppl):65–75
21. Volianskis A, France G, Jensen MS, Bortolotto ZA, Jane DE, Collingridge GL (2015) Long-term potentiation and the role of N-methyl-D-aspartate receptors. Brain Res 1621:5–16. doi:10.1016/j.brainres.2015.01.016
22. Zorumski CF, Izumi Y (2012) NMDA receptors and metaplasticity: mechanisms and possible roles in neuropsychiatric disorders. Neurosci Biobehav Rev 36(3):989–1000. doi:10.1016/j.neubiorev.2011.12.011
23. Lau CG, Zukin RS (2007) NMDA receptor trafficking in synaptic plasticity and neuropsychiatric disorders. Nat Rev Neurosci 8 (6):413–426. doi:10.1038/nrn2153
24. Holtmaat A, Svoboda K (2009) Experience-dependent structural synaptic plasticity in the mammalian brain. Nat Rev Neurosci 10 (9):647–658. doi:10.1038/nrn2699
25. Granger AJ, Nicoll RA (2014) Expression mechanisms underlying long-term potentiation: a postsynaptic view, 10 years on. Philos Trans R Soc Lond B Biol Sci 369(1633):20130136. doi:10.1098/rstb.2013.0136
26. Citri A, Malenka RC (2008) Synaptic plasticity: multiple forms, functions, and mechanisms. Neuropsychopharmacology 33 (1):18–41. doi:10.1038/sj.npp.1301559
27. Morris RG (2013) NMDA receptors and memory encoding. Neuropharmacology 74:32–40. doi:10.1016/j.neuropharm.2013.04.014
28. Hunt DL, Castillo PE (2012) Synaptic plasticity of NMDA receptors: mechanisms and functional implications. Curr Opin Neurobiol 22(3):496–508. doi:10.1016/j.conb.2012.01.007
29. Kleckner NW, Dingledine R (1988) Requirement for glycine in activation of NMDA-receptors expressed in Xenopus oocytes. Science 241(4867):835–837

30. Benveniste M, Mayer ML (1991) Kinetic analysis of antagonist action at N-methyl-D-aspartic acid receptors. Two binding sites each for glutamate and glycine. Biophys J 59 (3):560–573
31. Clements JD, Westbrook GL (1994) Kinetics of AP5 dissociation from NMDA receptors: evidence for two identical cooperative binding sites. J Neurophysiol 71(6):2566–2569
32. Anson LC, Chen PE, Wyllie DJA, Colquhoun D, Schoepfer R (1998) Identification of amino acid residues of the NR2A subunit that control glutamate potency in recombinant NR1/NR2A NMDA receptors. J Neurosci 18(2):581–589
33. Williams K, Chao J, Kashiwagi K, Masuko T, Igarashi K (1996) Activation of N-methyl-D-aspartate receptors by glycine: role of an aspartate residue in the M3-M4 loop of the NR1 subunit. Mol Pharmacol 50(4):701–708
34. Johnson JW, Ascher P (1987) Glycine potentiates the NMDA response in cultured mouse brain neurons. Nature 325 (6104):529–531. doi:10.1038/325529a0
35. Clements JD, Westbrook GL (1991) Activation kinetics reveal the number of glutamate and glycine binding sites on the N-methyl-D-aspartate receptor. Neuron 7(4):605–613
36. Oliet SH, Mothet JP (2009) Regulation of N-methyl-D-aspartate receptors by astrocytic D-serine. Neuroscience 158(1):275–283. doi:10.1016/j.neuroscience.2008.01.071
37. Wolosker H (2007) NMDA receptor regulation by D-serine: new findings and perspectives. Mol Neurobiol 36(2):152–164. doi:10.1007/s12035-007-0038-6
38. Mothet JP, Le Bail M, Billard JM (2015) Time and space profiling of NMDA receptor co-agonist functions. J Neurochem 135 (2):210–225. doi:10.1111/jnc.13204
39. Papouin T, Ladepeche L, Ruel J, Sacchi S, Labasque M, Hanini M, Groc L, Pollegioni L, Mothet JP, Oliet SH (2012) Synaptic and extrasynaptic NMDA receptors are gated by different endogenous coagonists. Cell 150 (3):633–646. doi:10.1016/j.cell.2012.06.029
40. Bergeron R, Meyer TM, Coyle JT, Greene RW (1998) Modulation of N-methyl-D-aspartate receptor function by glycine transport. Proc Natl Acad Sci U S A 95 (26):15730–15734
41. Billups D, Attwell D (2003) Active release of glycine or D-serine saturates the glycine site of NMDA receptors at the cerebellar mossy fibre to granule cell synapse. Eur J Neurosci 18 (11):2975–2980
42. Parsons MP, Raymond LA (2014) Extrasynaptic NMDA receptor involvement in central nervous system disorders. Neuron 82 (2):279–293. doi:10.1016/j.neuron.2014.03.030
43. Choi DW, Koh JY, Peters S (1988) Pharmacology of glutamate neurotoxicity in cortical cell culture: attenuation by NMDA antagonists. J Neurosci 8(1):185–196
44. Ellren K, Lehmann A (1989) Calcium dependency of N-methyl-D-aspartate toxicity in slices from the immature rat hippocampus. Neuroscience 32(2):371–379
45. Mody I, MacDonald JF (1995) NMDA receptor-dependent excitotoxicity: the role of intracellular Ca2+ release. Trends Pharmacol Sci 16(10):356–359
46. Wroge CM, Hogins J, Eisenman L, Mennerick S (2012) Synaptic NMDA receptors mediate hypoxic excitotoxic death. J Neurosci 32 (19):6732–6742. doi:10.1523/JNEUROSCI.6371-11.2012
47. Olney JW, Sharpe LG (1969) Brain lesions in an infant rhesus monkey treated with monsodium glutamate. Science 166(3903): 386–388
48. Surmeier DJ, Schumacker PT (2013) Calcium, bioenergetics, and neuronal vulnerability in Parkinson's disease. J Biol Chem 288 (15):10736–10741. doi:10.1074/jbc.R112.410530
49. Hallett PJ, Standaert DG (2004) Rationale for and use of NMDA receptor antagonists in Parkinson's disease. Pharmacol Ther 102 (2):155–174. doi:10.1016/j.pharmthera.2004.04.001
50. Low SJ, Roland CL (2004) Review of NMDA antagonist-induced neurotoxicity and implications for clinical development. Int J Clin Pharmacol Ther 42(1):1–14
51. Javitt DC (2007) Glutamate and schizophrenia: phencyclidine, N-methyl-D-aspartate receptors, and dopamine-glutamate interactions. Int Rev Neurobiol 78:69–108. doi:10.1016/S0074-7742(06)78003-5
52. Farber NB (2003) The NMDA receptor hypofunction model of psychosis. Ann N Y Acad Sci 1003:119–130
53. Moghaddam B, Krystal JH (2012) Capturing the angel in "angel dust": twenty years of translational neuroscience studies of NMDA receptor antagonists in animals and humans. Schizophr Bull 38(5):942–949. doi:10.1093/schbul/sbs075
54. Strong KL, Jing Y, Prosser AR, Traynelis SF, Liotta DC (2014) NMDA receptor modulators: an updated patent review (2013-2014).

Expert Opin Ther Pat 24(12):1349–1366. doi:10.1517/13543776.2014.972938

55. Ogden KK, Traynelis SF (2011) New advances in NMDA receptor pharmacology. Trends Pharmacol Sci 32(12):726–733. doi:10.1016/j.tips.2011.08.003
56. Zhu S, Paoletti P (2015) Allosteric modulators of NMDA receptors: multiple sites and mechanisms. Curr Opin Pharmacol 20:14–23. doi:10.1016/j.coph.2014.10.009
57. Divito CB, Underhill SM (2014) Excitatory amino acid transporters: roles in glutamatergic neurotransmission. Neurochem Int 73:172–180. doi:10.1016/j.neuint.2013.12.008
58. Clements JD, Lester RA, Tong G, Jahr CE, Westbrook GL (1992) The time course of glutamate in the synaptic cleft. Science 258 (5087):1498–1501
59. Lester RA, Clements JD, Westbrook GL, Jahr CE (1990) Channel kinetics determine the time course of NMDA receptor-mediated synaptic currents. Nature 346 (6284):565–567. doi:10.1038/346565a0
60. Erreger K, Dravid SM, Banke TG, Wyllie DJ, Traynelis SF (2005) Subunit-specific gating controls rat NR1/NR2A and NR1/NR2B NMDA channel kinetics and synaptic signalling profiles. J Physiol 563(Pt 2):345–358. doi:10.1113/jphysiol.2004.080028
61. Lester RA, Jahr CE (1992) NMDA channel behavior depends on agonist affinity. J Neurosci 12(2):635–643
62. Vicini S, Wang JF, Li JH, Zhu WJ, Wang YH, Luo JAH, Wolfe BB, Grayson DR (1998) Functional and pharmacological differences between recombinant N-methyl-D-aspartate receptors. J Neurophysiol 79(2):555–566
63. Monyer H, Burnashev N, Laurie DJ, Sakmann B, Seeburg PH (1994) Developmental and regional expression in the rat brain and functional properties of four NMDA receptors. Neuron 12(3):529–540
64. Monyer H, Sprengel R, Schoepfer R, Herb A, Higuchi M, Lomeli H, Burnashev N, Sakmann B, Seeburg PH (1992) Heteromeric NMDA receptors: molecular and functional distinction of subtypes. Science 256 (5060):1217–1221
65. Ulbrich MH, Isacoff EY (2007) Subunit counting in membrane-bound proteins. Nat Methods 4(4):319–321. doi:10.1038/nmeth1024
66. Lee CH, Lu W, Michel JC, Goehring A, Du J, Song X, Gouaux E (2014) NMDA receptor structures reveal subunit arrangement and pore architecture. Nature 511 (7508):191–197. doi:10.1038/nature13548
67. Karakas E, Furukawa H (2014) Crystal structure of a heterotetrameric NMDA receptor ion channel. Science 344(6187):992–997. doi:10.1126/science.1251915
68. Kehoe LA, Bernardinelli Y, Muller D (2013) GluN3A: an NMDA receptor subunit with exquisite properties and functions. Neural Plast 2013:145387. doi:10.1155/2013/145387
69. Pachernegg S, Strutz-Seebohm N, Hollmann M (2012) GluN3 subunit-containing NMDA receptors: not just one-trick ponies. Trends Neurosci 35(4):240–249. doi:10.1016/j.tins.2011.11.010
70. Henson MA, Roberts AC, Perez-Otano I, Philpot BD (2010) Influence of the NR3A subunit on NMDA receptor functions. Prog Neurobiol 91(1):23–37. doi:10.1016/j.pneurobio.2010.01.004
71. Cavara NA, Hollmann M (2008) Shuffling the deck anew: how NR3 tweaks NMDA receptor function. Mol Neurobiol 38(1):16–26. doi:10.1007/s12035-008-8029-9
72. Low CM, Wee KS (2010) New insights into the not-so-new NR3 subunits of N-methyl-D-aspartate receptor: localization, structure, and function. Mol Pharmacol 78(1):1–11. doi:10.1124/mol.110.064006
73. Hollmann M, Boulter J, Maron C, Beasley L, Sullivan J, Pecht G, Heinemann S (1993) Zinc potentiates agonist-induced currents at certain splice variants of the NMDA receptor. Neuron 10(5):943–954
74. Durand GM, Gregor P, Zheng X, Bennett MV, Uhl GR, Zukin RS (1992) Cloning of an apparent splice variant of the rat N-methyl-D-aspartate receptor NMDAR1 with altered sensitivity to polyamines and activators of protein kinase C. Proc Natl Acad Sci U S A 89 (19):9359–9363
75. Nakanishi N, Axel R, Shneider NA (1992) Alternative splicing generates functionally distinct N-methyl-D-aspartate receptors. Proc Natl Acad Sci U S A 89(18):8552–8556
76. Sugihara H, Moriyoshi K, Ishii T, Masu M, Nakanishi S (1992) Structures and properties of seven isoforms of the NMDA receptor generated by alternative splicing. Biochem Biophys Res Commun 185(3):826–832
77. Laurie DJ, Seeburg PH (1994) Regional and developmental heterogeneity in splicing of the rat brain NMDAR1 mRNA. J Neurosci 14(5 Pt 2):3180–3194
78. Paupard MC, Friedman LK, Zukin RS (1997) Developmental regulation and cell-specific

expression of N-methyl-D-aspartate receptor splice variants in rat hippocampus. Neuroscience 79(2):399–409

79. Zhong J, Carrozza DP, Williams K, Pritchett DB, Molinoff PB (1995) Expression of mRNAs encoding subunits of the NMDA receptor in developing rat brain. J Neurochem 64(2):531–539
80. Traynelis SF, Burgess MF, Zheng F, Lyuboslavsky P, Powers JL (1998) Control of voltage-independent zinc inhibition of NMDA receptors by the NR1 subunit. J Neurosci 18(16):6163–6175
81. Traynelis SF, Hartley M, Heinemann SF (1995) Control of proton sensitivity of the NMDA receptor by RNA splicing and polyamines. Science 268(5212):873–876
82. Rumbaugh G, Prybylowski K, Wang JF, Vicini S (2000) Exon 5 and spermine regulate deactivation of NMDA receptor subtypes. J Neurophysiol 83(3):1300–1306
83. Vance KM, Hansen KB, Traynelis SF (2012) GluN1 splice variant control of GluN1/GluN2D NMDA receptors. J Physiol 590 (16):3857–3875. doi:10.1113/jphysiol.2012.234062
84. Swanger SA, Vance KM, Pare JF, Sotty F, Fog K, Smith Y, Traynelis SF (2015) NMDA receptors containing the GluN2D subunit control neuronal function in the subthalamic nucleus. J Neurosci 35(48):15971–15983. doi:10.1523/JNEUROSCI.1702-15.2015
85. Mott DD, Doherty JJ, Zhang S, Washburn MS, Fendley MJ, Lyuboslavsky P, Traynelis SF, Dingledine R (1998) Phenylethanolamines inhibit NMDA receptors by enhancing proton inhibition. Nat Neurosci 1 (8):659–667. doi:10.1038/3661
86. Pahk AJ, Williams K (1997) Influence of extracellular pH on inhibition by ifenprodil at N-methyl-D-aspartate receptors in Xenopus oocytes. Neurosci Lett 225(1):29–32
87. Durand GM, Bennett MV, Zukin RS (1993) Splice variants of the N-methyl-D-aspartate receptor NR1 identify domains involved in regulation by polyamines and protein kinase C. Proc Natl Acad Sci U S A 90 (14):6731–6735
88. Zhang L, Zheng X, Paupard MC, Wang AP, Santchi L, Friedman LK, Zukin RS, Bennett MV (1994) Spermine potentiation of recombinant N-methyl-D-aspartate receptors is affected by subunit composition. Proc Natl Acad Sci U S A 91(23):10883–10887
89. Scott DB, Blanpied TA, Swanson GT, Zhang C, Ehlers MD (2001) An NMDA receptor ER retention signal regulated by phosphorylation and alternative splicing. J Neurosci 21 (9):3063–3072
90. Mu Y, Otsuka T, Horton AC, Scott DB, Ehlers MD (2003) Activity-dependent mRNA splicing controls ER export and synaptic delivery of NMDA receptors. Neuron 40(3):581–594
91. Scott DB, Blanpied TA, Ehlers MD (2003) Coordinated PKA and PKC phosphorylation suppresses RXR-mediated ER retention and regulates the surface delivery of NMDA receptors. Neuropharmacology 45 (6):755–767
92. Akazawa C, Shigemoto R, Bessho Y, Nakanishi S, Mizuno N (1994) Differential expression of five N-methyl-D-aspartate receptor subunit mRNAs in the cerebellum of developing and adult rats. J Comp Neurol 347 (1):150–160. doi:10.1002/cne.903470112
93. Watanabe M, Inoue Y, Sakimura K, Mishina M (1992) Developmental changes in distribution of NMDA receptor channel subunit mRNAs. Neuroreport 3(12):1138–1140
94. Yuan H, Hansen KB, Vance KM, Ogden KK, Traynelis SF (2009) Control of NMDA receptor function by the NR2 subunit amino-terminal domain. J Neurosci 29 (39):12045–12058. doi:10.1523/JNEUROSCI.1365-09.2009
95. Ishii T, Moriyoshi K, Sugihara H, Sakurada K, Kadotani H, Yokoi M, Akazawa C, Shigemoto R, Mizuno N, Masu M et al (1993) Molecular characterization of the family of the N-methyl-D-aspartate receptor subunits. J Biol Chem 268(4):2836–2843
96. Hackos DH, Hanson JE (2016) Diverse modes of NMDA receptor positive allosteric modulation: mechanisms and consequences. Neuropharmacology 112:34. doi:10.1016/j.neuropharm.2016.07.037
97. Vyklicky V, Korinek M, Smejkalova T, Balik A, Krausova B, Kaniakova M, Lichnerova K, Cerny J, Krusek J, Dittert I, Horak M, Vyklicky L (2014) Structure, function, and pharmacology of NMDA receptor channels. Physiol Res 63(Suppl 1):S191–S203
98. Qian A, Buller AL, Johnson JW (2005) NR2 subunit-dependence of NMDA receptor channel block by external Mg2+. J Physiol 562(Pt 2):319–331
99. Kuner T, Schoepfer R (1996) Multiple structural elements determine subunit specificity of Mg2+ block in NMDA receptor channels. J Neurosci 16(11):3549–3558
100. Paoletti P, Ascher P, Neyton J (1997) High-affinity zinc inhibition of NMDA NR1-NR2A receptors. J Neurosci 17(15):5711–5725

101. Chen PE, Geballe MT, Katz E, Erreger K, Livesey MR, O'Toole KK, Le P, Lee CJ, Snyder JP, Traynelis SF, Wyllie DJ (2008) Modulation of glycine potency in rat recombinant NMDA receptors containing chimeric NR2A/2D subunits expressed in Xenopus laevis oocytes. J Physiol 586(1):227–245. doi:10.1113/jphysiol.2007.143172
102. Erreger K, Geballe MT, Kristensen A, Chen PE, Hansen KB, Lee CJ, Yuan H, Le P, Lyuboslavsky PN, Micale N, Jorgensen L, Clausen RP, Wyllie DJ, Snyder JP, Traynelis SF (2007) Subunit-specific agonist activity at NR2A-, NR2B-, NR2C-, and NR2D-containing N-methyl-D-aspartate glutamate receptors. Mol Pharmacol 72(4):907–920. doi:10.1124/mol.107.037333
103. Hansen KB, Brauner-Osborne H, Egebjerg J (2008) Pharmacological characterization of ligands at recombinant NMDA receptor subtypes by electrophysiological recordings and intracellular calcium measurements. Comb Chem High Throughput Screen 11 (4):304–315. doi:10.2174/13862070878 4246040
104. Erreger K, Chen PE, Wyllie DJ, Traynelis SF (2004) Glutamate receptor gating. Crit Rev Neurobiol 16(3):187–224
105. Wyllie DJ, Livesey MR, Hardingham GE (2013) Influence of GluN2 subunit identity on NMDA receptor function. Neuropharmacology 74:4–17. doi:10.1016/j.neuropharm. 2013.01.016
106. Glasgow NG, Siegler Retchless B, Johnson JW (2015) Molecular bases of NMDA receptor subtype-dependent properties. J Physiol 593(1):83–95. doi:10.1113/jphysiol.2014. 273763
107. Eriksson M, Nilsson A, Samuelsson H, Samuelsson EB, Mo L, Akesson E, Benedikz E, Sundstrom E (2007) On the role of NR3A in human NMDA receptors. Physiol Behav 92 (1-2):54–59. doi:10.1016/j.physbeh.2007. 05.026
108. Yao Y, Harrison CB, Freddolino PL, Schulten K, Mayer ML (2008) Molecular mechanism of ligand recognition by NR3 subtype glutamate receptors. EMBO J 27(15):2158–2170. doi:10.1038/emboj.2008.140
109. Yao Y, Mayer ML (2006) Characterization of a soluble ligand binding domain of the NMDA receptor regulatory subunit NR3A. J Neurosci 26(17):4559–4566. doi:10. 1523/JNEUROSCI.0560-06.2006
110. Nilsson A, Duan J, Mo-Boquist LL, Benedikz E, Sundstrom E (2007) Characterisation of the human NMDA receptor subunit NR3A glycine binding site. Neuropharmacology 52 (4):1151–1159. doi:10.1016/j.neuropharm. 2006.12.002
111. Nilsson A, Eriksson M, Muly EC, Akesson E, Samuelsson EB, Bogdanovic N, Benedikz E, Sundstrom E (2007) Analysis of NR3A receptor subunits in human native NMDA receptors. Brain Res 1186:102–112. doi:10.1016/ j.brainres.2007.09.008
112. Pilli J, Kumar SS (2012) Triheteromeric N-methyl-D-aspartate receptors differentiate synaptic inputs onto pyramidal neurons in somatosensory cortex: involvement of the GluN3A subunit. Neuroscience 222:75–88. doi:10.1016/j.neuroscience.2012.07.020
113. Das S, Sasaki YF, Rothe T, Premkumar LS, Takasu M, Crandall JE, Dikkes P, Conner DA, Rayudu PV, Cheung W, Chen HS, Lipton SA, Nakanishi N (1998) Increased NMDA current and spine density in mice lacking the NMDA receptor subunit NR3A. Nature 393(6683):377–381. doi:10.1038/ 30748
114. Perez-Otano I, Schulteis CT, Contractor A, Lipton SA, Trimmer JS, Sucher NJ, Heinemann SF (2001) Assembly with the NR1 subunit is required for surface expression of NR3A-containing NMDA receptors. J Neurosci 21(4):1228–1237
115. Larsen RS, Corlew RJ, Henson MA, Roberts AC, Mishina M, Watanabe M, Lipton SA, Nakanishi N, Perez-Otano I, Weinberg RJ, Philpot BD (2011) NR3A-containing NMDARs promote neurotransmitter release and spike timing-dependent plasticity. Nat Neurosci 14(3):338–344. doi:10.1038/nn. 2750
116. Al-Hallaq RA, Jarabek BR, Fu Z, Vicini S, Wolfe BB, Yasuda RP (2002) Association of NR3A with the N-methyl-D-aspartate receptor NR1 and NR2 subunits. Mol Pharmacol 62(5):1119–1127
117. Matsuda K, Fletcher M, Kamiya Y, Yuzaki M (2003) Specific assembly with the NMDA receptor 3B subunit controls surface expression and calcium permeability of NMDA receptors. J Neurosci 23(31):10064–10073
118. Tong G, Takahashi H, Tu S, Shin Y, Talantova M, Zago W, Xia P, Nie Z, Goetz T, Zhang D, Lipton SA, Nakanishi N (2008) Modulation of NMDA receptor properties and synaptic transmission by the NR3A subunit in mouse hippocampal and cerebrocortical neurons. J Neurophysiol 99(1):122–132. doi:10.1152/ jn.01044.2006
119. Matsuda K, Kamiya Y, Matsuda S, Yuzaki M (2002) Cloning and characterization of a novel NMDA receptor subunit NR3B: a dominant subunit that reduces calcium

permeability. Mol Brain Res 100(1-2):43–52. doi:10.1016/S0169-328x(02)00173-0. pii S0169-328x(02)00173-0

120. Perez-Otano I, Lujan R, Tavalin SJ, Plomann M, Modregger J, Liu XB, Jones EG, Heinemann SF, Lo DC, Ehlers MD (2006) Endocytosis and synaptic removal of NR3A-containing NMDA receptors by PACSIN1/syndapin1. Nat Neurosci 9(5):611–621. doi:10.1038/nn1680

121. Chowdhury D, Marco S, Brooks IM, Zandueta A, Rao Y, Haucke V, Wesseling JF, Tavalin SJ, Perez-Otano I (2013) Tyrosine phosphorylation regulates the endocytosis and surface expression of GluN3A-containing NMDA receptors. J Neurosci 33 (9):4151–4164. doi:10.1523/JNEUROSCI.2721-12.2013

122. Yuan T, Mameli M, O'Connor EC, Dey PN, Verpelli C, Sala C, Perez-Otano I, Luscher C, Bellone C (2013) Expression of cocaine-evoked synaptic plasticity by GluN3A-containing NMDA receptors. Neuron 80 (4):1025–1038. doi:10.1016/j.neuron.2013.07.050

123. Chatterton JE, Awobuluyi M, Premkumar LS, Takahashi H, Talantova M, Shin Y, Cui J, Tu S, Sevarino KA, Nakanishi N, Tong G, Lipton SA, Zhang D (2002) Excitatory glycine receptors containing the NR3 family of NMDA receptor subunits. Nature 415 (6873):793–798. doi:10.1038/nature715

124. Ulbrich MH, Isacoff EY (2008) Rules of engagement for NMDA receptor subunits. Proc Natl Acad Sci U S A 105 (37):14163–14168. doi:10.1073/pnas.0802075105

125. Smothers CT, Woodward JJ (2007) Pharmacological characterization of glycine-activated currents in HEK 293 cells expressing N-methyl-D-aspartate NR1 and NR3 subunits. J Pharmacol Exp Ther 322(2):739–748. doi:10.1124/jpet.107.123836

126. Sasaki YF, Rothe T, Premkumar LS, Das S, Cui J, Talantova MV, Wong HK, Gong X, Chan SF, Zhang D, Nakanishi N, Sucher NJ, Lipton SA (2002) Characterization and comparison of the NR3A subunit of the NMDA receptor in recombinant systems and primary cortical neurons. J Neurophysiol 87 (4):2052–2063. doi:10.1152/jn.00531.2001

127. Madry C, Betz H, Geiger JR, Laube B (2010) Potentiation of glycine-gated NR1/NR3A NMDA receptors relieves Ca-dependent outward rectification. Front Mol Neurosci 3:6. doi:10.3389/fnmol.2010.00006

128. Madry C, Mesic I, Bartholomaus I, Nicke A, Betz H, Laube B (2007) Principal role of NR3 subunits in NR1/NR3 excitatory glycine receptor function. Biochem Biophys Res Commun 354(1):102–108. doi:10.1016/j.bbrc.2006.12.153

129. Awobuluyi M, Yang J, Ye Y, Chatterton JE, Godzik A, Lipton SA, Zhang D (2007) Subunit-specific roles of glycine-binding domains in activation of NR1/NR3 N-methyl-D-aspartate receptors. Mol Pharmacol 71(1):112–122. doi:10.1124/mol.106.030700

130. Kvist T, Greenwood JR, Hansen KB, Traynelis SF, Brauner-Osborne H (2013) Structure-based discovery of antagonists for GluN3-containing N-methyl-D-aspartate receptors. Neuropharmacology 75:324–336. doi:10.1016/j.neuropharm.2013.08.003

131. Al-Hallaq RA, Conrads TP, Veenstra TD, Wenthold RJ (2007) NMDA di-heteromeric receptor populations and associated proteins in rat hippocampus. J Neurosci 27 (31):8334–8343. doi:10.1523/JNEUROSCI.2155-07.2007

132. Rauner C, Kohr G (2011) Triheteromeric NR1/NR2A/NR2B receptors constitute the major N-methyl-D-aspartate receptor population in adult hippocampal synapses. J Biol Chem 286(9):7558–7566. doi:10.1074/jbc.M110.182600

133. Tovar KR, McGinley MJ, Westbrook GL (2013) Triheteromeric NMDA receptors at hippocampal synapses. J Neurosci 33 (21):9150–9160. doi:10.1523/JNEUROSCI.0829-13.2013

134. Luo JH, Wang YH, Yasuda RP, Dunah AW, Wolfe BB (1997) The majority of N-methyl-D-aspartate receptor complexes in adult rat cerebral cortex contain at least three different subunits (NR1/NR2A/NR2B). Mol Pharmacol 51(1):79–86

135. Sheng M, Cummings J, Roldan LA, Jan YN, Jan LY (1994) Changing subunit composition of heteromeric NMDA receptors during development of rat cortex. Nature 368 (6467):144–147. doi:10.1038/368144a0

136. Chazot PL, Stephenson FA (1997) Molecular dissection of native mammalian forebrain NMDA receptors containing the NR1 C2 exon: direct demonstration of NMDA receptors comprising NR1, NR2A, and NR2B subunits within the same complex. J Neurochem 69(5):2138–2144

137. Brickley SG, Misra C, Mok MH, Mishina M, Cull-Candy SG (2003) NR2B and NR2D subunits coassemble in cerebellar Golgi cells to form a distinct NMDA receptor subtype

restricted to extrasynaptic sites. J Neurosci 23 (12):4958–4966

138. Pina-Crespo JC, Gibb AJ (2002) Subtypes of NMDA receptors in new-born rat hippocampal granule cells. J Physiol 541(Pt 1):41–64
139. Jones S, Gibb AJ (2005) Functional NR2B- and NR2D-containing NMDA receptor channels in rat substantia nigra dopaminergic neurones. J Physiol 569(Pt 1):209–221. doi:10.1113/jphysiol.2005.095554
140. Chazot PL, Coleman SK, Cik M, Stephenson FA (1994) Molecular characterization of N-methyl-D-aspartate receptors expressed in mammalian cells yields evidence for the coexistence of three subunit types within a discrete receptor molecule. J Biol Chem 269 (39):24403–24409
141. Cathala L, Misra C, Cull-Candy S (2000) Developmental profile of the changing properties of NMDA receptors at cerebellar mossy fiber-granule cell synapses. J Neurosci 20 (16):5899–5905
142. Huang Z, Gibb AJ (2014) Mg2+ block properties of triheteromeric GluN1-GluN2B-GluN2D NMDA receptors on neonatal rat substantia nigra pars compacta dopaminergic neurones. J Physiol 592(10):2059–2078. doi:10.1113/jphysiol.2013.267864
143. Brothwell SL, Barber JL, Monaghan DT, Jane DE, Gibb AJ, Jones S (2008) NR2B- and NR2D-containing synaptic NMDA receptors in developing rat substantia nigra pars compacta dopaminergic neurones. J Physiol 586 (3):739–750. doi:10.1113/jphysiol.2007.144618
144. Sundstrom E, Whittemore S, Mo LL, Seiger A (1997) Analysis of NMDA receptors in the human spinal cord. Exp Neurol 148 (2):407–413
145. Dunah AW, Luo JH, Wang YH, Yasuda RP, Wolfe BB (1998) Subunit composition of N-methyl-D-aspartate receptors in the central nervous system that contain the NR2D subunit. Mol Pharmacol 53(3):429–437
146. Dunah AW, Standaert DG (2003) Subcellular segregation of distinct heteromeric NMDA glutamate receptors in the striatum. J Neurochem 85(4):935–943
147. Lu CY, Fu ZY, Karavanov I, Yasuda RP, Wolfe BB, Buonanno A, Vicini S (2006) NMDA receptor subtypes at autaptic synapses of cerebellar granule neurons. J Neurophysiol 96 (5):2282–2294. doi:10.1152/jn.00078.2006
148. Tovar KR, Westbrook GL (1999) The incorporation of NMDA receptors with a distinct subunit composition at nascent hippocampal synapses in vitro. J Neurosci 19 (10):4180–4188
149. Gray JA, Shi Y, Usui H, During MJ, Sakimura K, Nicoll RA (2011) Distinct modes of AMPA receptor suppression at developing synapses by GluN2A and GluN2B: single-cell NMDA receptor subunit deletion in vivo. Neuron 71(6):1085–1101. doi:10.1016/j.neuron.2011.08.007
150. Hatton CJ, Paoletti P (2005) Modulation of triheteromeric NMDA receptors by N-terminal domain ligands. Neuron 46 (2):261–274. doi:10.1016/j.neuron.2005.03.005
151. Hansen KB, Ogden KK, Yuan H, Traynelis SF (2014) Distinct functional and pharmacological properties of Triheteromeric GluN1/GluN2A/GluN2B NMDA receptors. Neuron 81(5):1084–1096. doi:10.1016/j.neuron.2014.01.035
152. Tang TT, Badger JD 2nd, Roche PA, Roche KW (2010) Novel approach to probe subunit-specific contributions to N-methyl-D-aspartate (NMDA) receptor trafficking reveals a dominant role for NR2B in receptor recycling. J Biol Chem 285 (27):20975–20981. doi:10.1074/jbc.M110.102210
153. Stroebel D, Carvalho S, Grand T, Zhu S, Paoletti P (2014) Controlling NMDA receptor subunit composition using ectopic retention signals. J Neurosci 34 (50):16630–16636. doi:10.1523/JNEUROSCI.2736-14.2014
154. Cheriyan J, Balsara RD, Hansen KB, Castellino FJ (2016) Pharmacology of triheteromeric N-methyl-D-aspartate receptors. Neurosci Lett 617:240–246. doi:10.1016/j.neulet.2016.02.032
155. Yi F, Mou T-C, Dorsett KN, Volkmann RA, Menniti FS, Sprang SR, Hansen KB (2016) Structural basis for negative allosteric modulation of GluN2A-containing NMDA receptors. Neuron 91(6):1316–1329. doi:10.1016/j.neuron.2016.08.014
156. Serraz B, Grand T, Paoletti P (2016) Altered zinc sensitivity of NMDA receptors harboring clinically-relevant mutations. Neuropharmacology 109:196–204. doi:10.1016/j.neuropharm.2016.06.008
157. Hackos DH, Lupardus PJ, Grand T, Chen Y, Wang TM, Reynen P, Gustafson A, Wallweber HJ, Volgraf M, Sellers BD, Schwarz JB, Paoletti P, Sheng M, Zhou Q, Hanson JE (2016) Positive allosteric modulators of GluN2A-containing NMDARs with distinct modes of action and impacts on circuit function.

Neuron 89(5):983–999. doi:10.1016/j.neuron.2016.01.016

158. Yuan H, Hansen KB, Zhang J, Pierson TM, Markello TC, Fajardo KV, Holloman CM, Golas G, Adams DR, Boerkoel CF, Gahl WA, Traynelis SF (2014) Functional analysis of a de novo GRIN2A missense mutation associated with early-onset epileptic encephalopathy. Nat Commun 5:3251. doi:10.1038/ncomms4251

159. Khatri A, Burger PB, Swanger SA, Hansen KB, Zimmerman S, Karakas E, Liotta DC, Furukawa H, Snyder JP, Traynelis SF (2014) Structural determinants and mechanism of action of a GluN2C-selective NMDA receptor positive allosteric modulator. Mol Pharmacol 86(5):548–560. doi:10.1124/mol.114.094516

160. Tajima N, Karakas E, Grant T, Simorowski N, Diaz-Avalos R, Grigorieff N, Furukawa H (2016) Activation of NMDA receptors and the mechanism of inhibition by ifenprodil. Nature 534(7605):63–68. doi:10.1038/nature17679

161. Zhu S, Stein RA, Yoshioka C, Lee CH, Goehring A, McHaourab HS, Gouaux E (2016) Mechanism of NMDA receptor inhibition and activation. Cell 165(3):704–714. doi:10.1016/j.cell.2016.03.028

162. Furukawa H, Gouaux E (2003) Mechanisms of activation, inhibition and specificity: crystal structures of the NMDA receptor NR1 ligand-binding core. EMBO J 22 (12):2873–2885. doi:10.1093/emboj/cdg303

163. Furukawa H, Singh SK, Mancusso R, Gouaux E (2005) Subunit arrangement and function in NMDA receptors. Nature 438 (7065):185–192. doi:10.1038/nature04089

164. Inanobe A, Furukawa H, Gouaux E (2005) Mechanism of partial agonist action at the NR1 subunit of NMDA receptors. Neuron 47(1):71–84. doi:10.1016/j.neuron.2005.05.022

165. Jespersen A, Tajima N, Fernandez-Cuervo G, Garnier-Amblard EC, Furukawa H (2014) Structural insights into competitive antagonism in NMDA receptors. Neuron 81 (2):366–378. doi:10.1016/j.neuron.2013.11.033

166. Hansen KB, Tajima N, Risgaard R, Perszyk RE, Jorgensen L, Vance KM, Ogden KK, Clausen RP, Furukawa H, Traynelis SF (2013) Structural determinants of agonist efficacy at the glutamate binding site of N-methyl-D-aspartate receptors. Mol Pharmacol 84(1):114–127. doi:10.1124/mol.113.085803

167. Yao Y, Belcher J, Berger AJ, Mayer ML, Lau AY (2013) Conformational analysis of NMDA receptor GluN1, GluN2, and GluN3 ligand-binding domains reveals subtype-specific characteristics. Structure 21 (10):1788–1799. doi:10.1016/j.str.2013.07.011

168. Vance KM, Simorowski N, Traynelis SF, Furukawa H (2011) Ligand-specific deactivation time course of GluN1/GluN2D NMDA receptors. Nat Commun 2:294. doi:10.1038/ncomms1295

169. Volgraf M, Sellers BD, Jiang Y, Wu G, Ly CQ, Villemure E, Pastor RM, Yuen PW, Lu A, Luo X, Liu M, Zhang S, Sun L, Fu Y, Lupardus PJ, Wallweber HJ, Liederer BM, Deshmukh G, Plise E, Tay S, Reynen P, Herrington J, Gustafson A, Liu Y, Dirksen A, Dietz MG, Liu Y, Wang TM, Hanson JE, Hackos D, Scearce-Levie K, Schwarz JB (2016) Discovery of GluN2A-selective NMDA receptor positive allosteric modulators (PAMs): tuning deactivation kinetics via structure-based design. J Med Chem 59(6):2760–2779. doi:10.1021/acs.jmedchem.5b02010

170. Kvist T, Steffensen TB, Greenwood JR, Mehrzad Tabrizi F, Hansen KB, Gajhede M, Pickering DS, Traynelis SF, Kastrup JS, Brauner-Osborne H (2013) Crystal structure and pharmacological characterization of a novel N-methyl-D-aspartate (NMDA) receptor antagonist at the GluN1 glycine binding site. J Biol Chem 288(46):33124–33135. doi:10.1074/jbc.M113.480210

171. Pohlsgaard J, Frydenvang K, Madsen U, Kastrup JS (2011) Lessons from more than 80 structures of the GluA2 ligand-binding domain in complex with agonists, antagonists and allosteric modulators. Neuropharmacology 60(1):135–150. doi:10.1016/j.neuropharm.2010.08.004

172. Kumar J, Mayer ML (2013) Functional insights from glutamate receptor ion channel structures. Annu Rev Physiol 75:313–337. doi:10.1146/annurev-physiol-030212-183711

173. Karakas E, Regan MC, Furukawa H (2015) Emerging structural insights into the function of ionotropic glutamate receptors. Trends Biochem Sci 40(6):328–337. doi:10.1016/j.tibs.2015.04.002

174. Borschel WF, Cummings KA, Tindell LK, Popescu GK (2015) Kinetic contributions to gating by interactions unique to N-methyl-D-aspartate (NMDA) receptors. J Biol Chem 290(44):26846–26855. doi:10.1074/jbc.M115.678656

175. Aizenman E, Lipton SA, Loring RH (1989) Selective modulation of NMDA responses by reduction and oxidation. Neuron 2 (3):1257–1263
176. Kohr G, Eckardt S, Luddens H, Monyer H, Seeburg PH (1994) NMDA receptor channels: subunit-specific potentiation by reducing agents. Neuron 12(5):1031–1040
177. Choi YB, Lipton SA (2000) Redox modulation of the NMDA receptor. Cell Mol Life Sci 57(11):1535–1541
178. Tang LH, Aizenman E (1993) The modulation of N-methyl-D-aspartate receptors by redox and alkylating reagents in rat cortical neurones in vitro. J Physiol 465:303–323
179. Sullivan JM, Traynelis SF, Chen HS, Escobar W, Heinemann SF, Lipton SA (1994) Identification of two cysteine residues that are required for redox modulation of the NMDA subtype of glutamate receptor. Neuron 13(4):929–936
180. Choi Y, Chen HV, Lipton SA (2001) Three pairs of cysteine residues mediate both redox and zn2+ modulation of the nmda receptor. J Neurosci 21(2):392–400
181. Talukder I, Kazi R, Wollmuth LP (2011) GluN1-specific redox effects on the kinetic mechanism of NMDA receptor activation. Biophys J 101(10):2389–2398. doi:10.1016/j.bpj.2011.10.015
182. Paganelli MA, Kussius CL, Popescu GK (2013) Role of cross-cleft contacts in NMDA receptor gating. PLoS One 8(11): e80953. doi:10.1371/journal.pone.0080953
183. Kalbaugh TL, VanDongen HM, VanDongen AM (2004) Ligand-binding residues integrate affinity and efficacy in the NMDA receptor. Mol Pharmacol 66(2):209–219. doi:10.1124/mol.66.2.209
184. Dolino DM, Cooper D, Ramaswamy S, Jaurich H, Landes CF, Jayaraman V (2015) Structural dynamics of the glycine-binding domain of the N-methyl-D-aspartate receptor. J Biol Chem 290(2):797–804. doi:10.1074/jbc.M114.605436
185. Dolino DM, Rezaei Adariani S, Shaikh SA, Jayaraman V, Sanabria H (2016) Conformational selection and submillisecond dynamics of the ligand-binding domain of the N-methyl-d-aspartate receptor. J Biol Chem 291(31):16175–16185. doi:10.1074/jbc.M116.721274
186. Dai J, Wollmuth LP, Zhou HX (2015) Mechanism-based mathematical model for gating of ionotropic glutamate receptors. J Phys Chem B 119(34):10934–10940. doi:10.1021/acs.jpcb.5b00521
187. Dai J, Zhou HX (2015) Reduced curvature of ligand-binding domain free-energy surface underlies partial agonism at NMDA receptors. Structure 23(1):228–236. doi:10.1016/j.str.2014.11.012
188. Regan MC, Romero-Hernandez A, Furukawa H (2015) A structural biology perspective on NMDA receptor pharmacology and function. Curr Opin Struct Biol 33:68–75. doi:10.1016/j.sbi.2015.07.012
189. Krupp JJ, Vissel B, Heinemann SF, Westbrook GL (1998) N-terminal domains in the NR2 subunit control desensitization of NMDA receptors. Neuron 20(2):317–327
190. Ogden KK, Traynelis SF (2013) Contribution of the M1 transmembrane helix and pre-M1 region to positive allosteric modulation and gating of N-methyl-D-aspartate receptors. Mol Pharmacol 83(5):1045–1056. doi:10.1124/mol.113.085209
191. Villarroel A, Regalado MP, Lerma J (1998) Glycine-independent NMDA receptor desensitization: localization of structural determinants. Neuron 20(2):329–339
192. Ren H, Honse Y, Karp BJ, Lipsky RH, Peoples RW (2003) A site in the fourth membrane-associated domain of the N-methyl-D-aspartate receptor regulates desensitization and ion channel gating. J Biol Chem 278(1):276–283
193. Schneggenburger R, Ascher P (1997) Coupling of permeation and gating in an NMDA-channel pore mutant. Neuron 18 (1):167–177
194. Alsaloum M, Kazi R, Gan Q, Amin J, Wollmuth LP (2016) A molecular determinant of subtype-specific desensitization in ionotropic glutamate receptors. J Neurosci 36 (9):2617–2622. doi:10.1523/JNEUROSCI.2667-15.2016
195. Kazi R, Gan Q, Talukder I, Markowitz M, Salussolia CL, Wollmuth LP (2013) Asynchronous movements prior to pore opening in NMDA receptors. J Neurosci 33 (29):12052–12066. doi:10.1523/JNEUROSCI.5780-12.2013
196. Talukder I, Borker P, Wollmuth LP (2010) Specific sites within the ligand-binding domain and ion channel linkers modulate NMDA receptor gating. J Neurosci 30 (35):11792–11804
197. Chang HR, Kuo CC (2008) The activation gate and gating mechanism of the NMDA receptor. J Neurosci 28(7):1546–1556. doi:10.1523/JNEUROSCI.3485-07.2008

198. Yuan H, Erreger K, Dravid SM, Traynelis SF (2005) Conserved structural and functional control of N-methyl-D-aspartate receptor gating by transmembrane domain M3. J Biol Chem 280(33):29708–29716
199. Jones KS, VanDongen HMA, VanDongen AMJ (2002) The NMDA receptor M3 segment is a conserved transduction element coupling ligand binding to channel opening. J Neurosci 22(6):2044–2053
200. Sobolevsky AI, Beck C, Wollmuth LP (2002) Molecular rearrangements of the extracellular vestibule in NMDAR channels during gating. Neuron 33(1):75–85
201. Beck C, Wollmuth LP, Seeburg PH, Sakmann B, Kuner T (1999) NMDAR channel segments forming the extracellular vestibule inferred from the accessibility of substituted cysteines. Neuron 22(3):559–570
202. Schorge S, Elenes S, Colquhoun D (2005) Maximum likelihood fitting of single channel NMDA activity with a mechanism composed of independent dimers of subunits. J Physiol 569(Pt 2):395–418
203. Popescu G, Robert A, Howe JR, Auerbach A (2004) Reaction mechanism determines NMDA receptor response to repetitive stimulation. Nature 430(7001):790–793
204. Banke TG, Traynelis SF (2003) Activation of NR1/NR2B NMDA receptors. Nat Neurosci 6(2):144–152. doi:10.1038/nn1000
205. Zhou Y, Auerbach A (2005) Gating reaction mechanisms for NMDA receptor channels. J Neurosci 25(35):7914–7923. doi:10.1523/Jneurosci.1471-05.2005
206. Kussius CL, Popescu GK (2009) Kinetic basis of partial agonism at NMDA receptors. Nat Neurosci 12(9):1114–U1110. doi:10.1038/nn.2361
207. Dravid SM, Prakash A, Traynelis SF (2008) Activation of recombinant NR1/NR2C NMDA receptors. J Physiol 586 (18):4425–4439. doi:10.1113/jphysiol.2008.158634
208. Amico-Ruvio SA, Popescu GK (2010) Stationary gating of GluN1/GluN2B receptors in intact membrane patches. Biophys J 98 (7):1160–1169. doi:10.1016/j.bpj.2009.12.4276
209. Benveniste M, Clements J, Vyklicky L Jr, Mayer ML (1990) A kinetic analysis of the modulation of N-methyl-D-aspartic acid receptors by glycine in mouse cultured hippocampal neurones. J Physiol 428:333–357
210. Banke TG, Dravid SM, Traynelis SF (2005) Protons trap NR1/NR2B NMDA receptors in a nonconducting state. J Neurosci 25 (1):42–51. doi:10.1523/JNEUROSCI.3154-04.2005
211. Erreger K, Traynelis SF (2008) Zinc inhibition of rat NR1/NR2A N-methyl-D-aspartate receptors. J Physiol 586(3):763–778. doi:10.1113/jphysiol.2007.143941
212. Amico-Ruvio SA, Murthy SE, Smith TP, Popescu GK (2011) Zinc effects on NMDA receptor gating kinetics. Biophys J 100 (8):1910–1918. doi:10.1016/j.bpj.2011.02.042
213. Kussius CL, Kaur N, Popescu GK (2009) Pregnanolone sulfate promotes desensitization of activated NMDA receptors. J Neurosci 29(21):6819–6827. doi:10.1523/JNEUROSCI.0281-09.2009
214. Paganelli MA, Popescu GK (2015) Actions of bupivacaine, a widely used local anesthetic, on NMDA receptor responses. J Neurosci 35 (2):831–842. doi:10.1523/JNEUROSCI.3578-14.2015
215. Dravid SM, Erreger K, Yuan H, Nicholson K, Le P, Lyuboslavsky P, Almonte A, Murray E, Mosely C, Barber J, French A, Balster R, Murray TF, Traynelis SF (2007) Subunit-specific mechanisms and proton sensitivity of NMDA receptor channel block. J Physiol 581 (Pt 1):107–128. doi:10.1113/jphysiol.2006.124958
216. Blanpied TA, Boeckman FA, Aizenman E, Johnson JW (1997) Trapping channel block of NMDA-activated responses by amantadine and memantine. J Neurophysiol 77 (1):309–323
217. Blanpied TA, Clarke RJ, Johnson JW (2005) Amantadine inhibits NMDA receptors by accelerating channel closure during channel block. J Neurosci 25(13):3312–3322. doi:10.1523/JNEUROSCI.4262-04.2005
218. Huettner JE, Bean BP (1988) Block of N-methyl-D-aspartate-activated current by the anticonvulsant MK-801: selective binding to open channels. Proc Natl Acad Sci U S A 85 (4):1307–1311
219. Rosenmund C, Stern-Bach Y, Stevens CF (1998) The tetrameric structure of a glutamate receptor channel. Science 280 (5369):1596–1599
220. Jin R, Banke TG, Mayer ML, Traynelis SF, Gouaux E (2003) Structural basis for partial agonist action at ionotropic glutamate receptors. Nat Neurosci 6(8):803–810. doi:10.1038/nn1091
221. Kristensen AS, Jenkins MA, Banke TG, Schousboe A, Makino Y, Johnson RC, Huganir R, Traynelis SF (2011) Mechanism of Ca2+/calmodulin-dependent kinase II regulation

of AMPA receptor gating. Nat Neurosci 14 (6):727–735. doi:10.1038/nn.2804

222. Erreger K, Geballe MT, Dravid SM, Snyder JP, Wyllie DJA, Traynelis SF (2005) Mechanism of partial agonism at NMDA receptors for a conformationally restricted glutamate analog. J Neurosci 25(34):7858–7866. doi:10.1523/Jneurosci.1613-05.2005
223. Wollmuth LP, Sobolevsky AI (2004) Structure and gating of the glutamate receptor ion channel. Trends Neurosci 27 (6):321–328. doi:10.1016/j.tins.2004.04.005
224. Sobolevsky AI, Rooney L, Wollmuth LP (2002) Staggering of subunits in NMDAR channels. Biophys J 83(6):3304–3314
225. Wollmuth LP, Kuner T, Seeburg PH, Sakmann B (1996) Differential contribution of the NR1- and NR2A-subunits to the selectivity filter of recombinant NMDA receptor channels. J Physiol 491(Pt 3):779–797
226. Burnashev N, Schoepfer R, Monyer H, Ruppersberg JP, Gunther W, Seeburg PH, Sakmann B (1992) Control by asparagine residues of calcium permeability and magnesium blockade in the NMDA receptor. Science 257(5075):1415–1419
227. Wollmuth LP, Kuner T, Sakmann B (1998) Adjacent asparagines in the NR2-subunit of the NMDA receptor channel control the voltage-dependent block by extracellular Mg2+. J Physiol 506(Pt 1):13–32
228. Sharma G, Stevens CF (1996) Interactions between two divalent ion binding sites in N-methyl-D-aspartate receptor channels. Proc Natl Acad Sci U S A 93(24):14170–14175
229. Schneggenburger R (1996) Simultaneous measurement of Ca2+ influx and reversal potentials in recombinant N-methyl-D-aspartate receptor channels. Biophys J 70 (5):2165–2174. doi:10.1016/S0006-3495(96)79782-5
230. Schneggenburger R (1998) Altered voltage dependence of fractional Ca2+ current in N-methyl-D-aspartate channel pore mutants with a decreased Ca2+ permeability. Biophys J 74(4):1790–1794. doi:10.1016/S0006-3495(98)77889-0
231. Burnashev N, Zhou Z, Neher E, Sakmann B (1995) Fractional calcium currents through recombinant Glur channels of the Nmda, Ampa and kainate receptor subtypes. J Physiol 485(2):403–418
232. Jatzke C, Watanabe J, Wollmuth LP (2002) Voltage and concentration dependence of Ca (2+) permeability in recombinant glutamate receptor subtypes. J Physiol 538(Pt 1):25–39
233. Premkumar LS, Auerbach A (1996) Identification of a high affinity divalent cation binding site near the entrance of the NMDA receptor channel. Neuron 16(4):869–880
234. Premkumar LS, Qin F, Auerbach A (1997) Subconductance states of a mutant NMDA receptor channel kinetics, calcium, and voltage dependence. J Gen Physiol 109 (2):181–189
235. Wyllie DJ, Behe P, Nassar M, Schoepfer R, Colquhoun D (1996) Single-channel currents from recombinant NMDA NR1a/NR2D receptors expressed in Xenopus oocytes. Proc Biol Sci 263 (1373):1079–1086. doi:10.1098/rspb.1996.0159
236. Watanabe J, Beck C, Kuner T, Premkumar LS, Wollmuth LP (2002) DRPEER: a motif in the extracellular vestibule conferring high Ca2+ flux rates in NMDA receptor channels. J Neurosci 22(23):10209–10216
237. Wada A, Takahashi H, Lipton SA, Chen HS (2006) NR3A modulates the outer vestibule of the "NMDA" receptor channel. J Neurosci 26(51):13156–13166. doi:10.1523/JNEUROSCI.2552-06.2006
238. Kawajiri S, Dingledine R (1993) Multiple structural determinants of voltage-dependent magnesium block in recombinant NMDA receptors. Neuropharmacology 32 (11):1203–1211
239. Vogt K, Mellor J, Tong G, Nicoll R (2000) The actions of synaptically released zinc at hippocampal mossy fiber synapses. Neuron 26(1):187–196
240. Christine CW, Choi DW (1990) Effect of zinc on NMDA receptor-mediated channel currents in cortical neurons. J Neurosci 10 (1):108–116
241. Clarke RJ, Johnson JW (2006) NMDA receptor NR2 subunit dependence of the slow component of magnesium unblock. J Neurosci 26(21):5825–5834. doi:10.1523/JNEUROSCI.0577-06.2006
242. Siegler Retchless B, Gao W, Johnson JW (2012) A single GluN2 subunit residue controls NMDA receptor channel properties via intersubunit interaction. Nat Neurosci 15 (3):406–413., S401-402. doi:10.1038/nn.3025
243. Qian A, Johnson JW (2006) Permeant ion effects on external Mg2+ block of NR1/2D NMDA receptors. J Neurosci 26 (42):10899–10910
244. Antonov SM, Johnson JW (1999) Permeant ion regulation of N-methyl-D-aspartate

receptor channel block by Mg(2+). Proc Natl Acad Sci U S A 96(25):14571–14576

245. Qian A, Antonov SM, Johnson JW (2002) Modulation by permeant ions of Mg(2+) inhibition of NMDA-activated whole-cell currents in rat cortical neurons. J Physiol 538(Pt 1):65–77

246. Zhu Y, Auerbach A (2001) K(+) occupancy of the N-methyl-d-aspartate receptor channel probed by Mg(2+) block. J Gen Physiol 117 (3):287–298

247. Zhu Y, Auerbach A (2001) Na(+) occupancy and Mg(2+) block of the n-methyl-d-aspartate receptor channel. J Gen Physiol 117 (3):275–286

248. Brackley PT, Bell DR, Choi SK, Nakanishi K, Usherwood PN (1993) Selective antagonism of native and cloned kainate and NMDA receptors by polyamine-containing toxins. J Pharmacol Exp Ther 266(3):1573–1580

249. Parsons CG, Quack G, Bresink I, Baran L, Przegalinski E, Kostowski W, Krzascik P, Hartmann S, Danysz W (1995) Comparison of the potency, kinetics and voltage-dependency of a series of uncompetitive NMDA receptor antagonists in vitro with anticonvulsive and motor impairment activity in vivo. Neuropharmacology 34 (10):1239–1258

250. Sobolevsky AI (1999) Two-component blocking kinetics of open NMDA channels by organic cations. Biochim Biophys Acta 1416(1-2):69–91

251. Barygin OI, Gmiro VE, Kim K, Magazanik LG, Tikhonov DB (2009) Blockade of NMDA receptor channels by 9-aminoacridine and its derivatives. Neurosci Lett 451 (1):29–33. doi:10.1016/j.neulet.2008.12.036

252. Bolshakov KV, Gmiro VE, Tikhonov DB, Magazanik LG (2003) Determinants of trapping block of N-methyl-d-aspartate receptor channels. J Neurochem 87(1):56–65

253. Benveniste M, Mayer ML (1995) Trapping of glutamate and glycine during open channel block of rat hippocampal neuron NMDA receptors by 9-aminoacridine. J Physiol 483 (Pt 2):367–384

254. Johnson JW, Glasgow NG, Povysheva NV (2015) Recent insights into the mode of action of memantine and ketamine. Curr Opin Pharmacol 20:54–63. doi:10.1016/j.coph.2014.11.006

255. Chen HS, Lipton SA (1997) Mechanism of memantine block of NMDA-activated channels in rat retinal ganglion cells: uncompetitive antagonism. J Physiol 499(Pt 1):27–46

256. Mealing GA, Lanthorn TH, Murray CL, Small DL, Morley P (1999) Differences in degree of trapping of low-affinity uncompetitive N-methyl-D-aspartic acid receptor antagonists with similar kinetics of block. J Pharmacol Exp Ther 288(1):204–210

257. Kotermanski SE, Wood JT, Johnson JW (2009) Memantine binding to a superficial site on NMDA receptors contributes to partial trapping. J Physiol 587(Pt 19):4589–4604. doi:10.1113/jphysiol.2009.176297

258. Sobolevsky AI, Yelshansky MV (2000) The trapping block of NMDA receptor channels in acutely isolated rat hippocampal neurones. J Physiol 526(Pt 3):493–506

259. Kotermanski SE, Johnson JW (2009) Mg2+ imparts NMDA receptor subtype selectivity to the Alzheimer's drug memantine. J Neurosci 29(9):2774–2779. doi:10.1523/JNEUROSCI.3703-08.2009

260. Parsons CG, Gruner R, Rozental J, Millar J, Lodge D (1993) Patch clamp studies on the kinetics and selectivity of N-methyl-D-aspartate receptor antagonism by memantine (1-amino-3,5-dimethyladamantan). Neuropharmacology 32(12):1337–1350

261. Chen HS, Lipton SA (2006) The chemical biology of clinically tolerated NMDA receptor antagonists. J Neurochem 97 (6):1611–1626. doi:10.1111/j.1471-4159.2006.03991.x

262. Karakas E, Simorowski N, Furukawa H (2009) Structure of the zinc-bound amino-terminal domain of the NMDA receptor NR2B subunit. EMBO J 28 (24):3910–3920. doi:10.1038/emboj.2009.338

263. Karakas E, Simorowski N, Furukawa H (2011) Subunit arrangement and phenylethanolamine binding in GluN1/GluN2B NMDA receptors. Nature 475 (7355):249–253. doi:10.1038/nature10180

264. Sobolevsky AI (2015) Structure and gating of tetrameric glutamate receptors. J Physiol 593 (1):29–38. doi:10.1113/jphysiol.2013.264911

265. Meyerson JR, Kumar J, Chittori S, Rao P, Pierson J, Bartesaghi A, Mayer ML, Subramaniam S (2014) Structural mechanism of glutamate receptor activation and desensitization. Nature 514(7522):328–334. doi:10.1038/nature13603

266. Hansen KB, Furukawa H, Traynelis SF (2010) Control of assembly and function of glutamate receptors by the amino-terminal

domain. Mol Pharmacol 78(4):535–549. doi:10.1124/mol.110.067157

267. Dutta A, Shrivastava IH, Sukumaran M, Greger IH, Bahar I (2012) Comparative dynamics of NMDA- and AMPA-glutamate receptor N-terminal domains. Structure 20 (11):1838–1849. doi:10.1016/j.str.2012.08.012
268. Sukumaran M, Rossmann M, Shrivastava I, Dutta A, Bahar I, Greger IH (2011) Dynamics and allosteric potential of the AMPA receptor N-terminal domain. EMBO J 30 (5):972–982. doi:10.1038/emboj.2011.17
269. Pasternack A, Coleman SK, Jouppila A, Mottershead DG, Lindfors M, Pasternack M, Keinanen K (2002) Alpha-amino-3-hydroxy-5-methyl-4-isoxazolepropionic acid (AMPA) receptor channels lacking the N-terminal domain. J Biol Chem 277 (51):49662–49667. doi:10.1074/jbc.M208349200
270. Gielen M, Siegler Retchless B, Mony L, Johnson JW, Paoletti P (2009) Mechanism of differential control of NMDA receptor activity by NR2 subunits. Nature 459 (7247):703–707. doi:10.1038/nature07993
271. Zhu S, Stroebel D, Yao CA, Taly A, Paoletti P (2013) Allosteric signaling and dynamics of the clamshell-like NMDA receptor GluN1 N-terminal domain. Nat Struct Mol Biol 20 (4):477–485. doi:10.1038/nsmb.2522
272. Gielen M, Le Goff A, Stroebel D, Johnson JW, Neyton J, Paoletti P (2008) Structural rearrangements of NR1/NR2A NMDA receptors during allosteric inhibition. Neuron 57(1):80–93. doi:10.1016/j.neuron.2007.11.021
273. Zhu S, Riou M, Yao CA, Carvalho S, Rodriguez PC, Bensaude O, Paoletti P, Ye S (2014) Genetically encoding a light switch in an ionotropic glutamate receptor reveals subunit-specific interfaces. Proc Natl Acad Sci U S A 111(16):6081–6086. doi:10.1073/pnas.1318808111
274. Choi YB, Lipton SA (1999) Identification and mechanism of action of two histidine residues underlying high-affinity Zn2+ inhibition of the NMDA receptor. Neuron 23 (1):171–180
275. Fayyazuddin A, Villarroel A, Le Goff A, Lerma J, Neyton J (2000) Four residues of the extracellular N-terminal domain of the NR2A subunit control high-affinity Zn2+ binding to NMDA receptors. Neuron 25 (3):683–694
276. Low CM, Zheng F, Lyuboslavsky P, Traynelis SF (2000) Molecular determinants of coordinated proton and zinc inhibition of N-methyl-D-aspartate NR1/NR2A receptors. Proc Natl Acad Sci U S A 97 (20):11062–11067. doi:10.1073/pnas.180307497
277. Williams K (1996) Separating dual effects of zinc at recombinant N-methyl-D-aspartate receptors. Neurosci Lett 215(1):9–12
278. Rachline J, Perin-Dureau F, Le Goff A, Neyton J, Paoletti P (2005) The micromolar zinc-binding domain on the NMDA receptor subunit NR2B. J Neurosci 25(2):308–317. doi:10.1523/JNEUROSCI.3967-04.2005
279. Paoletti P, Perin-Dureau F, Fayyazuddin A, Le Goff A, Callebaut I, Neyton J (2000) Molecular organization of a zinc binding n-terminal modulatory domain in a NMDA receptor subunit. Neuron 28(3):911–925
280. Stroebel D, Buhl DL, Knafels JD, Chanda PK, Green M, Sciabola S, Mony L, Paoletti P, Pandit J (2016) A novel binding mode reveals two distinct classes of NMDA receptor GluN2B-selective antagonists. Mol Pharmacol 89:541. doi:10.1124/mol.115.103036
281. Burger PB, Yuan H, Karakas E, Geballe M, Furukawa H, Liotta DC, Snyder JP, Traynelis SF (2012) Mapping the binding of GluN2B-selective N-methyl-D-aspartate receptor negative allosteric modulators. Mol Pharmacol 82(2):344–359. doi:10.1124/mol.112.078568
282. Mony L, Zhu S, Carvalho S, Paoletti P (2011) Molecular basis of positive allosteric modulation of GluN2B NMDA receptors by polyamines. EMBO J 30(15):3134–3146. doi:10.1038/emboj.2011.203
283. Sirrieh RE, MacLean DM, Jayaraman V (2015) Subtype-dependent N-methyl-D-aspartate receptor amino-terminal domain conformations and modulation by spermine. J Biol Chem 290(20):12812–12820. doi:10.1074/jbc.M115.649723
284. Paoletti P, Neyton J, Ascher P (1995) Glycine-independent and subunit-specific potentiation of NMDA responses by extracellular Mg2+. Neuron 15(5):1109–1120
285. Meddows E, Le Bourdelles B, Grimwood S, Wafford K, Sandhu S, Whiting P, McIlhinney RAJ (2001) Identification of molecular determinants that are important in the assembly of N-methyl-D-aspartate receptors. J Biol Chem 276(22):18795–18803. doi:10.1074/jbc.M101382200
286. Schorge S, Colquhoun D (2003) Studies of NMDA receptor function and stoichiometry with truncated and tandem subunits. J Neurosci 23(4):1151–1158

287. Papadakis M, Hawkins LM, Stephenson FA (2004) Appropriate NR1-NR1 disulfide-linked homodimer formation is requisite for efficient expression of functional, cell surface N-methyl-D-aspartate NR1/NR2 receptors. J Biol Chem 279(15):14703–14712
288. Qiu S, Hua YL, Yang F, Chen YZ, Luo JH (2005) Subunit assembly of N-methyl-d-aspartate receptors analyzed by fluorescence resonance energy transfer. J Biol Chem 280 (26):24923–24930
289. Schuler T, Mesic I, Madry C, Bartholomaus I, Laube B (2008) Formation of NR1/NR2 and NR1/NR3 heterodimers constitutes the initial step in N-methyl-D-aspartate receptor assembly. J Biol Chem 283(1):37–46. doi:10.1074/jbc.M703539200
290. Atlason PT, Garside ML, Meddows E, Whiting P, McIlhinney RA (2007) N-Methyl-D-aspartate (NMDA) receptor subunit NR1 forms the substrate for oligomeric assembly of the NMDA receptor. J Biol Chem 282 (35):25299–25307. doi:10.1074/jbc.M702778200
291. Farina AN, Blain KY, Maruo T, Kwiatkowski W, Choe S, Nakagawa T (2011) Separation of domain contacts is required for heterotetrameric assembly of functional NMDA receptors. J Neurosci 31(10):3565–3579. doi:10.1523/JNEUROSCI.6041-10.2011
292. Greger IH, Ziff EB, Penn AC (2007) Molecular determinants of AMPA receptor subunit assembly. Trends Neurosci 30(8):407–416. doi:10.1016/j.tins.2007.06.005
293. Sukumaran M, Penn AC, Greger IH (2012) AMPA receptor assembly: atomic determinants and built-in modulators. Adv Exp Med Biol 970:241–264. doi:10.1007/978-3-7091-0932-8_11
294. Qiu S, Zhang XM, Cao JY, Yang W, Yan YG, Shan L, Zheng J, Luo JH (2009) An endoplasmic reticulum retention signal located in the extracellular amino-terminal domain of the NR2A subunit of N-Methyl-D-aspartate receptors. J Biol Chem 284 (30):20285–20298
295. Mayer ML, Vyklicky L Jr, Clements J (1989) Regulation of NMDA receptor desensitization in mouse hippocampal neurons by glycine. Nature 338(6214):425–427. doi:10.1038/338425a0
296. Lester RAJ, Tong G, Jahr CE (1993) Interactions between the glycine and glutamate binding-sites of the Nmda receptor. J Neurosci 13(3):1088–1096
297. Chen N, Moshaver A, Raymond LA (1997) Differential sensitivity of recombinant N-methyl-D-aspartate receptor subtypes to zinc inhibition. Mol Pharmacol 51(6):1015–1023
298. Erreger K, Traynelis SF (2005) Allosteric interaction between zinc and glutamate binding domains on NR2A causes desensitization of NMDA receptors. J Physiol 569 (2):381–393. doi:10.1113/jphysiol.2005.095497
299. Zheng F, Erreger K, Low CM, Banke T, Lee CJ, Conn PJ, Traynelis SF (2001) Allosteric interaction between the amino terminal domain and the ligand binding domain of NR2A. Nat Neurosci 4(9):894–901
300. Clark GD, Clifford DB, Zorumski CF (1990) The effect of agonist concentration, membrane voltage and calcium on N-methyl-D-aspartate receptor desensitization. Neuroscience 39(3):787–797
301. Legendre P, Rosenmund C, Westbrook GL (1993) Inactivation of NMDA channels in cultured hippocampal neurons by intracellular calcium. J Neurosci 13(2):674–684
302. Rosenmund C, Feltz A, Westbrook GL (1995) Calcium-dependent inactivation of synaptic NMDA receptors in hippocampal neurons. J Neurophysiol 73(1):427–430
303. Vyklicky L Jr (1993) Calcium-mediated modulation of N-methyl-D-aspartate (NMDA) responses in cultured rat hippocampal neurones. J Physiol 470:575–600
304. Krupp JJ, Vissel B, Heinemann SF, Westbrook GL (1996) Calcium-dependent inactivation of recombinant N-methyl-D-aspartate receptors is NR2 subunit specific. Mol Pharmacol 50(6):1680–1688
305. Medina I, Filippova N, Charton G, Rougeole S, Ben-Ari Y, Khrestchatisky M, Bregestovski P (1995) Calcium-dependent inactivation of heteromeric NMDA receptor-channels expressed in human embryonic kidney cells. J Physiol 482(Pt 3):567–573
306. Rosenmund C, Westbrook GL (1993) Calcium-induced actin depolymerization reduces NMDA channel activity. Neuron 10 (5):805–814
307. Ehlers MD, Fung ET, O'Brien RJ, Huganir RL (1998) Splice variant-specific interaction of the NMDA receptor subunit NR1 with neuronal intermediate filaments. J Neurosci 18(2):720–730
308. Ehlers MD, Zhang S, Bernhadt JP, Huganir RL (1996) Inactivation of NMDA receptors by direct interaction of calmodulin with the NR1 subunit. Cell 84(5):745–755
309. Krupp JJ, Vissel B, Thomas CG, Heinemann SF, Westbrook GL (1999) Interactions of calmodulin and alpha-actinin with the NR1

subunit modulate Ca2+-dependent inactivation of NMDA receptors. J Neurosci 19 (4):1165–1178

310. Zhang S, Ehlers MD, Bernhardt JP, Su CT, Huganir RL (1998) Calmodulin mediates calcium-dependent inactivation of N-methyl-D-aspartate receptors. Neuron 21 (2):443–453

311. Sather W, Dieudonne S, MacDonald JF, Ascher P (1992) Activation and desensitization of N-methyl-D-aspartate receptors in nucleated outside-out patches from mouse neurones. J Physiol 450:643–672

312. Sather W, Johnson JW, Henderson G, Ascher P (1990) Glycine-insensitive desensitization of NMDA responses in cultured mouse embryonic neurons. Neuron 4(5):725–731

313. Chen NS, Li B, Murphy TH, Raymond LA (2004) Site within N-methyl-D-aspartate receptor pore modulates channel Gating. Mol Pharmacol 65(1):157–164. doi:10.1124/mol.65.1.157

314. Hu B, Zheng F (2005) Molecular determinants of glycine-independent desensitization of NR1/NR2A receptors. J Pharmacol Exp Ther 313(2):563–569. doi:10.1124/jpet.104.080168

315. Giffard RG, Monyer H, Christine CW, Choi DW (1990) Acidosis reduces NMDA receptor activation, glutamate neurotoxicity, and oxygen-glucose deprivation neuronal injury in cortical cultures. Brain Res 506 (2):339–342

316. Traynelis SF, Cull-Candy SG (1990) Proton inhibition of N-methyl-D-aspartate receptors in cerebellar neurons. Nature 345 (6273):347–350

317. Traynelis SF, Cull-Candy SG (1991) Pharmacological properties and H+ sensitivity of excitatory amino acid receptor channels in rat cerebellar granule neurones. J Physiol 433:727–763

318. Vyklicky L Jr, Vlachova V, Krusek J (1990) The effect of external pH changes on responses to excitatory amino acids in mouse hippocampal neurones. J Physiol 430:497–517

319. Chesler M (2003) Regulation and modulation of pH in the brain. Physiol Rev 83 (4):1183–1221. doi:10.1152/physrev.00010.2003

320. Low CM, Lyuboslavsky P, French A, Le P, Wyatte K, Thiel WH, Marchan EM, Igarashi K, Kashiwagi K, Gernert K, Williams K, Traynelis SF, Zheng F (2003) Molecular determinants of proton-sensitive N-methyl-D-aspartate receptor gating. Mol Pharmacol 63 (6):1212–1222. doi:10.1124/mol.63.6.1212

321. Kashiwagi K, Fukuchi J, Chao J, Igarashi K, Williams K (1996) An aspartate residue in the extracellular loop of the N-methyl-D-aspartate receptor controls sensitivity to spermine and protons. Mol Pharmacol 49 (6):1131–1141

322. Kashiwagi K, Pahk AJ, Masuko T, Igarashi K, Williams K (1997) Block and modulation of N-methyl-D-aspartate receptors by polyamines and protons: role of amino acid residues in the transmembrane and pore-forming regions of NR1 and NR2 subunits. Mol Pharmacol 52(4):701–713

323. Bhatt JM, Prakash A, Suryavanshi PS, Dravid SM (2013) Effect of ifenprodil on GluN1/GluN2B N-methyl-D-aspartate receptor gating. Mol Pharmacol 83(1):9–21. doi:10.1124/mol.112.080952

324. Anderson CT, Radford RJ, Zastrow ML, Zhang DY, Apfel UP, Lippard SJ, Tzounopoulos T (2015) Modulation of extrasynaptic NMDA receptors by synaptic and tonic zinc. Proc Natl Acad Sci U S A 112(20): E2705–E2714. doi:10.1073/pnas.1503348112

325. Sanz-Clemente A, Nicoll RA, Roche KW (2013) Diversity in NMDA receptor composition: many regulators, many consequences. Neuroscientist 19(1):62–75. doi:10.1177/1073858411435129

326. Lussier MP, Sanz-Clemente A, Roche KW (2015) Dynamic Regulation of N-Methyl-d-aspartate (NMDA) and alpha-Amino-3-hydroxy-5-methyl-4-isoxazolepropionic Acid (AMPA) Receptors by Posttranslational Modifications. J Biol Chem 290 (48):28596–28603. doi:10.1074/jbc.R115.652750

327. Sans N, Petralia RS, Wang YX, Blahos J, Hell JW, Wenthold RJ (2000) A developmental change in NMDA receptor-associated proteins at hippocampal synapses. J Neurosci 20 (3):1260–1271

328. Groc L, Heine M, Cousins SL, Stephenson FA, Lounis B, Cognet L, Choquet D (2006) NMDA receptor surface mobility depends on NR2A-2B subunits. Proc Natl Acad Sci U S A 103(49):18769–18774. doi:10.1073/pnas.0605238103

329. Steigerwald F, Schulz TW, Schenker LT, Kennedy MB, Seeburg PH, Kohr G (2000) C-Terminal truncation of NR2A subunits impairs synaptic but not extrasynaptic localization of NMDA receptors. J Neurosci 20 (12):4573–4581

330. Martel MA, Wyllie DJ, Hardingham GE (2009) In developing hippocampal neurons, NR2B-containing N-methyl-D-aspartate receptors (NMDARs) can mediate signaling to neuronal survival and synaptic potentiation, as well as neuronal death. Neuroscience 158(1):334–343. doi:10.1016/j.neuroscience.2008.01.080
331. Omkumar RV, Kiely MJ, Rosenstein AJ, Min KT, Kennedy MB (1996) Identification of a phosphorylation site for calcium/calmodulin-dependent protein kinase II in the NR2B subunit of the N-methyl-D-aspartate receptor. J Biol Chem 271(49):31670–31678
332. Raveendran R, Devi Suma Priya S, Mayadevi M, Steephan M, Santhoshkumar TR, Cheriyan J, Sanalkumar R, Pradeep KK, James J, Omkumar RV (2009) Phosphorylation status of the NR2B subunit of NMDA receptor regulates its interaction with calcium/calmodulin-dependent protein kinase II. J Neurochem 110(1):92–105. doi:10.1111/j.1471-4159.2009.06108.x
333. Rellos P, Pike AC, Niesen FH, Salah E, Lee WH, von Delft F, Knapp S (2010) Structure of the CaMKIIdelta/calmodulin complex reveals the molecular mechanism of CaMKII kinase activation. PLoS Biol 8(7):e1000426. doi:10.1371/journal.pbio.1000426
334. Strack S, Colbran RJ (1998) Autophosphorylation-dependent targeting of calcium/calmodulin-dependent protein kinase II by the NR2B subunit of the N-methyl-D-aspartate receptor. J Biol Chem 273(33):20689–20692
335. Bayer KU, De Koninck P, Leonard AS, Hell JW, Schulman H (2001) Interaction with the NMDA receptor locks CaMKII in an active conformation. Nature 411(6839):801–805. doi:10.1038/35081080
336. Bayer KU, LeBel E, McDonald GL, O'Leary H, Schulman H, De Koninck P (2006) Transition from reversible to persistent binding of CaMKII to postsynaptic sites and NR2B. J Neurosci 26(4):1164–1174. doi:10.1523/JNEUROSCI.3116-05.2006
337. Otmakhov N, Tao-Cheng JH, Carpenter S, Asrican B, Dosemeci A, Reese TS, Lisman J (2004) Persistent accumulation of calcium/calmodulin-dependent protein kinase II in dendritic spines after induction of NMDA receptor-dependent chemical long-term potentiation. J Neurosci 24(42):9324–9331. doi:10.1523/JNEUROSCI.2350-04.2004
338. Lisman J, Yasuda R, Raghavachari S (2012) Mechanisms of CaMKII action in long-term potentiation. Nat Rev Neurosci 13(3):169–182. doi:10.1038/nrn3192
339. Chen BS, Roche KW (2007) Regulation of NMDA receptors by phosphorylation. Neuropharmacology 53(3):362–368. doi:10.1016/j.neuropharm.2007.05.018
340. Li S, Tian X, Hartley DM, Feig LA (2006) Distinct roles for Ras-guanine nucleotide-releasing factor 1 (Ras-GRF1) and Ras-GRF2 in the induction of long-term potentiation and long-term depression. J Neurosci 26(6):1721–1729. doi:10.1523/JNEUROSCI.3990-05.2006
341. Tian X, Gotoh T, Tsuji K, Lo EH, Huang S, Feig LA (2004) Developmentally regulated role for Ras-GRFs in coupling NMDA glutamate receptors to Ras, Erk and CREB. EMBO J 23(7):1567–1575. doi:10.1038/sj.emboj.7600151
342. Krapivinsky G, Krapivinsky L, Manasian Y, Ivanov A, Tyzio R, Pellegrino C, Ben-Ari Y, Clapham DE, Medina I (2003) The NMDA receptor is coupled to the ERK pathway by a direct interaction between NR2B and RasGRF1. Neuron 40(4):775–784
343. Dore K, Aow J, Malinow R (2016) The emergence of NMDA receptor metabotropic function: insights from imaging. Front Synaptic Neurosci 8:20. doi:10.3389/fnsyn.2016.00020
344. Gray JA, Zito K, Hell JW (2016) Non-ionotropic signaling by the NMDA receptor: controversy and opportunity. F1000Res 5. doi:10.12688/f1000research.8366.1
345. Nong Y, Huang YQ, Ju W, Kalia LV, Ahmadian G, Wang YT, Salter MW (2003) Glycine binding primes NMDA receptor internalization. Nature 422(6929):302–307. doi:10.1038/nature01497
346. Nabavi S, Kessels HW, Alfonso S, Aow J, Fox R, Malinow R (2013) Metabotropic NMDA receptor function is required for NMDA receptor-dependent long-term depression. Proc Natl Acad Sci U S A 110(10):4027–4032. doi:10.1073/pnas.1219454110
347. Aow J, Dore K, Malinow R (2015) Conformational signaling required for synaptic plasticity by the NMDA receptor complex. Proc Natl Acad Sci U S A 112(47):14711–14716. doi:10.1073/pnas.1520029112
348. Williams K (1993) Ifenprodil discriminates subtypes of the N-methyl-D-aspartate receptor - selectivity and mechanisms at recombinant heteromeric receptors. Mol Pharmacol 44(4):851–859
349. Santangelo RM, Acker TM, Zimmerman SS, Katzman BM, Strong KL, Traynelis SF, Liotta DC (2012) Novel NMDA receptor

modulators: an update. Expert Opin Ther Pat 22(11):1337–1352. doi:10.1517/13543776.2012.728587

350. Auberson YP, Allgeier H, Bischoff S, Lingenhoehl K, Moretti R, Schmutz M (2002) 5-phosphonomethylquinoxalinediones as competitive NMDA receptor antagonists with a preference for the human 1A/2A, rather than 1A/2B receptor composition. Bioorg Med Chem Lett 12(7):1099–1102. doi:10.1016/S0960-894x(02)00074-4. pii S0960-894x(02)00074-4
351. Frizelle PA, Chen PE, Wyllie DJ (2006) Equilibrium constants for (R)-[(S)-1-(4-bromophenyl)-ethylamino]-(2,3-dioxo-1,2,3,4-tetrahydroquino xalin-5-yl)-methyl]-phosphonic acid (NVP-AAM077) acting at recombinant NR1/NR2A and NR1/NR2B N-methyl-D-aspartate receptors: implications for studies of synaptic transmission. Mol Pharmacol 70 (3):1022–1032
352. Neyton J, Paoletti P (2006) Relating NMDA receptor function to receptor subunit composition: limitations of the pharmacological approach. J Neurosci 26(5):1331–1333. doi:10.1523/JNEUROSCI.5242-05.2006
353. Bettini E, Sava A, Griffante C, Carignani C, Buson A, Capelli AM, Negri M, Andreetta F, Senar-Sancho SA, Guiral L, Cardullo F (2010) Identification and characterization of novel NMDA receptor antagonists selective for NR2A- over NR2B-containing receptors. J Pharmacol Exp Ther 335(3):636–644. doi:10.1124/jpet.110.172544
354. Edman S, McKay S, Macdonald LJ, Samadi M, Livesey MR, Hardingham GE, Wyllie DJ (2012) TCN 201 selectively blocks GluN2A-containing NMDARs in a GluN1 co-agonist dependent but non-competitive manner. Neuropharmacology 63(3):441–449. doi:10.1016/j.neuropharm.2012.04.027
355. Hansen KB, Ogden KK, Traynelis SF (2012) Subunit-selective allosteric inhibition of glycine binding to NMDA receptors. J Neurosci 32(18):6197–6208. doi:10.1523/JNEUROSCI.5757-11.2012
356. Volkmann RA, Fanger CM, Anderson DR, Sirivolu VR, Paschetto K, Gordon E, Virginio C, Gleyzes M, Buisson B, Steidl E, Mierau SB, Fagiolini M, Menniti FS (2016) MPX-004 and MPX-007: new pharmacological tools to study the physiology of NMDA receptors containing the GluN2A subunit. PLoS One 11(2):e0148129. doi:10.1371/journal.pone.0148129
357. McKay S, Griffiths NH, Butters PA, Thubron EB, Hardingham GE, Wyllie DJ (2012) Direct pharmacological monitoring of the developmental switch in NMDA receptor subunit composition using TCN 213, a GluN2A-selective, glycine-dependent antagonist. Br J Pharmacol 166(3):924–937. doi:10.1111/j.1476-5381.2011.01748.x
358. Izumi Y, Zorumski CF (2015) Sensitivity of N-methyl-D-aspartate receptor-mediated excitatory postsynaptic potentials and synaptic plasticity to TCN 201 and TCN 213 in rat hippocampal slices. J Pharmacol Exp Ther 352(2):267–273. doi:10.1124/jpet.114.220582
359. Gipson CD, Reissner KJ, Kupchik YM, Smith AC, Stankeviciute N, Hensley-Simon ME, Kalivas PW (2013) Reinstatement of nicotine seeking is mediated by glutamatergic plasticity. Proc Natl Acad Sci U S A 110 (22):9124–9129. doi:10.1073/pnas.1220591110
360. Bu F, Du R, Li Y, Quinn JP, Wang M (2016) NR2A contributes to genesis and propagation of cortical spreading depression in rats. Sci Rep 6:23576. doi:10.1038/srep23576
361. Masuko T, Kashiwagi K, Kuno T, Nguyen ND, Pahk AJ, Fukuchi J, Igarashi K, Williams K (1999) A regulatory domain (R1-R2) in the amino terminus of the N-methyl-D-aspartate receptor: effects of spermine, protons, and ifenprodil, and structural similarity to bacterial leucine/isoleucine/valine binding protein. Mol Pharmacol 55(6):957–969
362. Kew JN, Trube G, Kemp JA (1996) A novel mechanism of activity-dependent NMDA receptor antagonism describes the effect of ifenprodil in rat cultured cortical neurones. J Physiol 497(Pt 3):761–772
363. Chenard BL, Bordner J, Butler TW, Chambers LK, Collins MA, De Costa DL, Ducat MF, Dumont ML, Fox CB, Mena EE et al (1995) (1S,2S)-1-(4-hydroxyphenyl)-2-(4-hydroxy-4-phenylpiperidino)-1-propanol: a potent new neuroprotectant which blocks N-methyl-D-aspartate responses. J Med Chem 38(16):3138–3145
364. Fischer G, Mutel V, Trube G, Malherbe P, Kew JN, Mohacsi E, Heitz MP, Kemp JA (1997) Ro 25-6981, a highly potent and selective blocker of N-methyl-D-aspartate receptors containing the NR2B subunit. Characterization in vitro. J Pharmacol Exp Ther 283(3):1285–1292
365. Yuan H, Myers SJ, Wells G, Nicholson KL, Swanger SA, Lyuboslavsky P, Tahirovic YA, Menaldino DS, Ganesh T, Wilson LJ, Liotta DC, Snyder JP, Traynelis SF (2015) Context-Dependent GluN2B-Selective Inhibitors of NMDA Receptor Function Are Neuroprotective with Minimal Side Effects. Neuron 85

(6):1305–1318. doi:10.1016/j.neuron.2015.02.008

366. Hashimoto K, Malchow B, Falkai P, Schmitt A (2013) Glutamate modulators as potential therapeutic drugs in schizophrenia and affective disorders. Eur Arch Psychiatry Clin Neurosci 263(5):367–377. doi:10.1007/s00406-013-0399-y
367. Shipton OA (1633) Paulsen O (2014) GluN2A and GluN2B subunit-containing NMDA receptors in hippocampal plasticity. Philos Trans R Soc Lond B Biol Sci 369:20130163. doi:10.1098/rstb.2013.0163
368. Lai TW, Zhang S, Wang YT (2014) Excitotoxicity and stroke: identifying novel targets for neuroprotection. Prog Neurobiol 115:157–188. doi:10.1016/j.pneurobio.2013.11.006
369. Yurkewicz L, Weaver J, Bullock MR, Marshall LF (2005) The effect of the selective NMDA receptor antagonist traxoprodil in the treatment of traumatic brain injury. J Neurotrauma 22(12):1428–1443. doi:10.1089/neu.2005.22.1428
370. Preskorn S, Macaluso M, Mehra DO, Zammit G, Moskal JR, Burch RM, Group G-CS (2015) Randomized proof of concept trial of GLYX-13, an N-methyl-D-aspartate receptor glycine site partial agonist, in major depressive disorder nonresponsive to a previous antidepressant agent. J Psychiatr Pract 21 (2):140–149. doi:10.1097/01.pra.0000462606.17725.93
371. Farin A, Marshall LF (2004) Lessons from epidemiologic studies in clinical trials of traumatic brain injury. Acta Neurochir Suppl 89:101–107
372. Feng B, Morley RM, Jane DE, Monaghan DT (2005) The effect of competitive antagonist chain length on NMDA receptor subunit selectivity. Neuropharmacology 48 (3):354–359. doi:10.1016/j.neuropharm.2004.11.004
373. Feng B, Tse HW, Skifter DA, Morley R, Jane DE, Monaghan DT (2004) Structure-activity analysis of a novel NR2C/NR2D-preferring NMDA receptor antagonist: 1-(phenanthrene-2-carbonyl) piperazine-2,3-dicarboxylic acid. Br J Pharmacol 141(3):508–516. doi:10.1038/sj.bjp.0705644
374. Buller AL, Monaghan DT (1997) Pharmacological heterogeneity of NMDA receptors: characterization of NR1a/NR2D heteromers expressed in Xenopus oocytes. Eur J Pharmacol 320(1):87–94
375. Morley RM, Tse HW, Feng B, Miller JC, Monaghan DT, Jane DE (2005) Synthesis and pharmacology of N1-substituted piperazine-2,3-dicarboxylic acid derivatives acting as NMDA receptor antagonists. J Med Chem 48(7):2627–2637
376. Costa BM, Feng B, Tsintsadze TS, Morley RM, Irvine MW, Tsintsadze V, Lozovaya NA, Jane DE, Monaghan DT (2009) N-methyl-D-aspartate (NMDA) receptor NR2 subunit selectivity of a series of novel piperazine-2,3-dicarboxylate derivatives: preferential blockade of extrasynaptic NMDA receptors in the rat hippocampal CA3-CA1 synapse. J Pharmacol Exp Ther 331 (2):618–626. doi:10.1124/jpet.109.156752
377. Suarez F, Zhao Q, Monaghan DT, Jane DE, Jones S, Gibb AJ (2010) Functional heterogeneity of NMDA receptors in rat substantia nigra pars compacta and reticulata neurones. Eur J Neurosci 32(3):359–367. doi:10.1111/j.1460-9568.2010.07298.x
378. Lozovaya NA, Grebenyuk SE, Tsintsadze T, Feng B, Monaghan DT, Krishtal OA (2004) Extrasynaptic NR2B and NR2D subunits of NMDA receptors shape 'superslow' afterburst EPSC in rat hippocampus. J Physiol 558(Pt 2):451–463. doi:10.1113/jphysiol.2004.063792
379. Harney SC, Jane DE, Anwyl R (2008) Extrasynaptic NR2D-containing NMDARs are recruited to the synapse during LTP of NMDAR-EPSCs. J Neurosci 28 (45):11685–11694. doi:10.1523/JNEUROSCI.3035-08.2008
380. Costa BM, Irvine MW, Fang G, Eaves RJ, Mayo-Martin MB, Skifter DA, Jane DE, Monaghan DT (2010) A novel family of negative and positive allosteric modulators of NMDA receptors. J Pharmacol Exp Ther 335(3):614–621. doi:10.1124/jpet.110.174144
381. Costa BM, Irvine MW, Fang G, Eaves RJ, Mayo-Martin MB, Laube B, Jane DE, Monaghan DT (2012) Structure-activity relationships for allosteric NMDA receptor inhibitors based on 2-naphthoic acid. Neuropharmacology 62(4):1730–1736. doi:10.1016/j.neuropharm.2011.11.019
382. Mosley CA, Acker TM, Hansen KB, Mullasseril P, Andersen KT, Le P, Vellano KM, Brauner-Osborne H, Liotta DC, Traynelis SF (2010) Quinazolin-4-one derivatives: a novel class of noncompetitive NR2C/D subunit-selective N-methyl-D-aspartate receptor antagonists. J Med Chem 53 (15):5476–5490. doi:10.1021/jm100027p

383. Hansen KB, Traynelis SF (2011) Structural and mechanistic determinants of a novel site for noncompetitive inhibition of GluN2D-containing NMDA receptors. J Neurosci 31 (10):3650–3661. doi:10.1523/JNEUROSCI.5565-10.2011

384. Acker TM, Yuan H, Hansen KB, Vance KM, Ogden KK, Jensen HS, Burger PB, Mullasseril P, Snyder JP, Liotta DC, Traynelis SF (2011) Mechanism for noncompetitive inhibition by novel GluN2C/D N-methyl-D-aspartate receptor subunit-selective modulators. Mol Pharmacol 80(5):782–795. doi:10.1124/mol.111.073239

385. Acker TM, Khatri A, Vance KM, Slabber C, Bacsa J, Snyder JP, Traynelis SF, Liotta DC (2013) Structure-activity relationships and pharmacophore model of a noncompetitive pyrazoline containing class of GluN2C/GluN2D selective antagonists. J Med Chem 56(16):6434–6456. doi:10.1021/jm400652r

386. Hildebrand ME, Pitcher GM, Harding EK, Li H, Beggs S, Salter MW (2014) GluN2B and GluN2D NMDARs dominate synaptic responses in the adult spinal cord. Sci Rep 4:4094. doi:10.1038/srep04094

387. Pearlstein E, Gouty-Colomer LA, Michel FJ, Cloarec R, Hammond C (2015) Glutamatergic synaptic currents of nigral dopaminergic neurons follow a postnatal developmental sequence. Front Cell Neurosci 9:210. doi:10.3389/fncel.2015.00210

388. Ashhad S, Narayanan R (2016) Active dendrites regulate the impact of gliotransmission on rat hippocampal pyramidal neurons. Proc Natl Acad Sci U S A 113(23):E3280–E3289. doi:10.1073/pnas.1522180113

389. Mullasseril P, Hansen KB, Vance KM, Ogden KK, Yuan H, Kurtkaya NL, Santangelo R, Orr AG, Le P, Vellano KM, Liotta DC, Traynelis SF (2010) A subunit-selective potentiator of NR2C- and NR2D-containing NMDA receptors. Nat Commun 1:90. doi:10.1038/ncomms1085

390. Santangelo Freel RM, Ogden KK, Strong KL, Khatri A, Chepiga KM, Jensen HS, Traynelis SF, Liotta DC (2013) Synthesis and structure activity relationship of tetrahydroisoquinoline-based potentiators of GluN2C and GluN2D containing N-methyl-D-aspartate receptors. J Med Chem 56 (13):5351–5381. doi:10.1021/jm400177t

391. Santangelo Freel RM, Ogden KK, Strong KL, Khatri A, Chepiga KM, Jensen HS, Traynelis SF, Liotta DC (2014) Correction to Synthesis and Structure Activity Relationship of Tetrahydroisoquinoline-Based Potentiators of GluN2C and GluN2D Containing N-Methyl-d-aspartate Receptors. J Med Chem 57(11):4975. doi:10.1021/jm500710w

392. Perszyk RE, DiRaddo JO, Strong KL, Low CM, Ogden KK, Khatri A, Vargish GA, Pelkey KA, Tricoire L, Liotta DC, Smith Y, McBain CJ, Traynelis SF (2016) GluN2D-containing NMDA receptors mediate synaptic transmission in hippocampal interneurons and regulate interneuron activity. Mol Pharmacol 90:689. doi:10.1124/mol.116.105130

393. Ogden KK, Khatri A, Traynelis SF, Heldt SA (2014) Potentiation of GluN2C/D NMDA receptor subtypes in the amygdala facilitates the retention of fear and extinction learning in mice. Neuropsychopharmacology 39 (3):625–637. doi:10.1038/npp.2013.241

394. Suryavanshi PS, Ugale RR, Yilmazer-Hanke D, Stairs DJ, Dravid SM (2014) GluN2C/GluN2D subunit-selective NMDA receptor potentiator CIQ reverses MK-801-induced impairment in prepulse inhibition and working memory in Y-maze test in mice. Br J Pharmacol 171(3):799–809. doi:10.1111/bph.12518

395. Yamamoto H, Kamegaya E, Sawada W, Hasegawa R, Yamamoto T, Hagino Y, Takamatsu Y, Imai K, Koga H, Mishina M, Ikeda K (2013) Involvement of the N-methyl-D-aspartate receptor GluN2D subunit in phencyclidine-induced motor impairment, gene expression, and increased Fos immunoreactivity. Mol Brain 6:56. doi:10.1186/1756-6606-6-56

396. Zhang X, Feng ZJ, Chergui K (2014) Allosteric modulation of GluN2C/GluN2D-containing NMDA receptors bidirectionally modulates dopamine release: implication for Parkinson's disease. Br J Pharmacol 171 (16):3938–3945. doi:10.1111/bph.12758

397. Zimmerman SS, Khatri A, Garnier-Amblard EC, Mullasseril P, Kurtkaya NL, Gyoneva S, Hansen KB, Traynelis SF, Liotta DC (2014) Design, synthesis, and structure-activity relationship of a novel series of GluN2C-selective potentiators. J Med Chem 57(6):2334–2356. doi:10.1021/jm401695d

398. Watanabe M, Inoue Y, Sakimura K, Mishina M (1993) Distinct spatio-temporal distributions of the NMDA receptor channel subunit mRNAs in the brain. Ann N Y Acad Sci 707:463–466

399. Wenzel A, Fritschy JM, Mohler H, Benke D (1997) NMDA receptor heterogeneity during postnatal development of the rat brain: differential expression of the NR2A, NR2B,

and NR2C subunit proteins. J Neurochem 68 (2):469–478

400. Karavanova I, Vasudevan K, Cheng J, Buonanno A (2007) Novel regional and developmental NMDA receptor expression patterns uncovered in NR2C subunit-beta-galactosidase knock-in mice. Mol Cell Neurosci 34 (3):468–480. doi:10.1016/j.mcn.2006.12.001
401. Daggett LP, Johnson EC, Varney MA, Lin FF, Hess SD, Deal CR, Jachec C, Lu CC, Kerner JA, Landwehrmeyer GB, Standaert DG, Young AB, Harpold MM, Velicelebi G (1998) The human N-methyl-D-aspartate receptor 2C subunit: genomic analysis, distribution in human brain, and functional expression. J Neurochem 71(5):1953–1968
402. Wenzel A, Villa M, Mohler H, Benke D (1996) Developmental and regional expression of NMDA receptor subtypes containing the NR2D subunit in rat brain. J Neurochem 66(3):1240–1248
403. Standaert DG, Bernhard Landwehrmeyer G, Kerner JA, Penney JB, Young AB (1996) Expression of NMDAR2D glutamate receptor subunit mRNA in neurochemically identified interneurons in the rat neostriatum, neocortex and hippocampus. Mol Brain Res 42(1):89–102. doi:10.1016/s0169-328x(96)00117-9
404. Binshtok AM, Fleidervish IA, Sprengel R, Gutnick MJ (2006) NMDA receptors in layer 4 spiny stellate cells of the mouse barrel cortex contain the NR2C subunit. J Neurosci 26(2):708–715. doi:10.1523/JNEUROSCI.4409-05.2006
405. Misra C, Brickley SG, Farrant M, Cull-Candy SG (2000) Identification of subunits contributing to synaptic and extrasynaptic NMDA receptors in Golgi cells of the rat cerebellum. J Physiol 524(Pt 1):147–162
406. Ciabarra AM, Sullivan JM, Gahn LG, Pecht G, Heinemann S, Sevarino KA (1995) Cloning and characterization of chi-1: a developmentally regulated member of a novel class of the ionotropic glutamate receptor family. J Neurosci 15(10):6498–6508
407. Sucher NJ, Akbarian S, Chi CL, Leclerc CL, Awobuluyi M, Deitcher DL, Wu MK, Yuan JP, Jones EG, Lipton SA (1995) Developmental and regional expression pattern of a novel NMDA receptor-like subunit (NMDAR-L) in the rodent brain. J Neurosci 15(10):6509–6520
408. Wong HK, Liu XB, Matos MF, Chan SF, Perez-Otano I, Boysen M, Cui J, Nakanishi N, Trimmer JS, Jones EG, Lipton SA, Sucher NJ (2002) Temporal and regional expression of NMDA receptor subunit NR3A in the mammalian brain. J Comp Neurol 450 (4):303–317. doi:10.1002/cne.10314
409. Goebel DJ, Poosch MS (1999) NMDA receptor subunit gene expression in the rat brain: a quantitative analysis of endogenous mRNA levels of NR1Com, NR2A, NR2B, NR2C, NR2D and NR3A. Brain Res Mol Brain Res 69(2):164–170
410. Henson MA, Roberts AC, Salimi K, Vadlamudi S, Hamer RM, Gilmore JH, Jarskog LF, Philpot BD (2008) Developmental regulation of the NMDA receptor subunits, NR3A and NR1, in human prefrontal cortex. Cereb Cortex 18(11):2560–2573. doi:10.1093/cercor/bhn017
411. Fukaya M, Hayashi Y, Watanabe M (2005) NR2 to NR3B subunit switchover of NMDA receptors in early postnatal motoneurons. Eur J Neurosci 21(5):1432–1436. doi:10.1111/j.1460-9568.2005.03957.x
412. Prithviraj R, Inglis FM (2008) Expression of the N-methyl-D-aspartate receptor subunit NR3B regulates dendrite morphogenesis in spinal motor neurons. Neuroscience 155 (1):145–153. doi:10.1016/j.neuroscience.2008.03.089
413. Wee KS, Zhang Y, Khanna S, Low CM (2008) Immunolocalization of NMDA receptor subunit NR3B in selected structures in the rat forebrain, cerebellum, and lumbar spinal cord. J Comp Neurol 509(1):118–135. doi:10.1002/cne.21747
414. Wee KS, Tan FC, Cheong YP, Khanna S, Low CM (2016) Ontogenic profile and synaptic distribution of GluN3 proteins in the rat brain and hippocampal neurons. Neurochem Res 41(1-2):290–297. doi:10.1007/s11064-015-1794-8
415. Eriksson M, Nilsson A, Froelich-Fabre S, Akesson E, Dunker J, Seiger A, Folkesson R, Benedikz E, Sundstrom E (2002) Cloning and expression of the human N-methyl-D-aspartate receptor subunit NR3A. Neurosci Lett 321(3):177–181
416. Mueller HT, Meador-Woodruff JH (2005) Distribution of the NMDA receptor NR3A subunit in the adult pig-tail macaque brain. J Chem Neuroanat 29(3):157–172. doi:10.1016/j.jchemneu.2004.11.002
417. Bendel O, Meijer B, Hurd Y, von Euler G (2005) Cloning and expression of the human NMDA receptor subunit NR3B in the adult human hippocampus. Neurosci Lett 377(1):31–36. doi:10.1016/j.neulet.2004.11.064

418. Barth AL, Malenka RC (2001) NMDAR EPSC kinetics do not regulate the critical period for LTP at thalamocortical synapses. Nat Neurosci 4(3):235–236. doi:10.1038/85070
419. Bellone C, Nicoll RA (2007) Rapid bidirectional switching of synaptic NMDA receptors. Neuron 55(5):779–785. doi:10.1016/j.neuron.2007.07.035
420. Flint AC, Maisch US, Weishaupt JH, Kriegstein AR, Monyer H (1997) NR2A subunit expression shortens NMDA receptor synaptic currents in developing neocortex. J Neurosci 17(7):2469–2476
421. Crair MC, Malenka RC (1995) A critical period for long-term potentiation at thalamocortical synapses. Nature 375 (6529):325–328. doi:10.1038/375325a0
422. Tang YP, Shimizu E, Dube GR, Rampon C, Kerchner GA, Zhuo M, Liu G, Tsien JZ (1999) Genetic enhancement of learning and memory in mice. Nature 401 (6748):63–69
423. Kirkwood A, Rioult MC, Bear MF (1996) Experience-dependent modification of synaptic plasticity in visual cortex. Nature 381 (6582):526–528. doi:10.1038/381526a0
424. Philpot BD, Sekhar AK, Shouval HZ, Bear MF (2001) Visual experience and deprivation bidirectionally modify the composition and function of NMDA receptors in visual cortex. Neuron 29(1):157–169
425. Yashiro K, Philpot BD (2008) Regulation of NMDA receptor subunit expression and its implications for LTD, LTP, and metaplasticity. Neuropharmacology 55(7):1081–1094. doi:10.1016/j.neuropharm.2008.07.046
426. Zhou Q, Sheng M (2013) NMDA receptors in nervous system diseases. Neuropharmacology 74:69–75. doi:10.1016/j.neuropharm.2013.03.030
427. Duman RS, Aghajanian GK (2012) Synaptic dysfunction in depression: potential therapeutic targets. Science 338(6103):68–72. doi:10.1126/science.1222939
428. Niciu MJ, Henter ID, Luckenbaugh DA, Zarate CA Jr, Charney DS (2014) Glutamate receptor antagonists as fast-acting therapeutic alternatives for the treatment of depression: ketamine and other compounds. Annu Rev Pharmacol Toxicol 54:119–139. doi:10.1146/annurev-pharmtox-011613-135950
429. Berman RM, Cappiello A, Anand A, Oren DA, Heninger GR, Charney DS, Krystal JH (2000) Antidepressant effects of ketamine in depressed patients. Biol Psychiatry 47 (4):351–354
430. Cornwell BR, Salvadore G, Furey M, Marquardt CA, Brutsche NE, Grillon C, Zarate CA Jr (2012) Synaptic potentiation is critical for rapid antidepressant response to ketamine in treatment-resistant major depression. Biol Psychiatry 72(7):555–561. doi:10.1016/j.biopsych.2012.03.029
431. Zarate CA Jr, Singh JB, Carlson PJ, Brutsche NE, Ameli R, Luckenbaugh DA, Charney DS, Manji HK (2006) A randomized trial of an N-methyl-D-aspartate antagonist in treatment-resistant major depression. Arch Gen Psychiatry 63(8):856–864. doi:10.1001/archpsyc.63.8.856
432. Murrough JW, Soleimani L, DeWilde KE, Collins KA, Lapidus KA, Iacoviello BM, Lener M, Kautz M, Kim J, Stern JB, Price RB, Perez AM, Brallier JW, Rodriguez GJ, Goodman WK, Iosifescu DV, Charney DS (2015) Ketamine for rapid reduction of suicidal ideation: a randomized controlled trial. Psychol Med 45(16):3571–3580. doi:10.1017/S0033291715001506
433. Price RB, Nock MK, Charney DS, Mathew SJ (2009) Effects of intravenous ketamine on explicit and implicit measures of suicidality in treatment-resistant depression. Biol Psychiatry 66(5):522–526. doi:10.1016/j.biopsych.2009.04.029
434. Zarate CA Jr, Brutsche NE, Ibrahim L, Franco-Chaves J, Diazgranados N, Cravchik A, Selter J, Marquardt CA, Liberty V, Luckenbaugh DA (2012) Replication of ketamine's antidepressant efficacy in bipolar depression: a randomized controlled add-on trial. Biol Psychiatry 71(11):939–946. doi:10.1016/j.biopsych.2011.12.010
435. Rodriguez CI, Kegeles LS, Levinson A, Feng T, Marcus SM, Vermes D, Flood P, Simpson HB (2013) Randomized controlled crossover trial of ketamine in obsessive-compulsive disorder: proof-of-concept. Neuropsychopharmacology 38(12):2475–2483. doi:10.1038/npp.2013.150
436. Feder A, Parides MK, Murrough JW, Perez AM, Morgan JE, Saxena S, Kirkwood K, Aan Het Rot M, Lapidus KA, Wan LB, Iosifescu D, Charney DS (2014) Efficacy of intravenous ketamine for treatment of chronic posttraumatic stress disorder: a randomized clinical trial. JAMA Psychiat 71(6):681–688. doi:10.1001/jamapsychiatry.2014.62
437. aan het Rot M, Collins KA, Murrough JW, Perez AM, Reich DL, Charney DS, Mathew SJ (2010) Safety and efficacy of repeated-dose intravenous ketamine for treatment-resistant depression. Biol Psychiatry 67(2):139–145. doi:10.1016/j.biopsych.2009.08.038

438. Diamond PR, Farmery AD, Atkinson S, Haldar J, Williams N, Cowen PJ, Geddes JR, McShane R (2014) Ketamine infusions for treatment resistant depression: a series of 28 patients treated weekly or twice weekly in an ECT clinic. J Psychopharmacol 28 (6):536–544. doi:10.1177/0269881114527361
439. Murrough JW, Perez AM, Pillemer S, Stern J, Parides MK, aan het Rot M, Collins KA, Mathew SJ, Charney DS, Iosifescu DV (2013) Rapid and longer-term antidepressant effects of repeated ketamine infusions in treatment-resistant major depression. Biol Psychiatry 74(4):250–256. doi:10.1016/j.biopsych.2012.06.022
440. Singh JB, Fedgchin M, Daly EJ, De Boer P, Cooper K, Lim P, Pinter C, Murrough JW, Sanacora G, Shelton RC, Kurian B, Winokur A, Fava M, Manji H, Drevets WC, Van Nueten L (2016) A double-blind, randomized, placebo-controlled, dose-frequency study of intravenous ketamine in patients with treatment-resistant depression. Am J Psychiatry 173(8):816–826. doi:10.1176/appi.ajp.2016.16010037
441. Wan LB, Levitch CF, Perez AM, Brallier JW, Iosifescu DV, Chang LC, Foulkes A, Mathew SJ, Charney DS, Murrough JW (2015) Ketamine safety and tolerability in clinical trials for treatment-resistant depression. J Clin Psychiatry 76(3):247–252. doi:10.4088/JCP.13m08852
442. Lapidus KA, Levitch CF, Perez AM, Brallier JW, Parides MK, Soleimani L, Feder A, Iosifescu DV, Charney DS, Murrough JW (2014) A randomized controlled trial of intranasal ketamine in major depressive disorder. Biol Psychiatry 76(12):970–976. doi:10.1016/j.biopsych.2014.03.026
443. Loo CK, Galvez V, O'Keefe E, Mitchell PB, Hadzi-Pavlovic D, Leyden J, Harper S, Somogyi AA, Lai R, Weickert CS, Glue P (2016) Placebo-controlled pilot trial testing dose titration and intravenous, intramuscular and subcutaneous routes for ketamine in depression. Acta Psychiatr Scand 134 (1):48–56. doi:10.1111/acps.12572
444. Sanacora G, Smith MA, Pathak S, Su HL, Boeijinga PH, McCarthy DJ, Quirk MC (2014) Lanicemine: a low-trapping NMDA channel blocker produces sustained antidepressant efficacy with minimal psychotomimetic adverse effects. Mol Psychiatry 19 (9):978–985. doi:10.1038/mp.2013.130
445. Zarate CA Jr, Mathews D, Ibrahim L, Chaves JF, Marquardt C, Ukoh I, Jolkovsky L, Brutsche NE, Smith MA, Luckenbaugh DA (2013) A randomized trial of a low-trapping nonselective N-methyl-D-aspartate channel blocker in major depression. Biol Psychiatry 74(4):257–264. doi:10.1016/j.biopsych.2012.10.019
446. Ibrahim L, Diaz Granados N, Jolkovsky L, Brutsche N, Luckenbaugh DA, Herring WJ, Potter WZ, Zarate CA Jr (2012) A Randomized, placebo-controlled, crossover pilot trial of the oral selective NR2B antagonist MK-0657 in patients with treatment-resistant major depressive disorder. J Clin Psychopharmacol 32(4):551–557. doi:10.1097/JCP.0b013e31825d70d6
447. Preskorn SH, Baker B, Kolluri S, Menniti FS, Krams M, Landen JW (2008) An innovative design to establish proof of concept of the antidepressant effects of the NR2B subunit selective N-methyl-D-aspartate antagonist, CP-101,606, in patients with treatment-refractory major depressive disorder. J Clin Psychopharmacol 28(6):631–637. doi:10.1097/JCP.0b013e31818a6cea
448. Abdallah CG, Sanacora G, Duman RS, Krystal JH (2015) Ketamine and rapid-acting antidepressants: a window into a new neurobiology for mood disorder therapeutics. Annu Rev Med 66:509–523. doi:10.1146/annurev-med-053013-062946
449. Abraham WC, Bear MF (1996) Metaplasticity: the plasticity of synaptic plasticity. Trends Neurosci 19(4):126–130
450. Cooper LN, Bear MF (2012) The BCM theory of synapse modification at 30: interaction of theory with experiment. Nat Rev Neurosci 13(11):798–810. doi:10.1038/nrn3353
451. Turrigiano GG (2008) The self-tuning neuron: synaptic scaling of excitatory synapses. Cell 135(3):422–435. doi:10.1016/j.cell.2008.10.008
452. Zanos P, Moaddel R, Morris PJ, Georgiou P, Fischell J, Elmer GI, Alkondon M, Yuan P, Pribut HJ, Singh NS, Dossou KS, Fang Y, Huang XP, Mayo CL, Wainer IW, Albuquerque EX, Thompson SM, Thomas CJ, Zarate CA Jr, Gould TD (2016) NMDAR inhibition-independent antidepressant actions of ketamine metabolites. Nature 533 (7604):481–486. doi:10.1038/nature17998
453. Duman RS, Li N, Liu RJ, Duric V, Aghajanian G (2012) Signaling pathways underlying the rapid antidepressant actions of ketamine. Neuropharmacology 62(1):35–41. doi:10.1016/j.neuropharm.2011.08.044
454. Li N, Lee B, Liu RJ, Banasr M, Dwyer JM, Iwata M, Li XY, Aghajanian G, Duman RS (2010) mTOR-dependent synapse formation underlies the rapid antidepressant effects of

NMDA antagonists. Science 329 (5994):959–964. doi:10.1126/science.1190287

455. Li N, Liu RJ, Dwyer JM, Banasr M, Lee B, Son H, Li XY, Aghajanian G, Duman RS (2011) Glutamate N-methyl-D-aspartate receptor antagonists rapidly reverse behavioral and synaptic deficits caused by chronic stress exposure. Biol Psychiatry 69 (8):754–761. doi:10.1016/j.biopsych.2010.12.015
456. Graef JD, Newberry K, Newton A, Pieschl R, Shields E, Luan FN, Simmermacher J, Luchetti D, Schaeffer E, Li YW, Kiss L, Bristow LJ (2015) Effect of acute NR2B antagonist treatment on long-term potentiation in the rat hippocampus. Brain Res 1609:31–39. doi:10.1016/j.brainres.2015.03.019
457. Nagy D, Stoiljkovic M, Menniti FS, Hajos M (2016) Differential effects of an NR2B NAM and ketamine on synaptic potentiation and gamma synchrony: relevance to rapid-onset antidepressant efficacy. Neuropsychopharmacology 41(6):1486–1494. doi:10.1038/npp.2015.298
458. Homayoun H, Moghaddam B (2007) NMDA receptor hypofunction produces opposite effects on prefrontal cortex interneurons and pyramidal neurons. J Neurosci 27(43):11496–11500. doi:10.1523/JNEUROSCI.2213-07.2007
459. Kocsis B (2012) Differential role of NR2A and NR2B subunits in N-methyl-D-aspartate receptor antagonist-induced aberrant cortical gamma oscillations. Biol Psychiatry 71 (11):987–995. doi:10.1016/j.biopsych.2011.10.002
460. Sivarao DV, Chen P, Yang Y, Li YW, Pieschl R, Ahlijanian MK (2014) NR2B Antagonist CP-101,606 Abolishes Pitch-Mediated Deviance Detection in Awake Rats. Front Psych 5:96. doi:10.3389/fpsyt.2014.00096
461. Keavy D, Bristow LJ, Sivarao DV, Batchelder M, King D, Thangathirupathy S, Macor JE, Weed MR (2016) The qEEG Signature of Selective NMDA NR2B Negative Allosteric Modulators; A Potential Translational Biomarker for Drug Development. PLoS One 11(4):e0152729. doi:10.1371/journal.pone.0152729
462. Matta JA, Pelkey KA, Craig MT, Chittajallu R, Jeffries BW, McBain CJ (2013) Developmental origin dictates interneuron AMPA and NMDA receptor subunit composition and plasticity. Nat Neurosci 16(8):1032–1041. doi:10.1038/nn.3459
463. Pfeffer CK, Xue MS, He M, Huang ZJ, Scanziani M (2013) Inhibition of inhibition in visual cortex: the logic of connections between molecularly distinct interneurons. Nat Neurosci 16(8):1068–U1130. doi:10.1038/nn.3446
464. Harris KD, Shepherd GM (2015) The neocortical circuit: themes and variations. Nat Neurosci 18(2):170–181. doi:10.1038/nn.3917
465. Wang M, Yang Y, Wang CJ, Gamo NJ, Jin LE, Mazer JA, Morrison JH, Wang XJ, Arnsten AF (2013) NMDA receptors subserve persistent neuronal firing during working memory in dorsolateral prefrontal cortex. Neuron 77 (4):736–749. doi:10.1016/j.neuron.2012.12.032
466. Hardingham GE, Bading H (2010) Synaptic versus extrasynaptic NMDA receptor signalling: implications for neurodegenerative disorders. Nat Rev Neurosci 11(10):682–696. doi:10.1038/nrn2911
467. Lee MC, Yasuda R, Ehlers MD (2010) Metaplasticity at single glutamatergic synapses. Neuron 66(6):859–870. doi:10.1016/j.neuron.2010.05.015
468. Chen WS, Bear MF (2007) Activity-dependent regulation of NR2B translation contributes to metaplasticity in mouse visual cortex. Neuropharmacology 52(1):200–214. doi:10.1016/j.neuropharm.2006.07.003
469. Stocca G, Vicini S (1998) Increased contribution of NR2A subunit to synaptic NMDA receptors in developing rat cortical neurons. J Physiol 507(Pt 1):13–24
470. Chiocchetti AG, Bour HS, Freitag CM (2014) Glutamatergic candidate genes in autism spectrum disorder: an overview. J Neural Transm (Vienna) 121 (9):1081–1106. doi:10.1007/s00702-014-1161-y
471. Greenwood TA, Lazzeroni LC, Murray SS, Cadenhead KS, Calkins ME, Dobie DJ, Green MF, Gur RE, Gur RC, Hardiman G, Kelsoe JR, Leonard S, Light GA, Nuechterlein KH, Olincy A, Radant AD, Schork NJ, Seidman LJ, Siever LJ, Silverman JM, Stone WS, Swerdlow NR, Tsuang DW, Tsuang MT, Turetsky BI, Freedman R, Braff DL (2011) Analysis of 94 candidate genes and 12 endophenotypes for schizophrenia from the Consortium on the Genetics of Schizophrenia. Am J Psychiatry 168(9):930–946. doi:10.1176/appi.ajp.2011.10050723
472. Hall J, Trent S, Thomas KL, O'Donovan MC, Owen MJ (2015) Genetic risk for schizophrenia: convergence on synaptic pathways involved in plasticity. Biol Psychiatry 77 (1):52–58. doi:10.1016/j.biopsych.2014.07.011

473. Menniti FS, Lindsley CW, Conn PJ, Pandit J, Zagouras P, Volkmann RA (2013) Allosteric modulators for the treatment of schizophrenia: targeting glutamatergic networks. Curr Top Med Chem 13(1):26–54
474. Domino EF, Luby ED (2012) Phencyclidine/schizophrenia: one view toward the past, the other to the future. Schizophr Bull 38(5):914–919. doi:10.1093/schbul/sbs011
475. Krystal JH, Karper LP, Seibyl JP, Freeman GK, Delaney R, Bremner JD, Heninger GR, Bowers MB Jr, Charney DS (1994) Subanesthetic effects of the noncompetitive NMDA antagonist, ketamine, in humans. Psychotomimetic, perceptual, cognitive, and neuroendocrine responses. Arch Gen Psychiatry 51(3):199–214
476. Xu K, Krystal JH, Ning Y, Chen da C, He H, Wang D, Ke X, Zhang X, Ding Y, Liu Y, Gueorguieva R, Wang Z, Limoncelli D, Pietrzak RH, Petrakis IL, Zhang X, Fan N (2015) Preliminary analysis of positive and negative syndrome scale in ketamine-associated psychosis in comparison with schizophrenia. J Psychiatr Res 61:64–72. doi:10.1016/j.jpsychires.2014.12.012
477. Luby ED, Cohen BD, Rosenbaum G, Gottlieb JS, Kelley R (1959) Study of a new schizophrenomimetic drug; sernyl. AMA Arch Neurol Psychiatry 81(3):363–369
478. Shaffer CL, Osgood SM, Smith DL, Liu J, Trapa PE (2014) Enhancing ketamine translational pharmacology via receptor occupancy normalization. Neuropharmacology 86:174–180. doi:10.1016/j.neuropharm.2014.07.008
479. Javitt DC, Zukin SR (1990) The role of excitatory amino acids in neuropsychiatric illness. J Neuropsychiatry Clin Neurosci 2(1):44–52. doi:10.1176/jnp.2.1.44
480. Javitt DC, Zukin SR, Heresco-Levy U, Umbricht D (2012) Has an angel shown the way? Etiological and therapeutic implications of the PCP/NMDA model of schizophrenia. Schizophr Bull 38(5):958–966. doi:10.1093/schbul/sbs069
481. Krystal JH, D'Souza DC, Mathalon D, Perry E, Belger A, Hoffman R (2003) NMDA receptor antagonist effects, cortical glutamatergic function, and schizophrenia: toward a paradigm shift in medication development. Psychopharmacology (Berl) 169(3-4):215–233. doi:10.1007/s00213-003-1582-z
482. Kantrowitz ER, Javitt D (2011) Glutamate: new hope for schizophrenia therapy. Curr Psychiatry 10(4):69–74
483. Tsai GE, Lin PY (2010) Strategies to enhance N-methyl-D-aspartate receptor-mediated neurotransmission in schizophrenia, a critical review and meta-analysis. Curr Pharm Des 16(5):522–537
484. Buchanan RW, Javitt DC, Marder SR, Schooler NR, Gold JM, McMahon RP, Heresco-Levy U, Carpenter WT (2007) The Cognitive and Negative Symptoms in Schizophrenia Trial (CONSIST): the efficacy of glutamatergic agents for negative symptoms and cognitive impairments. Am J Psychiatry 164(10):1593–1602. doi:10.1176/appi.ajp.2007.06081358
485. Goff DC (2012) D-cycloserine: an evolving role in learning and neuroplasticity in schizophrenia. Schizophr Bull 38(5):936–941. doi:10.1093/schbul/sbs012
486. Goff DC (2017) D-cycloserine in schizophrenia: new strategies for improving clinical outcomes by enhancing plasticity. Curr Neuropharmacol 15(1):21–34
487. Umbricht D, Alberati D, Martin-Facklam M, Borroni E, Youssef EA, Ostland M, Wallace TL, Knoflach F, Dorflinger E, Wettstein JG, Bausch A, Garibaldi G, Santarelli L (2014) Effect of bitopertin, a glycine reuptake inhibitor, on negative symptoms of schizophrenia: a randomized, double-blind, proof-of-concept study. JAMA Psychiat 71(6):637–646. doi:10.1001/jamapsychiatry.2014.163
488. Goff DC, Cather C, Gottlieb JD, Evins AE, Walsh J, Raeke L, Otto MW, Schoenfeld D, Green MF (2008) Once-weekly D-cycloserine effects on negative symptoms and cognition in schizophrenia: an exploratory study. Schizophr Res 106(2-3):320–327. doi:10.1016/j.schres.2008.08.012
489. Gottlieb JD, Cather C, Shanahan M, Creedon T, Macklin EA, Goff DC (2011) D-cycloserine facilitation of cognitive behavioral therapy for delusions in schizophrenia. Schizophr Res 131(1-3):69–74. doi:10.1016/j.schres.2011.05.029
490. Martina M, Gorfinkel Y, Halman S, Lowe JA, Periyalwar P, Schmidt CJ, Bergeron R (2004) Glycine transporter type 1 blockade changes NMDA receptor-mediated responses and LTP in hippocampal CA1 pyramidal cells by altering extracellular glycine levels. J Physiol 557(Pt 2):489–500. doi:10.1113/jphysiol.2004.063321
491. Sheinin A, Shavit S, Benveniste M (2001) Subunit specificity and mechanism of action of NMDA partial agonist D-cycloserine. Neuropharmacology 41(2):151–158
492. Dravid SM, Burger PB, Prakash A, Geballe MT, Yadav R, Le P, Vellano K, Snyder JP,

Traynelis SF (2010) Structural determinants of D-cycloserine efficacy at the NR1/NR2C NMDA receptors. J Neurosci 30 (7):2741–2754. doi:10.1523/JNEUROSCI.5390-09.2010

493. Standaert DG, Testa CM, Young AB, Penney JB Jr (1994) Organization of N-methyl-D-aspartate glutamate receptor gene expression in the basal ganglia of the rat. J Comp Neurol 343(1):1–16. doi:10.1002/cne.903430102

494. Zhang Y, Buonanno A, Vertes RP, Hoover WB, Lisman JE (2012) NR2C in the thalamic reticular nucleus; effects of the NR2C knockout. PLoS One 7(7):e41908. doi:10.1371/journal.pone.0041908

495. Merchant RE, Bullock MR, Carmack CA, Shah AK, Wilner KD, Ko G, Williams SA (1999) A double-blind, placebo-controlled study of the safety, tolerability and pharmacokinetics of CP-101,606 in patients with a mild or moderate traumatic brain injury. Ann N Y Acad Sci 890:42–50

496. Nutt JG, Gunzler SA, Kirchhoff T, Hogarth P, Weaver JL, Krams M, Jamerson B, Menniti FS, Landen JW (2008) Effects of a NR2B selective NMDA glutamate antagonist, CP-101,606, on dyskinesia and Parkinsonism. Mov Disord 23(13):1860–1866. doi:10.1002/mds.22169

497. Chaperon F, Muller W, Auberson YP, Tricklebank MD, Neijt HC (2003) Substitution for PCP, disruption of prepulse inhibition and hyperactivity induced by N-methyl-D-aspartate receptor antagonists: preferential involvement of the NR2B rather than NR2A subunit. Behav Pharmacol 14(5-6):477–487. doi:10.1097/01.fbp.0000091471.79060.ed

498. Nicholson KL, Mansbach RS, Menniti FS, Balster RL (2007) The phencyclidine-like discriminative stimulus effects and reinforcing properties of the NR2B-selective N-methyl-D-aspartate antagonist CP-101 606 in rats and rhesus monkeys. Behav Pharmacol 18 (8):731–743. doi:10.1097/FBP.0b013e3282f14ed6

499. Epi PMC (2015) A roadmap for precision medicine in the epilepsies. Lancet Neurol 14 (12):1219–1228. doi:10.1016/S1474-4422(15)00199-4

500. Carvill GL, Regan BM, Yendle SC, O'Roak BJ, Lozovaya N, Bruneau N, Burnashev N, Khan A, Cook J, Geraghty E, Sadleir LG, Turner SJ, Tsai MH, Webster R, Ouvrier R, Damiano JA, Berkovic SF, Shendure J, Hildebrand MS, Szepetowski P, Scheffer IE, Mefford HC (2013) GRIN2A mutations cause epilepsy-aphasia spectrum disorders. Nat Genet 45(9):1073–1076. doi:10.1038/ng.2727

501. Endele S, Rosenberger G, Geider K, Popp B, Tamer C, Stefanova I, Milh M, Kortum F, Fritsch A, Pientka FK, Hellenbroich Y, Kalscheuer VM, Kohlhase J, Moog U, Rappold G, Rauch A, Ropers HH, von Spiczak S, Tonnies H, Villeneuve N, Villard L, Zabel B, Zenker M, Laube B, Reis A, Wieczorek D, Van Maldergem L, Kutsche K (2010) Mutations in GRIN2A and GRIN2B encoding regulatory subunits of NMDA receptors cause variable neurodevelopmental phenotypes. Nat Genet 42(11):1021–1026. doi:10.1038/ng.677

502. Lemke JR, Lal D, Reinthaler EM, Steiner I, Nothnagel M, Alber M, Geider K, Laube B, Schwake M, Finsterwalder K, Franke A, Schilhabel M, Jahn JA, Muhle H, Boor R, Van Paesschen W, Caraballo R, Fejerman N, Weckhuysen S, De Jonghe P, Larsen J, Moller RS, Hjalgrim H, Addis L, Tang S, Hughes E, Pal DK, Veri K, Vaher U, Talvik T, Dimova P, Guerrero Lopez R, Serratosa JM, Linnankivi T, Lehesjoki AE, Ruf S, Wolff M, Buerki S, Wohlrab G, Kroell J, Datta AN, Fiedler B, Kurlemann G, Kluger G, Hahn A, Haberlandt DE, Kutzer C, Sperner J, Becker F, Weber YG, Feucht M, Steinbock H, Neophythou B, Ronen GM, Gruber-Sedlmayr U, Geldner J, Harvey RJ, Hoffmann P, Herms S, Altmuller J, Toliat MR, Thiele H, Nurnberg P, Wilhelm C, Stephani U, Helbig I, Lerche H, Zimprich F, Neubauer BA, Biskup S, von Spiczak S (2013) Mutations in GRIN2A cause idiopathic focal epilepsy with rolandic spikes. Nat Genet 45(9):1067–1072. doi:10.1038/ng.2728

503. Lesca G, Rudolf G, Bruneau N, Lozovaya N, Labalme A, Boutry-Kryza N, Salmi M, Tsintsadze T, Addis L, Motte J, Wright S, Tsintsadze V, Michel A, Doummar D, Lascelles K, Strug L, Waters P, de Bellescize J, Vrielynck P, de Saint MA, Ville D, Ryvlin P, Arzimanoglou A, Hirsch E, Vincent A, Pal D, Burnashev N, Sanlaville D, Szepetowski P (2013) GRIN2A mutations in acquired epileptic aphasia and related childhood focal epilepsies and encephalopathies with speech and language dysfunction. Nat Genet 45(9):1061–1066. doi:10.1038/ng.2726

504. Strehlow V, Heyne HO, Lemke JR (2015) The spectrum of GRIN2A-associated disorders. Epileptologie 32(3):147–151

505. Insel TR (2010) Rethinking schizophrenia. Nature 468(7321):187–193. doi:10.1038/nature09552

506. Yuan H, Low CM, Moody OA, Jenkins A, Traynelis SF (2015) Ionotropic GABA and glutamate receptor mutations and human neurologic diseases. Mol Pharmacol 88 (1):203–217. doi:10.1124/mol.115.097998
507. Burnashev N, Szepetowski P (2015) NMDA receptor subunit mutations in neurodevelopmental disorders. Curr Opin Pharmacol 20:73–82. doi:10.1016/j.coph.2014.11.008
508. Pierson TM, Yuan H, Marsh ED, Fuentes-Fajardo K, Adams DR, Markello T, Golas G, Simeonov DR, Holloman C, Tankovic A, Karamchandani MM, Schreiber JM, Mullikin JC, Ph DNCSP, Tifft CJ, Toro C, Boerkoel CF, Traynelis SF, Gahl WA (2014) GRIN2A mutation and early-onset epileptic encephalopathy: personalized therapy with memantine. Ann Clin Transl Neurol 1(3):190–198. doi:10.1002/acn3.39
509. Chahrour M, Zoghbi HY (2007) The story of Rett syndrome: from clinic to neurobiology. Neuron 56(3):422–437. doi:10.1016/j.neuron.2007.10.001
510. Robinson L, Guy J, McKay L, Brockett E, Spike RC, Selfridge J, De Sousa D, Merusi C, Riedel G, Bird A, Cobb SR (2012) Morphological and functional reversal of phenotypes in a mouse model of Rett syndrome. Brain 135(Pt 9):2699–2710. doi:10.1093/brain/aws096
511. Guy J, Gan J, Selfridge J, Cobb S, Bird A (2007) Reversal of neurological defects in a mouse model of Rett syndrome. Science 315 (5815):1143–1147. doi:10.1126/science.1138389
512. Katz DM, Bird A, Coenraads M, Gray SJ, Menon DU, Philpot BD, Tarquinio DC (2016) Rett syndrome: crossing the threshold to clinical translation. Trends Neurosci 39 (2):100–113. doi:10.1016/j.tins.2015.12.008
513. Blue ME, Kaufmann WE, Bressler J, Eyring C, O'Driscoll C, Naidu S, Johnston MV (2011) Temporal and regional alterations in NMDA receptor expression in Mecp2-null mice. Anat Rec (Hoboken) 294 (10):1624–1634. doi:10.1002/ar.21380
514. Durand S, Patrizi A, Quast KB, Hachigian L, Pavlyuk R, Saxena A, Carninci P, Hensch TK, Fagiolini M (2012) NMDA receptor regulation prevents regression of visual cortical function in the absence of Mecp2. Neuron 76(6):1078–1090. doi:10.1016/j.neuron.2012.12.004
515. Mierau SB, Patrizi A, Hensch TK, Fagiolini M (2016) Cell-specific regulation of N-methyl-D-aspartate receptor maturation by Mecp2 in cortical circuits. Biol Psychiatry 79 (9):746–754. doi:10.1016/j.biopsych.2015.05.018
516. Kron M, Howell CJ, Adams IT, Ransbottom M, Christian D, Ogier M, Katz DM (2012) Brain activity mapping in Mecp2 mutant mice reveals functional deficits in forebrain circuits, including key nodes in the default mode network, that are reversed with ketamine treatment. J Neurosci 32(40):13860–13872. doi:10.1523/JNEUROSCI.2159-12.2012
517. Patrizi A, Picard N, Simon AJ, Gunner G, Centofante E, Andrews NA, Fagiolini M (2016) Chronic administration of the N-methyl-D-aspartate receptor antagonist ketamine improves Rett syndrome phenotype. Biol Psychiatry 79(9):755–764. doi:10.1016/j.biopsych.2015.08.018
518. Bienenstock EL, Cooper LN, Munro PW (1982) Theory for the development of neuron selectivity: orientation specificity and binocular interaction in visual cortex. J Neurosci 2(1):32–48
519. Jones KA, Menniti FS, Sivarao DV (2015) Translational psychiatry–light at the end of the tunnel. Ann N Y Acad Sci 1344:1–11. doi:10.1111/nyas.12725
520. Monaghan DT, Irvine MW, Costa BM, Fang G, Jane DE (2012) Pharmacological modulation of NMDA receptor activity and the advent of negative and positive allosteric modulators. Neurochem Int 61(4):581–592. doi:10.1016/j.neuint.2012.01.004
521. Kohr G (2006) NMDA receptor function: subunit composition versus spatial distribution. Cell Tissue Res 326(2):439–446. doi:10.1007/s00441-006-0273-6
522. Kohr G (2007) NMDA receptor antagonists: tools in neuroscience with promise for treating CNS pathologies. J Physiol 581(Pt 1):1–2. doi:10.1113/jphysiol.2007.130732
523. Yuan H, Geballe MT, Hansen KB, Traynelis SF (2008) Structure and function relationship of the NMDA receptor. In: Hell JW, Ehlers MD (eds) Structural and functional organization of the synapse, 1st edn. Springer, New York, NY, pp 289–316
524. Hess SD, Daggett LP, Crona J, Deal C, Lu CC, Urrutia A, Chavez-Noriega L, Ellis SB, Johnson EC, Velicelebi G (1996) Cloning and functional characterization of human heteromeric N-methyl-D-aspartate receptors. J Pharmacol Exp Ther 278(2):808–816

Part I

NMDAR Subunits Expression and Mutations

Chapter 2

Quantification of NMDAR Subunit Genes Expression by qRT-PCR

Emilie Pallesi-Pocachard

Abstract

Transcription is the initial and generally the most sensitive step to cellular needs and environmental cues. Thus, it serves as a major mechanism controlling gene expression. Using reverse-transcription quantitative polymerase chain reaction technology (RT-qPCR), we will present how to quantify the transcriptional expression of NMDARs subunits during brain development and in both healthy and pathological conditions.

Key words RNA, RT-qPCR, Reverse transcription, cDNA, Relative quantification

1 Introduction

N-Methyl-D-aspartate receptors (NMDARs) are glutamate-gated cation channels that are expressed throughout the brain and play an important role in proper brain functioning [1]. To date, seven different subunits of NMDARs have been characterized [2–4]:

- the GluN1 subunit required for the normal function of the NMDA ionophore [5–8];
- the four distinct GluN2 subunits (GluN2A, GluN2B, GluN2C, and GluN2D), which are encoded by four separate genes and appear to serve a modulatory role in NMDA channel function [7, 8, 9, 10];
- and two GluN3 subunits (GluN3A and GluN3B) encoded by two distinct genes [11–13].

NMDARs are generated by the heteromeric assembly of the GluN1 subunit with specific members of the GluN2 and/or GluN3 subunits. These different NMDARs subtypes have distinct physiological properties, which may produce synaptic plasticity by expression in specific neurons or other cells. When present, GluN3 subunits may function as dominant-negative subunits in a NMDA

Nail Burnashev and Pierre Szepetowski (eds.), *NMDA Receptors: Methods and Protocols*, Methods in Molecular Biology, vol. 1677, DOI 10.1007/978-1-4939-7321-7_2, © Springer Science+Business Media LLC 2017

receptor complex [12]. Therefore, transcription of each NMDAR subunit gene in a given neuron must be coordinately controlled but also differentially responsive to cell type, developmental stage, and environmental signals to maintain healthy cellular function. In this section, we present how to quantify the transcriptional expression of NMDARs subunits during brain development using RT-qPCR. RT-qPCR utilizing primer–probe combinations directed to exons flanking the breakpoint junctional region, offers very high levels of both specificity and sensitivity, in a scalable, robust, and cost-effective assay.

2 Materials

2.1 RNA Extraction and Quantification

2.1.1 Columns Extraction Method

1. RNeasy Mini Kit.
2. 14.3 M β-mercaptoethanol.
3. 70% Ethanol.
4. 96–100% Ethanol.
5. 1 mL Syringes.
6. 18 G and 26 G needles.
7. Sterile, RNAse-free pipet tips.
8. 1.5 mL microcentrifuge capped tubes.
9. Microcentrifuge (with rotor for 1.5 and 2 mL tubes).

2.1.2 Trizol Extraction Method

1. TRIzol® RNA Isolation Reagent.
2. Chloroform.
3. Cooled Isopropanol.
4. Cooled 75% Ethanol.
5. Sterile RNAse-DNAse water.
6. 1 mL Syringes.
7. 18 G and 26 G needles.
8. Sterile, RNAse free pipet tips.
9. 1.5 mL microcentrifuge capped tubes.
10. Microcentrifuge (with rotor for 1.5 and 2 mL tubes).

2.1.3 RNA Quantification

1. Sterile water.
2. Sterile, RNAse free pipet tips.
3. Kimtech paper.
4. Nanophotometer.

2.2 cDNA Synthesis

1. QuantiTect® Reverse Transcription kit.
2. Thermocycler (SimpliAmp Thermal Cycler).

3. Micropipets.
4. Disposable plasticware (tubes, filters tips).

2.3 qPCR

1. Forward and reverse stock primers at the concentration of 100 μM.
2. Ready-to-use hot start reaction mix for SYBR Green I-based real-time PCR using the LightCycler®480 Instrument.
3. Sterile RNAse-DNAse water.
4. cDNA (Reverse transcripts and control reactions).
5. Optical 96 wells PCR plates.
6. Microseal film.
7. Plastics tubes.
8. Single repeater pipettes.
9. qPCR thermalcycler (LightCycler® 480 Instrument II) (*see* **Note 1**).

3 Methods

3.1 Extraction and Quantification of RNA

Whole dissected mouse brain or parts of the cerebral cortex are frozen in dry ice and directly stored at −80 °C. Immediate freezing of fresh tissue samples preserves good quality RNA for gene expression studies. Two alternative protocols are described depending on the weight of the available frozen tissue.

3.1.1 Extraction of RNA from Frozen Brain Tissue <30 mg (See Note 2)

The RNeasy Mini Kit provides a fast purification of high-quality RNA from brain tissues

1. Add 600 μL RLT containing 6 μL of β-mercaptoethanol.
2. Homogenize by passing the lysate at least ten times through a blunt 18 G needle fitted to an RNAse-free syringe.
3. Homogenize carefully by passing the lysate at least 20 times through a blunt 26 G needle fitted to the same RNAse-free syringe.
4. Centrifuge the lysate for 3 min at full speed.
5. Transfer the supernatant by pipetting to a new microcentrifuge tube and add 600 μL of 70% ethanol.
6. Transfer the mix to an RNeasy spin column placed in a 2 mL collection tube (*see* **Note 3**).
7. Centrifuge for 15 s at 15,000 × *g* and discard the flow-through.
8. Add 700 μL Buffer RW1 to the spin column and centrifuge for 15 s at 9000 × *g*, discard the flow-through, and replace the 2 mL collection tube (*see* **Note 4**).

9. Add 500 μL RPE wash buffer to the spin column and centrifuge for 15 s at 9000 × *g*, discarding the flow-through.
10. Add 500 μL RPE washing buffer to the spin column and centrifuge for 2 min at 9000 × *g*, then transfer the spin column to a new 2 mL collection tube and discard the old collection tube with the flow-through.
11. Centrifuge for 1 min at 15,000 × *g*.
12. Place the RNeasy Spin column in a new 1.5 mL collection tube.
13. Add 30 μL of RNAse-free water to the spin column and incubate for 1 min before elution of RNA by centrifuging for 1 min at 15,000 × *g*.
14. Repeat the previous step with 25 μL of RNAse-free water. The spin column can then be discarded.
15. Freeze RNA at −20 °C for one night before quantification.

3.1.2 Extraction of RNA from Frozen Brain Tissue >30 mg

This technique is a common method to extract total RNA from tissues. It takes slightly longer than column-based methods, but it has higher capacity and can yield more RNA.

1. Add 1 mL TRIzol Reagent per 50–100 mg of tissue sample.
2. Homogenize sample by passing the lysate at least ten times through a blunt 18 G needle fitted to an RNase-free syringe and at least 20 times through a blunt 26 G needle fitted to the same RNase-free syringe.
3. Incubate the homogenized sample for 5 min at room temperature.
4. Add 200 μL of chloroform per 1 mL of TRIzol Reagent and shake vigorously by hand for 15 s.
5. Incubate for 3 min at room temperature (20–22 °C).
6. Centrifuge at 10,000 × *g* for 15 min at 4 °C. The mixture will separate into three phases. The upper aqueous phase containing RNA represents approximatively half the total volume.
7. Transfer the aqueous phase to another new tube by pipetting carefully.
8. Add 1 volume of cooled isopropanol (−20 °C).
9. Incubate for 10 min at room temperature.
10. Centrifuge at 10,000 × *g* for 15 min at 4 °C.
11. Remove the supernatant and wash the pellet with 1 volume of cooled 75% ethanol.
12. Briefly vortex the sample and centrifuge at 7500 × *g* for 5 min at 4 °C. Discard the wash. Do not allow RNA to dry completely as the pellet might lose solubility. Partially dissolved RNA samples have an A260/280 ratio <1.6.

13. Resuspend the RNA pellet in RNAse-free water.
14. Store at −80 °C.

3.1.3 Quantification of RNA

The usual method for assessing RNA concentration and purity is UV spectroscopy. The absorbance of a diluted RNA sample is measured at 260 and 280 nm. The nucleic acid concentration is calculated using the Beer-Lambert law, which predicts a linear change in absorbance with concentration.

1. Defroze RNA on ice.
2. Quantify RNA using a nanodrop (nanophotometer) according to the manufacturer's instructions. An A260/A280 ratio of 1.8–2.1 indicates highly purified RNA.

3.2 cDNA Synthesis

1. Thaw template RNA, genomic DNA (gDNA) wipeout buffer and Quantiscript Reverse Transcriptase on ice. Thaw Quantiscript RT buffer, RT Primer Mix, and RNase-free water at room temperature.
2. Prepare the gDNA wipeout reaction on ice to a final total volume of 14 μL. The components are: 2 μL of gDNA Wipeout Buffer (7×), and 500 ng to 1 μg of RNA (*see* **Note 5**). One reaction tube, called minus-RT control, has to be prepared for reverse transcription but no reverse transcriptase enzyme will be added (*see* **Note 6**).
3. Incubate for 2 min at 42 °C in a thermal cycler. Then place immediately on ice.
4. Prepare the reverse-transcription master mix on ice. The mix contains 1 μL of Quantiscript reverse transcriptase, 4 μL de Quantiscript RT buffer, and 1 μL RT primer Mix per reaction.
5. Add reverse transcription master mix (6 μL) to each tube containing template RNA from **step 4** (14 μL). In the minus-RT control tube add 1 μL of DNAse-RNAse-free water, 4 μL of Quantiscript RT buffer, and 1 μL RT primer Mix per reaction.
6. Incubate for 20 min at 42 °C in a thermal cycler.
7. Incubate for 3 min at 95 °C to inactivate Quantiscript Reverse Transcriptase.
8. Store cDNA at −20 °C.

3.3 qPCR Primers

3.3.1 qPCR Primers Design [14]

1. Log on to www.universalprobelibrary.com/ (*see* **Note 7**).
2. On the assay design center screen, select your organism of interest from the drop down menu.
3. Specify the name or the sequence ID of the gene of interest.
4. Verify that the intron-spanning assay option is active and Click "design."

5. Select the right RefSeq sequence in the transcript table and click again on "Design."
6. Copy-paste the complete exon sequence in the text box. For real-time PCR purposes, we set "product size ranges" to "100–200," and "number to return" to "50." Short PCR products are typically amplified with higher efficiency than longer ones, but the PCR product should be at least 100 bp long to easily distinguish it from any primer-dimers that could potentially form.
7. Try to select two to three different primer pairs, preferably targeting different regions in the cDNA, except the untranslated regions. Annealing temperature must be between 58 °C and 61 °C and the percentage of Guanine cytosine (GC) contents between 45 and 55%. Primer length should be around 20 bp. Forward and reverse primers must span at least one intron within the gene of interest. Indeed, primers were designed to anneal to exons on both the sides of a given intron; this helps in discriminating between amplification of cDNA and amplification of potentially contaminating genomic DNA.
8. Check primer pairs for cross-hybridization with other genes using NCBI/Primer-BLAST at http://www.ncbi.nlm.nih.gov/tools/primer-blast. If there is cross-hybridization, go back to **step** 7. No mismatch in the primer set is acceptable.
9. To validate qPCR assay with the correct primers pair, at least two different pairs are ordered.
10. Primers are shipped in a dry form. Briefly spin the tube before you open the cap to avoid loss of DNA pellet. Dissolve the oligonucleotide in DNAse/RNAse-free water to make a primer stock at concentration 100 μM. Dilute from this stock 1:20 (in water) to make a working solution of mixing solution (forward and reverse) at 5 μM for use in setting up qPCR reactions.

3.3.2 Specificity (Melting Curve) and Validation (Standard Curve/Efficiency) of Primers

Each qPCR reaction must efficiently amplify a single product corresponding to a unique peak in post-amplification melting curve analysis, indicating that the design of the corresponding primers pair is correct. This analysis is frequently used as a diagnostic tool for assessing qPCR amplicon sequence and primer-dimers (*see* **Note 8**). The melting temperature of the peak correlates with the length and the GC content of the amplicon but not with the concentration template.

Amplification efficiency of these primers should be independent of cDNA template dilutions and should exhibit a twofold increase in copy numbers during each cycle of exponential amplification.

1. Make five serial dilutions of cDNA (could be any cDNA sample, but it is preferable to use a mixture of target samples to reduce heterogeneity). cDNA amount units of templates could be dilution values at 0.5, 0.1, 0.05, 0.01, and 0.002. Minus-RT and water will be used as control conditions.
2. Set up PCR reactions (*see* Subheading 3.4 for protocols). Do duplicated reactions for each diluted template. Add the appropriate controls (minus RT and water samples).
3. RT-qPCR conditions and instrument setup are described above (*see* Subheading 3.4 for protocols).
4. Data will be analyzed using LightCycler 480 SW 1.5.1 software and following the manufacturer's instructions.

3.4 qPCR (*See* Note 1)

We routinely use Sybrgreen-based application. SYBR® Green I is a fluorescent dye that binds to the minor groove of double-stranded DNA and fluoresces when bound to DNA and excited by a light source. It is useful as a screening tool to quickly assess the relative expression levels of a variety of genes in different sample types. All the procedures should be done on ice. It is advisable to run at least a triplicate for each sample.

1. Prepare one mix per reaction: 1 μL of primers solution (mixture of forward and reverse, each 5 μM) (*see* **Note 9**), 7.5 μL ready-to-use hot start reaction mix for SYBR Green I-based real-time PCR, and 2.75 μL water.
2. Pipet reaction mixture on a PCR plate, using a single channel repeater pipette.
3. Add 3.75 μL of 1/10 or 1/20 diluted cDNA per reaction (*see* **Note 10**).
4. Cover the PCR plate with optical seal.
5. Incubate the plate in the thermal cycler at the following temperatures: 10 min at 95 °C, then 50 cycles of 10 s at 95 °C, 20 s at 60 °C, and 10 s at 72 °C. At the end, a melting curve analysis needs to be performed. This is done after a denaturation step of 5 s at 95 °C followed by 1 min at 65 °C and increasing the temperature from 65 to 95 °C continuously. Five acquisitions are obtained per °C. The ramp rate is of 0.11 °C/s.
6. As standardization of the expression levels, reference housekeeping genes need to be taken along for qPCR (*see* **Note 11**).
7. Data will be analyzed using LightCycler 480 SW 1.5.1 software and following the manufacturer's instructions (*see* **Note 12**).

4 Notes

1. If using different instrument or fluorescent dye, the protocol needs to be adapted.
2. It is essential to use the correct amount of starting material to obtain optimal RNA yield and purity. A maximum amount of 30 mg fresh or frozen tissue can generally be processed. For most tissues, the RNA binding capacity of the RNeasy spin column and the lysing capacity of Buffer RLT will not be exceeded by these amounts.
3. As the volume exceeds 700 μL, centrifuge the mix two times in the same RNeasy spin column and discard the flow-through after each centrifugation.
4. After centrifugation, carefully remove the RNeasy spin column from the collection tube so that the column does not contact the flow-through. Be sure to empty the collection tube completely.
5. If using >1 μg RNA, scale up the reaction linearly to the appropriate volume.
6. The amount of amplification in the samples that is attributable to genomic DNA contamination can be evaluated by running this kind of control (Minus-RT controls) in a run of qPCR.
7. Specific primers for qPCR should be designed with the aid of primer design software to avoid primer-dimers and secondary structures. ProbeFinder is a web-based software tool based on Primer3 software using optimized settings as default.
8. If a dissociation curve gives multiple peaks (even though some are very weak), we recommend not using the corresponding primers pair for further experiments.
9. If the initial set of 300 nM primer does not work, optimization of primer concentration might be required. Lower primer concentrations decrease the accumulation of primer-dimer formation and nonspecific product formation, which is critical in using SYBR Green I dye in quantitative PCR.
10. Generally, the volume of reverse-transcription reaction added should not exceed 10% of the final PCR volume. Having a look at the Ct value gives a rough idea about how much RNA/cDNA is needed for further experiments. If Ct value is high (>30), it is better to increase the amount of template for further experiments.
11. Before selecting the most appropriate reference gene, test several housekeeping genes with preliminary experiments.
12. In some studies standards curves are not determined and target quantities are reported as a fraction of the reference, a technique

termed comparative quantitation. In other studies, it is assumed that amplification efficiencies of target and reference are optimal. However in the most accurate relative quantification techniques, amplification efficiencies of both the reference and target are measured and a correction factor is determined. This process, termed normalization, requires a sample containing known concentrations of both target and reference and the generation of two standard curves. We perform relative quantification of our samples. Indeed, the ratio between the amount of NMDA-R subunit transcripts and of housekeeping transcripts (such as *Ppia1*, *Hprt*, *Rpl13a* in rodents and *HPRT*, *β-Actin* in humans) can be calculated at different stages of development and in different regions of the brain in rodents and humans.

Acknowledgments

This work was supported by INSERM (Institut National de la Santé et de la Recherche Médicale) and by the European Union Seventh Framework Programme FP7/2007–2013 under the project DESIRE (grant agreement no. 602531).

References

1. Burnashev N, Szepetowski P (2015) NMDA receptor subunit mutations in neurodevelopmental disorders. Curr Opin Pharmacol 20:73–82. doi:10.1016/j.coph.2014.11.008
2. Paoletti P (2011) Molecular basis of NMDA receptor functional diversity. Eur J Neurosci 33(8):1351–1365. doi:10.1111/j.1460-9568.2011.07628.x
3. Paoletti P, Bellone C, Zhou Q (2013) NMDA receptor subunit diversity: impact on receptor properties, synaptic plasticity and disease. Nat Rev Neurosci 14(6):383–400. doi:10.1038/nrn3504
4. Traynelis SF, Wollmuth LP, McBain CJ, Menniti FS, Vance KM, Ogden KK, Hansen KB, Yuan H, Myers SJ, Dingledine R (2010) Glutamate receptor ion channels: structure, regulation, and function. Pharmacol Rev 62(3):405–496. doi:10.1124/pr.109.002451
5. Hollmann M, Heinemann S (1994) Cloned glutamate receptors. Annu Rev Neurosci 17:31–108. doi:10.1146/annurev.ne.17.030194.000335
6. McBain CJ, Mayer ML (1994) N-methyl-D-aspartic acid receptor structure and function. Physiol Rev 74(3):723–760
7. Moriyoshi K, Masu M, Ishii T, Shigemoto R, Mizuno N, Nakanishi S (1991) Molecular cloning and characterization of the rat NMDA receptor. Nature 354(6348):31–37. doi:10.1038/354031a0
8. Petralia RS, Yokotani N, Wenthold RJ (1994) Light and electron microscope distribution of the NMDA receptor subunit NMDAR1 in the rat nervous system using a selective anti-peptide antibody. J Neurosci 14(2):667–696
9. Anson LC, Chen PE, Wyllie DJ, Colquhoun D, Schoepfer R (1998) Identification of amino acid residues of the NR2A subunit that control glutamate potency in recombinant NR1/NR2A NMDA receptors. J Neurosci 18(2):581–589
10. Monyer H, Sprengel R, Schoepfer R, Herb A, Higuchi M, Lomeli H, Burnashev N, Sakmann B, Seeburg PH (1992) Heteromeric NMDA receptors: molecular and functional distinction of subtypes. Science 256(5060):1217–1221
11. Das S, Sasaki YF, Rothe T, Premkumar LS, Takasu M, Crandall JE, Dikkes P, Conner DA, Rayudu PV, Cheung W, Chen HS, Lipton SA, Nakanishi N (1998) Increased NMDA current and spine density in mice lacking the NMDA receptor subunit NR3A. Nature 393(6683):377–381. doi:10.1038/30748

12. Nishi M, Hinds H, Lu HP, Kawata M, Hayashi Y (2001) Motoneuron-specific expression of NR3B, a novel NMDA-type glutamate receptor subunit that works in a dominant-negative manner. J Neurosci 21(23):RC185
13. Pachernegg S, Strutz-Seebohm N, Hollmann M (2012) GluN3 subunit-containing NMDA receptors: not just one-trick ponies. Trends Neurosci 35(4):240–249. doi:10.1016/j.tins.2011.11.010
14. Lozovaya N, Gataullina S, Tsintsadze T, Tsintsadze V, Pallesi-Pocachard E, Minlebaev M, Goriounova NA, Buhler E, Watrin F, Shityakov S, Becker AJ, Bordey A, Milh M, Scavarda D, Bulteau C, Dorfmuller G, Delalande O, Represa A, Cardoso C, Dulac O, Ben-Ari Y, Burnashev N (2014) Selective suppression of excessive GluN2C expression rescues early epilepsy in a tuberous sclerosis murine model. Nat Commun 5:4563. doi:10.1038/ncomms5563

Chapter 3

Genetic and Functional Analysis of *GRIN2A* in Tumor Samples

Todd D. Prickett, Jared J. Gartner, and Yardena Samuels

Abstract

Ionotropic glutamate receptors (iGluRs) are large integral membrane multi-protein complexes that create ion channels in plasma membranes. Upon binding of receptor specific ligands (e.g., glutamate), increased efflux or influx of mono- or divalent cations (e.g., Ca^{2+}) promotes synaptic transmission, cellular migration, and survival. Three classes of iGluRs were originally defined after their respective agonists: AMPA, kainate, and NMDA receptors (NMDARs). Recently, we examined iGluR families at the genetic level using Next-Generation Sequencing (NGS) (whole-exome sequencing (WES)) and discovered a high prevalence of somatic mutations within the gene for one of the NMDAR subunits, *GRIN2A*, specifically in malignant melanoma. Following confirmation of the somatic mutations, we focused on functional characterization of a subset of the *GRIN2A* mutants that demonstrated a loss of NMDAR functionality. We used gene expression and protein biochemistry to examine complex formation between GluN1 subunit (encoded by *GRIN1*) and GluN2A subunit (encoded by *GRIN2A*), anchorage-independent growth in soft agar and cellular migration. Furthermore, we used shRNA depletion of endogenous *GRIN2A* in melanoma cells expressing either wild-type *GRIN2A* or mutant *GRIN2A* and measured cellular proliferation compared to negative controls. Our data show that somatic mutation of certain residues in GluN2A results in increased survival and is the first such report to demonstrate the functional importance of *GRIN2A* mutations in melanoma and the significance ionotropic glutamate receptor signaling plays in malignant melanoma.

Key words NMDA receptor, Melanoma, Sequencing, Somatic mutation, Ion channel

1 Introduction

N-methyl-D-aspartate receptors (NMDAR) play major roles in many neuronal processes and disease etiologies such as learning, memory, neurodegeneration (i.e., Schizophrenia), epilepsy, and cancer [1–6]. The NMDARs are heterotetrameric complexes consisting of two NR1 (GRIN1) subunits and two NR2 (GRIN2 (A–D)), or a mixture or GRIN2 and GRIN3 subunits predominantly expressed in neuronal cells. Upon binding of its cognate ligands (e.g., glutamate), NMDAR permits Ca^{2+} influx resulting in increased intracellular calcium levels leading to activation of calcium-dependent signal transduction [7]. To understand how

Nail Burnashev and Pierre Szepetowski (eds.), *NMDA Receptors: Methods and Protocols*, Methods in Molecular Biology, vol. 1677, DOI 10.1007/978-1-4939-7321-7_3, © Springer Science+Business Media LLC 2017

NMDA receptors might be connected to the etiology of melanoma, whole-exome sequencing and functional biochemical and cell biology assays can be applied.

To study the mechanisms that NMDARs complexes employ in the functioning of cells multiple genomic and proteomic approaches can be used. Utilizing whole-exome sequencing, a high prevalence of somatic mutations in *GRIN2A* in malignant melanoma was discovered [3] revealing the unexpected discovery that *GRIN2A* was somatically mutated in ~25% of melanoma cases. Employing techniques such as coexpression of both NMDAR subunits in transformed human cells allows for biochemical analysis of complex formation, ion channel influx, as well as functional studies allow for detection of mutation effect on gene [2]. To examine the roles that the normal versus mutated *GRIN2A* plays in melanoma cells, anchorage-independent (soft agar colony formation) and migration assays were performed to help delineate whether the wild-type *GRIN2A* gene product might be a proto-oncogene or a tumor suppressor. Mutated *GRIN2A* versions of proto-oncogenes would have magnitudes of growth (colony formation) and increased migratory potential over the wild-type gene products. Conversely, wild-type *GRIN2A* versions of tumor suppressors would show little to no soft agar colonies nor migratory cells compared to mutated versions of tumor suppressors. Furthermore, using cellular proliferation assays and shRNA-mediated knockdown of *GRIN2A* (wild-type vs. mutants) should help in determining what the tumorigenic potential is for mutated versions of *GRIN2A* in melanoma. Taken together, the genomic and proteomic analyses will show how these mutations affect NMDAR function, giving a better understanding of how NMDA receptors work in melanomagenesis.

2 Materials

2.1 Sequencing Materials

1. DEPC H_2O.
2. Hot start, high fidelity Taq polymerase 5 units/μL.
3. Exonuclease 20 units/μL.
4. Shrimp Alkaline Phosphatase 5 units/μL.
5. Sephadex G50 fine.
6. Filter plate millipore (0.45 μm, clear, nonsterile).
7. 5× sequencing buffer.
8. M13 forward primer.
9. dNTP mix. Stock of each dNTP at an initial concentration of 100 mM. Combine 1 mL of each dNTP + 6 mL DEPC H_2O

for a final concentration of 10 mM. Mix well, aliquot in DNase/RNase-free tubes, and store at −20 °C.

10. 10× PCR buffer: 16.6 mL of 2 M Ammonium Sulfate, 89 mL of 1.5 M Tris–HCl (pH 8.8), 13.4 mL of 1 M $MgCl_2$, 1.4 mL of 14.4 M of 2-Mercaptoethanol, 79.6 mL of dH2O. Combine ingredients mix well and store at −20 °C.

2.2 *GRIN2A* Functional Studies

All the solutions used in Subheading 3 use molecular biology-grade water (RNase-free, DNase-free, Protease-free, and Pyrogen-free) suitable for cloning, Western Blot (WB) and Immunoprecipitation (IP) analysis, stable expression, proliferation assays, migration, and soft agar assays, shRNA knockdown analysis and measurement of calcium uptake. A list of items used to determine, detect, and measure *GRIN2A* are shown below:

1. *GRIN1* and *GRIN2A* open reading frame (ORF) constructs for expression and *GRIN2A* shRNA for knockdown.
2. PCR 2× master mix.
3. HEK293T and 31T melanoma.
4. 1% lysis buffer: 1% NP-40, 50 mM Tris–HCl pH 7.5, 150 mM NaCl, Complete Protease Inhibitor tablet EDTA-free, 1 μM sodium orthovanadate, 1 mM sodium fluoride, and 0.1% β-mercaptoethanol.
5. Protein A/G beads.
6. Tween-20.
7. Antibodies: anti-GluN1 (sc162902), anti-GluN2A (sc1468), or anti-GluN2A (sc-136004) (Santa Cruz Biotechnology), anti-P-p38 MAPK, anti-p38 (Cell Signaling), and anti-GAPDH (Calbiochem-EMD Biosciences).
8. SYBR Green I.
9. 8.0 μm of BD Biocoat.
10. TCN-213 (dissolved in DMSO).
11. Hema 3 Stat Pack.
12. Bacto-Agar.
13. Calf intestine alkaline phosphatase (CIP).
14. LB broth/Super Broth medium.
15. Enhanced chemiluminesences reagent (ECL).
16. NotI (20 U/mL) and XbaI (10 U/mL).
17. Bovine serum albumin (BSA).
18. DNA ligation kit.

19. Puromycin (4 μg/mL) and Neomycin (~400 μg/mL).
20. Gel Extraction Kit, Plasmid DNA kit, RNA purification kit.
21. SuperScript III First Strand kit.
22. 35 mm glass-bottomed culture dishes.
23. Flou-3 AM (4-(6-Acetoxymethoxy-2,7-dichloro-3-oxo-9-xanthenyl)-4′-methyl-2,2′(ethylenedioxy)dianiline-N,N,N′,N′-tetraacetic acid tetrakis(acetoxymethyl) ester).
24. Laser scanning microscope.

3 Methods

3.1 Primer Design

Primers for targeted sequencing should be designed according to the capabilities of your sequencer. Recommendations below should be adapted if necessary:

1. Primers are designed to create amplicons in the size range of 300–600 bp covering all coding exons.
2. In event that exons are larger than amplicon size primers were tiles to overlap by 100 bases until entire exon was covered.
3. BLAST algorithm is used for all primers against the target genome to ensure amplification of only desired target with no other potential annealing locations.
4. An M13 sequence is added to all reverse primers so that all amplicons can be sequenced using the same sequencing primer.
5. There are several freely available online tools to help with primer design. PrimerTile (http://research.nhgri.nih.gov/tools/) and PCRTiler (http://pcrtiler.alaingervais.org:8080/PCRTiler/) both utilize Primer3 for primer design and perform searches for cross hybridization.
6. Example of primers designed to target *GRIN2A* is provided in Table 1.

3.2 Touchdown PCR

Touchdown PCR is used to generate the amplicons for sequencing. Touchdown uses higher annealing temperatures that are gradually lowered allowing for increased specificity as well as yield of desired target amplicon [8].

1. Run PCR reaction (10 μL total volume).
    - A master mix can be prepared without the primers or template DNA and used accordingly to perform multiple reactions in a high-throughput manner.

Table 1
Primers designed for Sanger sequencing of GRIN2A. An M13F sequence has been tagged onto the end of all reverse primers allowing for sequencing of all amplicons with same sequencing primer. Exoms that are represented by multiple primers are to necessitate the full coverage of the coding exon while still keeping the amplicon at appropriate length for sequencing platform

Primer_name	Gene	Exon	Oligo sequence	TM	Mol. weight
GRIN2A_R1	GRIN2A	Exon1_5'_forward	AGTTTCCGGCCTTACCTTGTC	57.25 °C	6339
GRIN2A_F1	GRIN2A	Exon1_5'_reverse	GTAAAACGACGGCCAGTCCTATCCTGCTGCTGAGTTCC	67.76 °C	11,615
GRIN2A_R2	GRIN2A	Exon1_3'_froward	GAGGCAAGACCTGGTTCTCAC	57.71 °C	6431
GRIN2A_F2	GRIN2A	Exon1_3'_reverse	GTAAAACGACGGCCAGTAGAGTGGGCTATTGGACCCTG	67.62 °C	11,793
GRIN2A_R3	GRIN2A	Exon2_5'_forward	GAACAGCCTCGTCTTTGGAAC	56.16 °C	6406
GRIN2A_F3	GRIN2A	Exon2_5'_reverse	GTAAAACGACGGCCAGTCTAGGACGCAGTTTGTGCTTC	67.04 °C	11,719
GRIN2A_R4	GRIN2A	Exon2_3'_forward	TCAGTGCGTATTTCCAACAATG	53.72 °C	6709
GRIN2A_F4	GRIN2A	Exon2_3"_reverse	GTAAAACGACGGCCAGTGGCTACAGGGAATTCATCAGC	67.12 °C	11,746
GRIN2A_R5	GRIN2A	Exon3_forward	AGAAAGAAGCACTGTGAGCCC	57.48 °C	6473
GRIN2A_F5	GRIN2A	Exon3_reverse	GTAAAACGACGGCCAGTGCAGAGAGGCTTCTTGTGATG	67.08 °C	11,808
GRIN2A_R6	GRIN2A	Exon4_forward	GCAAGTGTGGCACATCTCTAGG	57.90 °C	6775
GRIN2A_F6	GRIN2A	Exon4_reverse	GTAAAACGACGGCCAGTGGAAAGGATTTGCCTCTCCAG	67.15 °C	11,737
GRIN2A_R7	GRIN2A	Exon5_forward	CGTTGATAGACCACCTGGATG	55.10 °C	6446
GRIN2A_F7	GRIN2A	Exon5_reverse	GTAAAACGACGGCCAGTGTCCTTGGGAAAGCCACTTC	67.60 °C	11,383
GRIN2A_R8	GRIN2A	Exon6_forward	CATCCTCTGAAATATGCTGCC	53.60 °C	6341
GRIN2A_F8	GRIN2A	Exon6_reverse	GTAAAACGACGGCCAGTTCTTACTCAGACAAAGGGCCTG	66.64 °C	12,010
GRIN2A_R9	GRIN2A	Exon7_forward	TCCTGACCTCATGATCCACC	56.16 °C	5973
GRIN2A_F9	GRIN2A	Exon7_reverse	GTAAAACGACGGCCAGTATGTCTGGGCTTCCTGCTG	67.52 °C	11,092
GRIN2A_R10	GRIN2A	Exon8_forward	TCAATGAGAGGCACCTGAATC	54.63 °C	6439
GRIN2A_F10	GRIN2A	Exon8_reverse	GTAAAACGACGGCCAGTTTCCATCTTCTGGCAACCTTC	66.25 °C	11,574
GRIN2A_R11	GRIN2A	Exon9_forward	CATGCCGAGAGTCAATTTCTG	53.89 °C	6421

(continued)

Table 1 (continued)

Primer_name	Gene	Exon	Oligo sequence	TM	Mol. weight
GRIN2A_F11	GRIN2A	Exon9_reverse	GTAAAACGACGGCCAGTTTGTCATCCTGCCCTAATGC	66.55 °C	11,309
GRIN2A_R12	GRIN2A	Exon10_forward	ATGCAAAGATCCACTGGGAAG	55.21 °C	6488
GRIN2A_F12	GRIN2A	Exon10_reverse	GTAAAACGACGGCCAGTAAAGTGTGGGATGCTTTCAGG	66.12 °C	11,832
GRIN2A_R13	GRIN2A	Exon11_forward	TGAGACATCAAGAACCCAAGC	55.04 °C	6417
GRIN2A_F13	GRIN2A	Exon11_reverse	GTAAAACGACGGCCAGTGTCTGTTCCAAACCCAGAAGC	67.36 °C	11,666
GRIN2A_R14	GRIN2A	Exon12_forward_1	GAAACCATGTCCATGATGAGG	53.52 °C	6479
GRIN2A_F14	GRIN2A	Exon12_reverse_1	GTAAAACGACGGCCAGTCCCTATGCTTTGCAACTTGTC	66.32 °C	11,614
GRIN2A_R15	GRIN2A	Exon12_forward_2	GGAGTGGGTCCTATTCTCTGC	56.89 °C	6444
GRIN2A_F15	GRIN2A	Exon12_reverse_2	GTAAAACGACGGCCAGTCAGGCATCTACAGCTGCATTC	67.13 °C	11,657
GRIN2A_R16	GRIN2A	Exon12_forward_3	TGTTGTCAGGTTCCCTGTGG	57.29 °C	6146
GRIN2A_F16	GRIN2A	Exon12_reverse_3	GTAAAACGACGGCCAGTAGACAACAGGTCCTTTCAGGG	66.67 °C	11,746
GRIN2A_R17	GRIN2A	Exon12_forward_4	GTATCGCTCGCTGGTCTCAC	57.70 °C	6060
GRIN2A_F17	GRIN2A	Exon12_reverse_4	GTAAAACGACGGCCAGTATTCCATACGCCAGGATTCAC	65.74 °C	11,641
GRIN2A_R18	GRIN2A	Exon12_forward_5	GCCCAGTCCTGCTGGTAGAC	60.02 °C	6094
GRIN2A_F18	GRIN2A	Exon12_reverse_5	GTAAAACGACGGCCAGTAACGTGGACTTCCCGGAC	67.58 °C	10,775
GRIN2A_R19	GRIN2A	Exon12_forward_6	GAAAGGGTTATCGGAGGTGTG	55.33 °C	6606
GRIN2A_F19	GRIN2A	Exon12_reverse_6	GTAAAACGACGGCCAGTGTGAGACCAGCGAGCGATAC	68.09 °C	11,466
GRIN2A_R20	GRIN2A	Exon12_forward_7	TTTACCCTCCAGAACATTGGC	55.15 °C	6341
GRIN2A_F20	GRIN2A	Exon12_reverse_7	GTAAAACGACGGCCAGTGCTAGACCTTAGCAGGCCCTC	68.87 °C	11,658
GRIN2A_R21	GRIN2A	Exon12_forward_8	TGGACATCCAACATTTACCCTC	54.74 °C	6614
GRIN2A_F21	GRIN2A	Exon12_reverse_8	GTAAAACGACGGCCAGTTACAAACACTCGTTGCCATCC	66.31 °C	11,601

Components	Reaction volume
ddH_2O	3.3 μL
10× PCR buffer	1 μL
10 μM dNTP	1 μL
DMSO	0.6 μL
50 μM forward primer	1 μL
50 μM reverse primer	1 μL
Hot start Taq polymerase	0.6 μL
(3 ng/μL) template gDNA	2 μL

- Cycling conditions.

Stage	Cycles	Step	Temperature (°C)	Time (s)
1	1	1	94	120
2	4	1 2 3	94 67 68	10 30 50
3	4	1 2 3	94 64 68	10 30 50
4	4	1 2 3	94 61 68	10 30 50
5	30	1 2 3	94 58 68	10 30 50
6	1	1	68	180

2. Verify amplification of a single PCR product.

Run your PCR products on a gel to confirm amplification of only a single PCR product of the desired size can help to identify any problems that may have occurred. Multiple bands or a smear are usually the result of mispriming. These can sometimes be eliminated by raising the annealing temperature in **step 2** of the touchdown PCR by a few degrees for all stages.

3.3 Sanger Sequencing

3.3.1 Exonuclease (Exo)—Shrimp Alkaline Phosphatase (SAP) Treatment

1. Once a PCR product has been verified perform an exonuclease (Exo) and SAP treatment to remove single-strand DNA and free dNTPs.

   Single Exo/SAP reaction scale accordingly for multiple reactions:

Components	Reaction volume
dH_2O	1 μL
SAP buffer	0.3 μL
Exonuclease	0.3 μL
SAP	0.6 μL
PCR product	4 μL

Reaction conditions:

Temperature	Duration (Time)
37 °C	30 min
80 °C	15 min
4 °C	∞

2. After the EXO/SAP reaction store plates overnight at 4 °C or at −20 °C until ready.

3.3.2 Sequencing Reaction

Use chain-terminating chemistry to label sample.

1. Dilute EXO/SAP reactions 1:2 (add 6.5 μL dH_2O per well).
2. Place 2 μL DNA product into each well of a clean PCR plate or tube.
3. Master Mix recipe (1 rxn):

Components	Reaction volume
dH_2O	1.7 μL
5× sequencing buffer	1 μL
50 μM M13 seq primer	0.04 μL
Big dye	0.26 μL

4. Add 3 μL master mix and spin briefly.
5. Sequencing PCR conditions.

Stage	Cycles	Step	Temperature (°C)	Time (s)
1	1	1	94	120
2	25	1 2 3	94 50 64	30 15 4
3	1	1	10	Hold

3.3.3 Sephadex Purification

To remove free labeled nucleotides.

Single reactions can be diluted and spun through a G50 purification column. For more high-throughput screens it is often more useful to make your own Sephadex plates. These can be made ahead of time and stored (you will need to use a well for each reaction that is being performed, 96 reactions will use one 96-well plate).

1. Pour dry Sephadex over a column loader plate and scrape into all of the wells, returning excess to a container.
2. Place the filter plate upside down on a loader and flip so that the Sephadex falls into the wells.
3. Add 300 μL sterile water to each well.
4. Allow the plate to incubate at room temperature for at least 3 h, or overnight at 4 °C sealed in a plastic bag.
5. Before loading sequencing reactions, spin plates placed on the top of a collection container at 2100 × *g* for 5 min to remove H_2O and create columns.
6. Dilute sequencing reactions to a total volume of 20 μL with H_2O.
7. Load all 20 μL directly onto Sephadex columns in filter plate.
8. Spin sample through a Sephadex plate and into a collection plate at 2100 × *g* for 5 min.

3.3.4 Run Sample on Sequencer

Samples were run on an Applied Biosystems 3730 × L Sanger Sequencer. Samples in 96-well PCR plates were setup in plate assemblies making certain that the plate septas were correctly placed. Plates assemblies were then placed into the instrument, stacking up to 16 plate assemblies at a time. Samples were then sequenced using the run scheduler to insert each individual plates ID. Once samples were sequenced, the data was then taken and analyzed for somatic changes in *GRIN2A*.

3.4 Construction of Wild-Type and Mutant GRIN2A Expression Vectors

- Cloning of mouse *GRIN1* and mouse *GRIN2A* was done using PCR primers ((forward (5′-tctagagccaccatgagcacca tgcgcctgc-3′, reverse (5′-gcggccgctcagctctccctatgacg-3′)) for *GRIN1* and ((forward (5′-tctagagccaccat gggcagagtgggctattg-3′), reverse (5′-gcggccgcttaaacatcagattcgatactag-3′)) for *GRIN2A*.
- Full-length (wild-type and mutant) and truncated fragments were amplified from GRIN1 and GRIN2A clones using Hotstart PCR 2X Master mix.
- Use the standard PCR method to generate each fragment.

1. Prepare PCR reagents as follows:

Component	Amount per reaction
Hot start polymerase	25 µL
Forward primer (10 µM)	1 µL
Reverse primer (10 µM)	1 µL
DNA template (100 ng/µL)	1 µL
Distilled H_2O	22 µL
Total volume	50 µL

2. Perform PCR reaction as shown below:

Number of cycles	Temperature	Duration (Time)
1	95 °C	2 min
10–20	95 °C T_m-5 °C 72 °C	20 s 20 s 15–30 s
1	72 °C	3 min
1	4 °C	Infinite

3. Run PCR products on 0.9% agarose gels.
4. Purify PCR products from the gel using Gel Extraction kit and elute in 30 µL dH_2O.
5. Digest purified gene fragments, and pCDF1 and pCDH1 lentiviral expression vectorsPlasmid with XbaI (5′) and NotI (3′) using the following reaction setup:

Components	Reaction volume
dH_2O	Up to 50 µL
pCDF1 or pCDH1	3 µg
10× restriction buffer	5 µL
BSA (10 mg/mL)	1 µL
XbaI (20,000 U/mL)	2 µL
NotI (20,000 U/mL)	2 µL

6. Mix well and incubate at 37 °C for 2 h.
7. Add 2 µL calf intestine alkaline phosphatase (CIP-NEB) to vectors only, mix well, and incubate for an additional 30 min at 37 °C.
8. Run digested gene fragments and vectors on 0.8% agarose gel-EtBr and cut/purify using the Gel Extraction kit and elute in 30 µL of dH_2O.

9. Ligate purified vectors and gene products using the following reaction setup:

Components	Reaction volume
dH_2O	Up to 20 μL
Digested vector	1 μL
Digested gene fragment	2–4 μL
5× ligase buffer	4 μL
T4 DNA ligase	2 μL

10. Transform *E. coli* chemically competent bacteria with 3–5 μL of the ligation mixture and incubate on ice for 30 min.
11. Heat shock bacteria: ligation mix using 42 °C for 1 min, then put mixture back on ice for 2–3 min.
12. Add 400–500 μL Super Broth to each tube and incubate in a 37 °C shaker for 1–2 h.
13. Spread 150 μL of transformed bacteria on LB/Amp plates and incubate overnight in a 37 °C incubator.
14. Pick a few colonies and grow in 2× Super Broth media containing 100 μg/mL Amp overnight in a 37 °C bacterial shaker.
15. Purify plasmids from bacteria using Plasmid preparation kit and elute DNA in 40–50 μL of dH_2O. Sequence each construct to ensure proper sequence (*see* **Note 1**).

3.5 Cell Culture and Transient Expression

1. Grow HEK293T cells in complete RPMI-1640 medium supplemented with 10% Fetal Bovine Serum (FBS) and split 1–10 the day prior to transfection.
2. Grow Melanoma cells in RPMI-1640 and supplemented with 10% fetal bovine serum.
3. Maintain A375Cell, 501Mel, SK-Mel-2Cell cells (purchased from National Cancer Institute, Division of Cancer Treatment, Developmental Therapeutics Program, Frederick, MD) in RPMI-1640 and supplemented with 10% FBS.
4. Co-transfect HEK293T cells at a 6:1 ratio of reagent to constructs (μL:μg) using 2–6 μg of *GRIN1* and *GRIN2A* constructs, *or at* a 1 to 1 ratio for expression studies and functional analysis.
5. Incubate transfected cells overnight at 37 °C plus CO_2 in a humidified incubator.

3.6 Immunoprecipitation and Western Blotting

GluN1 and GluN2A expression were then tested by western blot (WB) analysis (*see* detailed protocol below). Primary antibodies used to interrogate specific protein complex formation for both normal phenotype and dominant negative phenotypes (Fig. 1a–c) and signal transduction pathway analysis (Fig. 1d) were anti-GluN1, anti-GluN2A, anti-GluN2A, anti-P-p38 MAPK, anti-p38, and anti-GAPDH.

1. Seed HEK293T cells in a T75 flask at 1×10^6 cells 24 h prior to transfection.
2. Transfect cells as described above to detect expression of GluN1 and GluN2A to help analyze NMDAR complex formation, phenotype and ionic influx of calcium.
3. Incubate cells overnight at 37 °C in 5% CO_2 and humidified incubator.
4. Gently wash transfected cells twice with 6 mL 1× PBS.
5. Lyse cells using 1.0 mL 1% NP-40 lysis buffer (1% NP-40, 50 mM Tris–HCl pH 7.5, 150 mM NaCl, Complete Protease Inhibitor tablet, EDTA-free, 1 μM sodium orthovanadate, 1 mM sodium fluoride, and 0.1% β-mercaptoethanol) per T-75 flask for 20 min on ice.
6. Scrap lysed cells using a cell scraper and then transfer into a 1.5 mL microcentrifuge tube.
7. Centrifuge cell extracts for 10 min at 14,000 × *g* at 4 °C.
8. Immunoprecipitate 800 μL of cell supernatant overnight using 20 μL of anti-GluN1 or GluN2A and 30 μL of 50% slurry of Protein A/G beads.
9. Wash the immunoprecipitates (IPs) three times using 800 μL per wash.
10. After the final wash, add 50 μL of 2× SDS Sample buffer with β-ME to each IP and heat at 95 °C for 5 min.
11. Centrifuge Heated IPs at 10,000 × *g* for 1 min prior to WB analysis.
12. Load 10 μL of IP samples onto 4–20% gradient SDS gels and run for approximately 35 min at 130 V.
13. Transfer material for western blot analysis using a semidry western blot system with 1× transfer buffer, two thick blot papers, and a nitrocellulose membrane.
14. Once transfer is done, wash membrane briefly before blocking with a 5% dry milk solution (w/v) dissolved in 1× Tris Buffered Saline with Tween-20 (TBST) (0.05% Tween-20 (1× TBS) for 30–60 min.
15. Wash membranes quickly 2–3× in 1× TBST prior to primary antibody incubation (*see* **Note 2**).

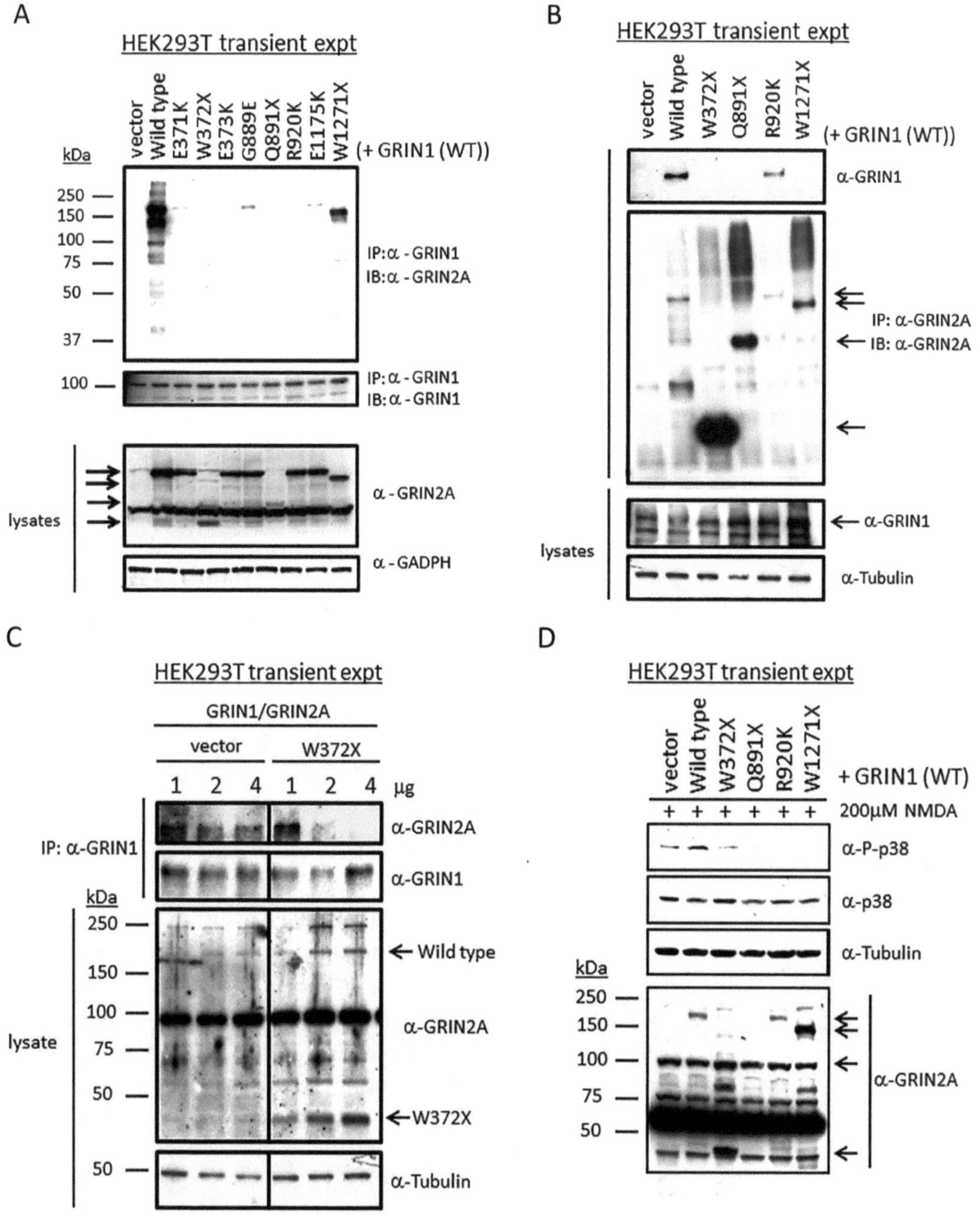

Fig. 1 Somatic mutations in GRIN2A have adverse effects on receptor function and formation. Mutant forms of GRIN2A binding GRIN1 with reduced affinity, thus causing decreased NMDAR complex formation. (**a**) HEK293T cells were transiently transfected with WT GRIN1 and GRIN2A (WT or mutants) or empty vectors as control and immunoprecipitated with anti-GRIN1. Immunoprecipitates were probed with anti-GRIN2A and anti-GRIN1 to confirm binding. Lysates were probed with anti-GRIN2A, and anti-GAPDH as a loading control. (**b**) HEK293T cells were transiently transfected with WT GRIN1 and GRIN2A (WT or mutants) or empty vectors as control and immunoprecipitated with anti-GRIN2A. Immunoprecipitates were probed with anti-GRIN1 and anti-GRIN2A to confirm binding. Lysates were probed with anti-GRIN1, and anti-GAPDH as a loading control. (**c**) HEK293T cells

3.7 Pooled Stable Expression

As per the methodology of cloning the *GRIN1* or *GRIN2A* genes, we used lentiviral constructs (either HIV-based or FIV-based) allowing for the use of two different antibiotic selections for dual-infection.

3.7.1 Lentiviral Production in HEK293 Cells

1. Day 0. Seed T-75 flasks with 2–3 million HEK293T cells per flask and incubate until they are >70% confluent.
2. Day 1. Add 8.6 μg of FIV-34 N plus 1.4 μg of pVSV-G plus 2 μg of expression/stable plasmid (pCDF1 expression vector) followed by the addition of 500 μL of Opti-MEMI.
3. Mix DNA and Opti-MEMI mixture well and then add 24 μL of Turbofect to mix and gently mix.
4. Incubate for 15–20 min at room temperature and then add DNA/Turbofect mixture to cells.
5. Incubate cells and mixture for 60+ h before harvesting medium.
6. Day 4. Carefully remove medium from flask and add to 15 or 50 mL conical tubes and centrifuge at 1000 × *g* for 5 min.
7. Carefully remove tubes from centrifuge and filter the medium through a 0.45 μm filter attached to a sterile 20 mL or 50 mL syringe (*see* **Note 3**).

3.7.2 Lentiviral Production in 31T or SK-Mel-2 Cells

Lentivirus for *GRIN1* and *GRIN2A* (wild-type or mutants) and empty vector control were used to sequentially infect 31T or SK-Mel-2 cells as described below:

1. Seed 31T or SK-Mel-2 cells at 1.5×10^6 cells per T75 flask 24 h prior to infection.
2. Day 0. Plate cells into T-75 flasks at a density of 2×10^6 cells so that the day of infection the cells are >80% confluent before transducing (infecting) with virus.
3. Day 1(2). Make infection cocktails of complete medium for your cells of interest plus harvested virus from Day 3 above plus polybrene (Hexadimethrine bromide). Tend to try two to three different viral ratios if a new cell line or a new batch of virus is made. Use a total of 8–10 mL of viral mixture at this ratio for transduction when using a T75 flask. All viral cocktails contain 8 μg/mL of

Fig. 1 (continued) were transiently co-transfected with WT GRIN1 and GRIN2A (WT or mutants) or empty vectors as control and stimulated with 200 μM NMDA in the presence of Ringer's solution for 60 min. Lysates were probed with anti-P-p38 MAPK, anti-p38 MAPK, and anti-GRIN2A to confirm expression. Lysates were probed with anti-Tubulin as a loading control. (**d**) HEK293T cells were transiently co-transfected with WT GRIN1 and WT GRIN2A and increasing amounts of empty vector or W372X and immunoprecipitated with anti-GRIN1. Immunoprecipitates were probed with anti-GRIN1 and anti-GRIN2A to confirm binding. Lysates were probed with anti-GRIN1, and anti-GAPDH as a loading control

polybrene final concentration (stock solution is at 8 mg/mL made up in water and filter sterilized). This is stored at 4 °C.

4. Day 2 (3). Change medium on infected cells to complete medium.
5. Day 3 (4). GFP lentivirus: keep incubating for an additional 48–72 h and then visually inspect cells using a fluorescent microscope.
6. Puromycin lentivirus: change medium to complete medium with the appropriate amount of puromycin as determined by kill curve (1–40 μg/mL) (*see* **Note 4**).
7. Days 5–8. Check that cells are dying off and that your control flask (no virus added to same cells infected) is dead or almost completely dead. After the control flask is completely dead then the cells that were infected can be harvested by either picking of individual clones or pooling all the cells together or both. These are your pooled clones to be used for functional assays.
8. Determine stable expression of GluN1 proteins (wild-type) and GluN2A (wild-type or mutants) by immunoprecipitation and SDS-PAGE analysis.
9. Then perform immunoblotting with anti-GluN1, anti-GluN2A, and anti-GAPDH to show equivalent expression among pools. (Described above) (*see* Fig. 2a–b).
10. Determine stable expression of *GRIN2A* (wild-type or mutants) by RT-PCR analysis of mRNA from 31T or SK-Mel-2 stable pooled clones using *GRIN2A*-specific primers and *GAPDH* primers as a loading control (data not shown).

3.8 Proliferation Assays

1. To examine growth potential of cells stably expressing *GRIN1*: *GRIN2A* (wild-type and mutants), seed pooled 31T or SK-Mel-2 pooled clones into 96-well plates at 300 cells per well in either 1%, 2.5% or 10% serum-containing medium.
2. Incubate cells for 13–17 days in 96-well plates covered loosely with saran wrap and a damp paper towel (reduces evaporation in wells).
3. Analyze samples every 48 h by lysing cells in 50 μL 0.2% SDS/well and incubating for 2 h at 37 °C.
4. Add 150 μL/well of SYBR Green I solution (1:750 SYBR Green I diluted in dH_2O).
5. Analyze plates using a fluorescence reader.
6. Plot data for proliferation assays using MicroSoft Excel to show growth or lack of potential for each stable pooled clone (Fig. 3a–b).

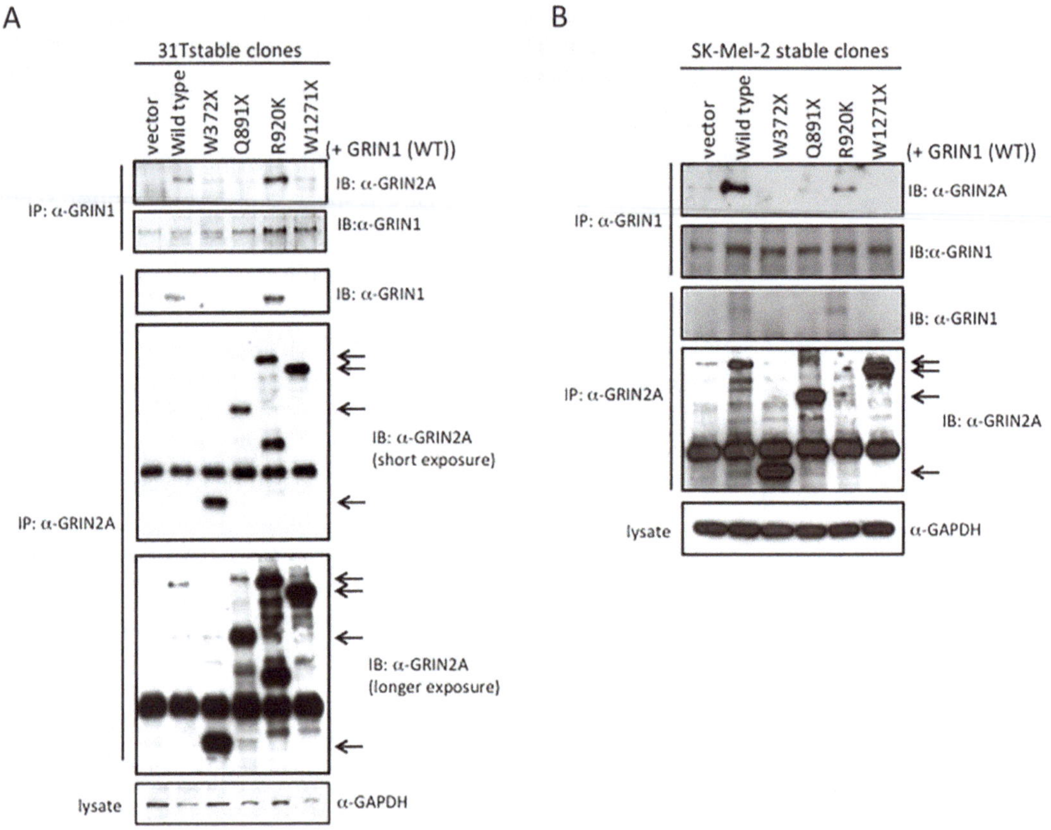

Fig. 2 Melanoma cells expressing mutant forms of GRIN2A have increased ability for anchorage-independent growth. Stable melanoma cell lines 31T and SK-Mel-2 expressing WT GRIN1 and GRIN2A (WT or mutants) or empty vector control and immunoprecipitated with anti-GRIN1 and anti-GRIN2A. Immunoprecipitates were probed with anti-GRIN2A and anti-GRIN1 to examine NMDAR complexes in 31T (**a**) or SK-Mel-2 (**b**)

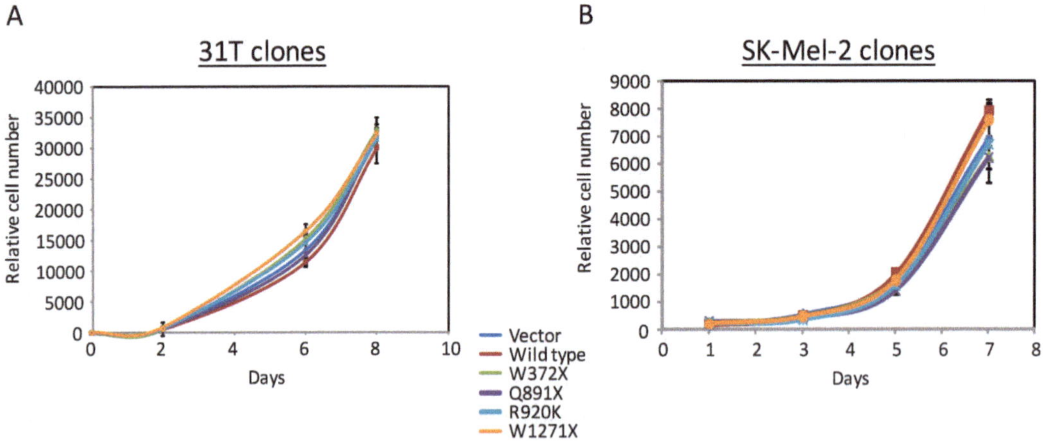

Fig. 3 Expression of GRIN2A mutants in melanoma cell lines results in similar proliferation. Proliferation assay of stable pooled clones 2359 (31T) or SK-Mel-2 cells expressing WT or mutant forms of GRIN2A and empty vector control. Cells seeded in a 96-well plate were measured for proliferation using dilute SYBR Green I over 7–8 days

3.9 Migration Assays

1. To pre-condition migration wells (8.0 μm), add 500–600 μL of warmed (37 °C) RPMI-1640 for 30–60 min at 37 °C.
2. Once pre-conditioned, carefully remove media from the top chamber and add 750 μL of RPMI-1640 supplemented with 10% FBS to the bottom chamber.
3. Seed 31T or SK-Mel-2 pooled clones into pre-conditioned migration wells at 30,000–100,000 cells per well in the serum-free medium in the top chamber and incubate for 24–48 h prior to harvesting (*see* **Note 5**).
4. For antagonist studies use a final concentration of 10 μM TCN-213 dissolved in DMSO in the top and bottom chambers of the Boyden chamber.
5. Use DMSO as a negative control for antagonist assays at <0.02% (vol/vol).
6. Once incubation times are up remove inserts and washed-differentially fix using a Hema 3 Stat Pack as per manufacturer's protocol.
7. Analyze inserts and count for cells migrated per field of view using ImageJ (NIH software) (*see* Fig. 4a–b).

3.10 Soft Agar Assay

1. Prior to starting a soft agar assay, preheat a clean water bath to 45 °C.
2. Prepare 2.5% Bacto-Agar (H_2O) (1.5 g/60 mL dH_2O) and autoclave to dissolve and sterilize. Cool to 45 °C before use (keep in 45 °C water bath throughout the procedure).
3. Make Nutrient Mix: (*see* **Note 6**): 60 mL of 2× RPMI media (1.2 g powdered media into 58.8 mL sterile dH_2O, plus 1.2 mL of 100× Glutamine), 30 mL of FBS, 150 mL of 1× RPMI.

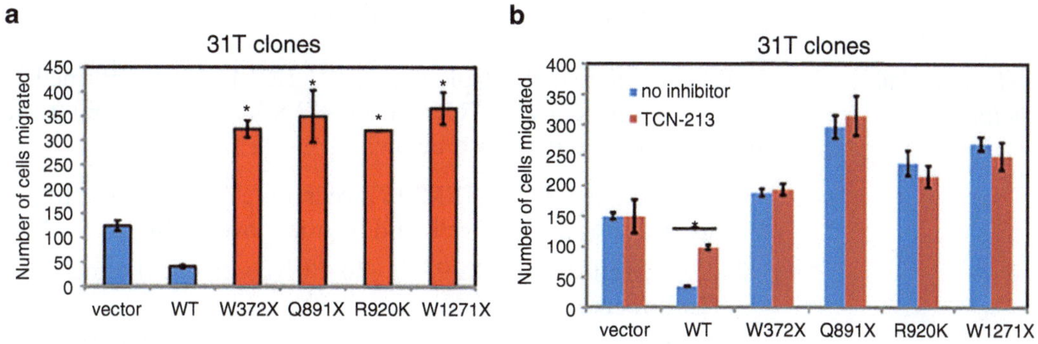

Fig. 4 Melanoma cells expressing mutant forms of GRIN2A have adverse effects on receptor function in cellular migration. Melanoma cell line 31T (**a**) stably expressing GRIN1 and GRIN2A mutants or empty vector were seeded into Boyden chambers in 0.5% serum and grown for 48 h before staining and counting. Melanoma cell line 31T (**b**) stably expressing GRIN1 and GRIN2A mutants or empty vector were seeded into Boyden chambers plus and minus 10 μM TCN-213 for 48 h before staining and counting. Error bars are representative of $n = 3$ (s.d.) ($*p < 0.01$ students t-test)

4. Mix Bacto-Agar mix into nutrient mix for a 0.5% agar concentration, and swirl gently (avoid making bubbles). **Keep this mix at 45 °C** or it will cool and harden. This volume (300 mL) will be enough for up to 200 wells in 24-well plates.
5. Pipet 1 mL nutrient-agar mix into each well (24-well plate) without bubbles—try to pop any that may get into the wells with sterile microtips.
6. Allow agar to cool/harden at room temperature (should only take ~10 min), and then store at 37 °C until cell mixes are ready.
7. Trypsinize and count cells. You will need to make up a mix for plating cell samples in duplicate.
8. Dilute the 0.5 mL cell suspension into 1 mL nutrient-agar mix (has to be a 1/3 dilution for a final agar concentration of 0.33%) and swirl gently to mix.
9. Immediately layer 0.5 mL of cell-agar mix into each well already containing the 0.5% nutrient-agar base.
10. Allow several minutes at room temperature to cool/harden, and then wrap plates in saran wrap and place at 37 °C to grow—check for colonies every few days starting 1 week after plating (*see* **Note** 7).
11. To stain the colonies, add 500 μL crystal violet staining solution (0.005% C.V. with 0.25% formalin) on the top of soft agar.
12. Incubate dishes for 1 h in a 37 °C incubator.
13. To test anchorage independence of 31T or SK-Mel-2 pooled clones, plate cells in triplicate at 1000 cells/well and in top plugs consisting of sterile 0.33% Bacto-Agar (BD, Sparks, MD) and 10% FBS in a 24-well plate. The lower plug contains sterile 0.5% Bacto-Agar and 10% fetal bovine serum.
14. After 2 weeks, photograph the colonies and count colonies by quantifying using ImageJ (NIH software) (*see* Fig. 5a–b).

3.11 Lentiviral shRNA

Constructs for stable depletion of *GRIN2A* were confirmed to efficiently knock down *GRIN2A* at the RNA level. Lentiviral stocks were prepared as described below:

1. Day 1. Seed fresh HEK293T cells in T75 flask/per shRNA lentivirus at 2×10^6 cells and incubate overnight.
2. Day2. Transfect HEK293T (~70–80% confluence) using either TurboFect of Arrest-In.
3. In Tube A add 3 μg of shRNA vector, 3 μg of pHRCMV-8.2ΔR, and 0.3 μg of pCMV-VSV-G and then add 500 μL of OptiMEM-I serum-free medium.
4. In Tube B mix 500 μL OptiMEM-I serum-free medium and 40 μL of TurboFect (or Arrest-In). Allow this to incubate for 5 min at room temperature.

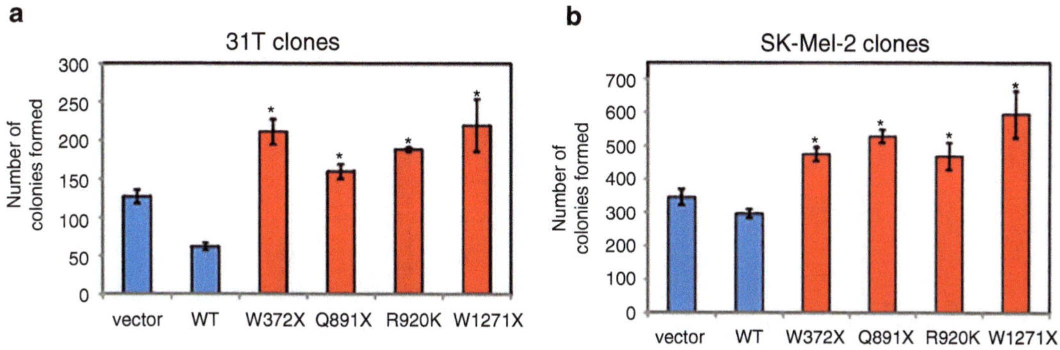

Fig. 5 Melanoma cells expressing mutant forms of GRIN2A have adverse effects on receptor function in anchor dependency. Melanoma cell line 31T (**a**) and SK-Mel-28 (**b**) stably expressing GRIN1 and GRIN2A mutants or empty vector tested for colony formation of 31T or SK-Mel-2 cells in soft agar, respectively. Error bars are representative of $n = 3$ (s.d.) (*$p < 0.01$ Student's *t*-test in quantitative graphs)

5. Add contents of Tube B to Tube A and incubate at room temperature for 20–40 min.
6. Add the mixture (~1 mL) to the HEK293T cells and incubate for 48 h.
7. Before adding the transfection mixture to the HEK293T cells you can change the media to lower serum containing media (from 10% to 2–3% serum) if you desire. This will allow the cells to still produce virus at a good rate but will stop them from getting to over-crowded.
8. Day4. To harvest the virus, carefully remove the medium from flask to 50 mL conical tubes.
9. Spin at 3000 × *g* for 10–15 min.
10. Carefully remove the supernatant and add to 20 mL syringe attached to sterile filter (0.45 μm).
11. Aliquot into sterile 1.50 mL tubes for small infections or sterile 15 mL conical tubes for larger infections.
12. Store at −80 °C.
13. Infect melanoma cell lines (31T, 39T, 501Mel, or 125T) with shRNA lentiviruses for each condition (vector and two-three different *GRIN2A* specific shRNAs).
14. Select stable pooled clones in the presence of 3 μg/mL puromycin containing normal medium for 3–5 days prior to determining knockdown efficiency.
15. Test stably infected pooled clones for proper knockdown of mRNA via qRT-PCR analysis (Fig. 6a–b).
16. Test stably infected pooled clones in functional assays, proliferation assay (*see* Fig. 6c–f).

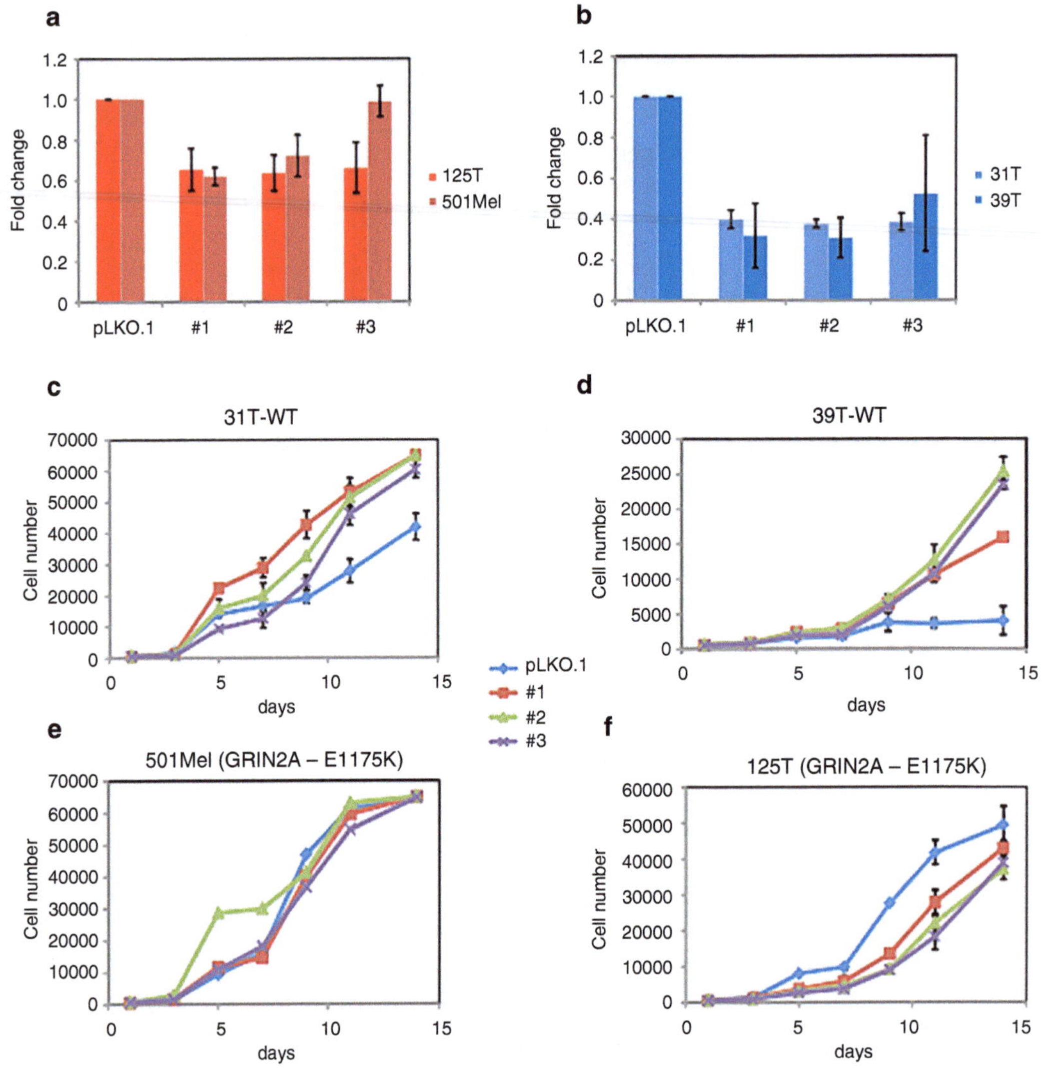

Fig. 6 GRIN2A functions as a tumor suppressor in melanoma cells. Stable depletion of endogenous GRIN2A in melanoma cells. Melanoma cells stably depleted of endogenous GRIN2A were confirmed via qRT-PCR of all clones. (**a–b**) Analysis of knockdown used specific GRIN2A primers and efficiency was compared to GAPDH levels in each sample, thus generating a fold change. Proliferation assay of wild-type 31T (**c**) or 39T (**d**) and mutant 501Mel (**e**) or 125T (**f**) expressing GRIN2A cell lines depleted of GRIN2A. Knockdown resulted in little to no change in proliferation for 125T or 501Mel melanoma cell lines expressing mutant GRIN2A but increased proliferation for both 31T and 39T cell lines expressing WT GRIN2A. Each cell line used pLKO.1 as an empty vector control

3.12 Reverse Transcription PCR

1. Extract total RNA from pooled clones of melanoma cells stably expressing wild-type or mutant *GRIN2A* forms or knocked down for endogenous *GRIN2A*.
2. Elute total RNA in 30 μL diethylpyrocarbonate (DEPC)-treated distilled H_2O.

3. To make cDNA use in qRT-PCR assay, aliquot 1 μg RNA into a labeled RNase-free tube on ice. Bring volume up to 8 μL with DEPC-H_2O.
4. Add 1 μL 10× buffer and 1 μL DNase to the RNA and tap gently to mix. Incubate at room temperature for 15 min. Add 1 μL of 25 mM EDTA to the RNA mix and incubate at 65 °C to heat inactivate the DNase I for 10 min.
5. Place on ice for 1 min and then proceed to cDNA synthesis protocol.
6. For cDNA Synthesis, combine the following in clean DNase/RNase-free tube: 7 μL of DEPC-H_2O, 1 μL of 10 mM dNTP, 1 μL of Primer (50 μM Oligo(dT) or 50 ng/μL random hexamers).
7. Tap mix well to combine and add 9 μL to the 11 μL RNA tube (*see* the DNase step above).
8. Incubate at 65 °C for 5 min, then place on ice for 1 min.
9. Prepare the cDNA synthesis mix. Add each component in the order shown as follows: 4 μL of 10× RT buffer, 8 μL of 25 mM $MgCl_2$, 4 μL of 0.1 M DTT, 1 mL of RNase OUT, 1 mL of RT enzyme, 2 μL of DEPC-H_2O.
10. Add 20 μL mix to each RNA mix and tap gently to combine.
11. For Oligo(dT)—incubate for 50 min at 50 °C.
12. For Random Hexamers—incubate for 10 min at room temperature, then 50 min at 50 °C.
13. Terminate reactions at 85 °C for 5 min—chill on ice.
14. Collect the reaction by a brief centrifugation, then add 1 μL RNase H to each tube and incubate at 37 °C for 20 min.
15. cDNA synthesis is complete: dilute product with 40 μL of sterile dH_2O prior to use.
16. Amplify cDNA using the oligo dT20 primer supplied in the kit.
17. To test for loss of *GRIN2A* message, use 1 μL of cDNA in the PCR reaction with either *GRIN2A* primers or *GAPDH* primers.

3.13 Proliferation Assays of Stable Knockdown Cells

1. Stable pooled knockdown clones for *GRIN2A* were seeded into 96-well plates at 500 cells per well in either 1%, 2.5% or 10% serum-containing medium.
2. Wrap plates loosely in saran wrap with a damp paper towel and incubate at 37 °C for 6–8 days.
3. Analyze samples every 2–3 days by lysing cells in 50 μL 0.2% SDS/well and incubating for 2 h at 37 °C.
4. Add 150 μL/well of SYBR Green I solution (1:750 SYBR Green I diluted in dH_2O).
5. Analyze plates using a fluorescence reader (Fig. 6c–f).

3.14 Calcium Assays

1. Seed HEK293T cells in 35 mm glass-bottomed culture dishes at a density of 1×10^5 cells per plate.
2. Transfect HEK293T cells transiently with GluN1 and either GluN2A WT or the following mutants (Q891X, R920K, W1271X).
3. Thirty six hours post transfection wash cells twice with Tyrode solution (5 mM HEPES, 136 mM NaCl, 2.7 mM KCl, 1 mM $MgCl_2$, 1.9 mM $CaCl_2$, 5.6 mM glucose, buffered to pH 7.4 with Tris base) prior to loading with Fluo-3Am.
4. Load cells with 10 μM Fluo-3 AM diluted in Tyrode solution supplemented with 0.1% BSA for 45 min in the dark.
5. Wash cells subsequently with Tyrode-BSA twice followed by two washes of Tyrode solution alone.
6. Leave cells for 15 min in the dark to ensure complete hydrolysis of the acetoxymethyl ester (AM) groups.
7. Take cell images every 5 s over a period of 15 min and collect images on a Zeiss LSM-700 laser-scanning microscope with a 63× oil immersion lens in a single-track mode using excitation 488 nm and emission BP 505-530 filter sets.
8. Analyze images to determine how mutant GluN2A molecules function compared to wild-type GluN2A in calcium influx (Fig. 7).

4 Notes

1. Send several constructs of the same type for sequencing and then analyze sequence to ensure that the fragments contain either full length wild-type or one of the specific somatic mutations found during whole exome sequencing.
2. Once the secondary antibody incubation and final washes are done rinse three to four times in TBS or dH_2O prior to the addition of ECL reagent. Use premixed ECL reagent on membranes for 1–5 min depending on the strength of the antibody. This is to be tested prior to utilizing it for experiments.
3. Aliquot this into 1.5 mL eppendorff tubes and store at −80 °C until ready to do infection.
4. A kill curve needs to be done for every cell that has not been tested before in this lab. I usually do a range from 1, 2, 4, 5, 7, 10, 15, 20, 30, 40, 50, 100 μg/mL. Using puromycin, the cells will be dead after 2 days or almost if they are sensitive.
5. When using tumor lines for the first time test which cell number works best to give you a fair number of cells that migrated

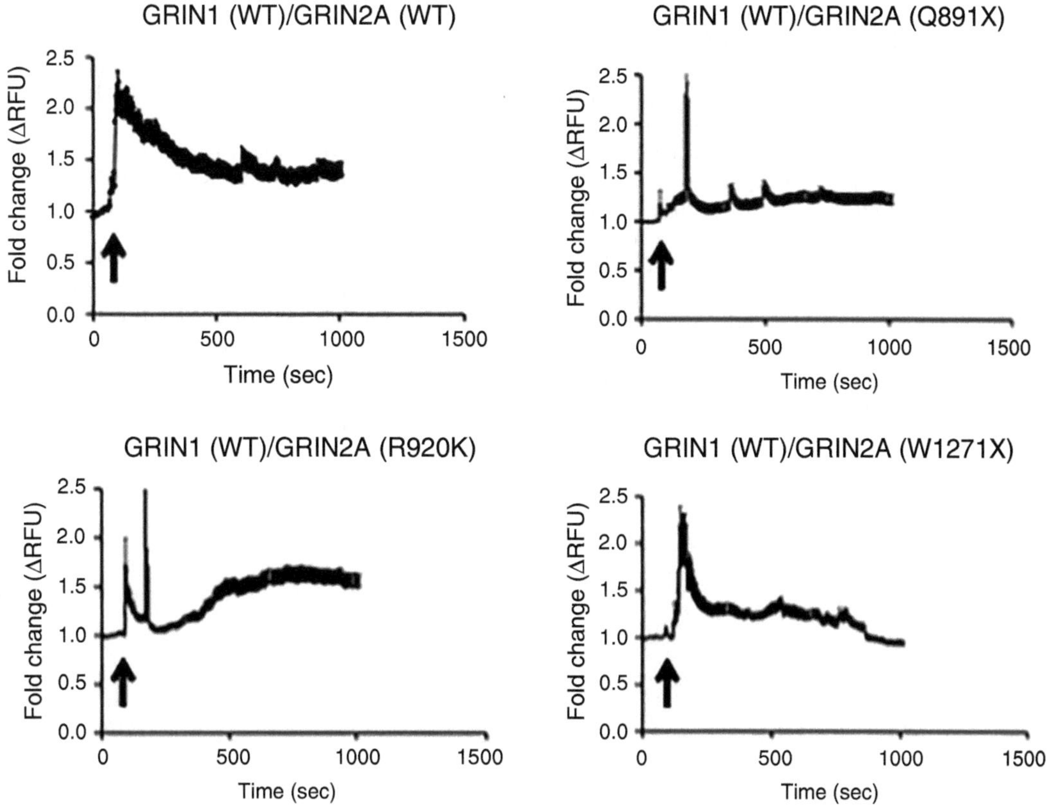

Fig. 7 GRIN2A mutations perturb calcium influx into cells. Influx of calcium upon NMDA stimulation of transient transfected HEK293T cells shows decreased calcium permeability in cells expressing mutant forms of GRIN2A. Cells were pretreated with Fluo-3/AM probe prior to imaging on Ziess confocal microscope. Images were taken every 5 s over a period of 15 min. NMDA (200 μM) was added to the dishes at the 100 s mark. Graphs are fold increase in fluorescent signal vs. time (s) (in this figure). Values are representative of 8–10 individual cells from the image view

within 24 h. Good numbers to start with range from 10,000 to 100,000 cells/well.

6. Make sure to filter components together and keep heated at 45 °C until you are ready to mix with preheated Bacto-Agar. Once you have made the Bacto-Agar nutrient mix keep heated at 45 °C so that it does not harden.
7. First time trials with new cells will require that you plate several different concentrations to optimize density for colony formation (500–5000 cells per well).

References

1. Lang UE, Puls I, Muller DJ, Strutz-Seebohm N, Gallinat J (2007) Molecular mechanisms of schizophrenia. Cell Physiol Biochem 20 (6):687–702. doi:10.1159/000110430
2. Prickett TD, Zerlanko BJ, Hill VK, Gartner JJ, Qutob N, Jiang J et al (2014) Somatic mutation of GRIN2A in malignant melanoma results in loss of tumor suppressor activity via aberrant

NMDAR complex formation. J Invest Dermatol 134(9):2390–2398. doi:10.1038/jid.2014.190
3. Wei X, Walia V, Lin JC, Teer JK, Prickett TD, Gartner J et al (2011) Exome sequencing identifies GRIN2A as frequently mutated in melanoma. Nat Genet 43(5):442–446. doi:10.1038/ng.810
4. Coban A, Gundogdu G, Poyraz M, Yegen G, Demirtas-Tatlidede A et al (2015) NMDA receptor encephalitis with cancer of unknown primary origin. Tumori 102(Suppl 2). doi:10.5301/tj.5000447
5. Carvill GL, Regan BM, Yendle SC, O'Roak BJ, Lozovaya N, Bruneau N et al (2013) GRIN2A mutations cause epilepsy-aphasia spectrum disorders. Nat Genet 45(9):1073–1076. doi:10.1038/ng.2727
6. Lemke JR, Lal D, Reinthaler EM, Steiner I, Nothnagel M, Alber M et al (2013) Mutations in GRIN2A cause idiopathic focal epilepsy with rolandic spikes. Nat Genet 45(9):1067–1072. doi:10.1038/ng.2728
7. Traynelis SF, Wollmuth LP, McBain CJ, Menniti FS, Vance KM, Ogden KK et al (2010) Glutamate receptor ion channels: structure, regulation, and function. Pharmacol Rev 62(3):405–496. doi:10.1124/pr.109.002451
8. Palavalli LH, Prickett TD, Wunderlich JR, Wei X, Burrell AS, Porter-Gill P et al (2009) Analysis of the matrix metalloproteinase family reveals that MMP8 is often mutated in melanoma. Nat Genet 41(5):518–520. doi:10.1038/ng.340

Chapter 4

Detection of NMDARs Antibodies in Encephalitis

Matteo Gastaldi, Patrick Waters, and Angela Vincent

Abstract

There is a range of diseases of the brain that are associated with rapidly developing inflammation. These conditions are called encephalitis, and can be caused by infections, or autoimmune disorders when the subject's immune system reacts against host proteins. The most common form of autoimmune encephalitis is associated with antibodies against the NR1 subunit of the N-Methyl-D-Aspartate Receptor (anti-NMDAR-encephalitis). Antibody detection is important for diagnosis, and can be performed only with techniques that preserve the full tertiary structure (conformation) of the antigen. Here, we describe the methods commonly used to detect NMDAR-antibodies with immunohistochemistry on rat brain slices and on live HEK cells transfected with components of the NMDAR.

Key words NMDAR, Encephalitis, Autoantibody

1 Introduction

N-Methyl-D-Aspartate receptors (NMDARs) are neuronal transmembrane ion channels implicated in several key functions of the central nervous system such as neuronal plasticity [1]. The structure of the channel on the surface of the neuron is heterotetrameric, composed of 2 GluN1 (also called NR1) subunits that express the binding site for glycine, and 2 GluN2 (or NR2) subunits that express the binding site for the excitatory neurotransmitter glutamate.

Antibodies directed against the NR1 subunit are the hallmark and probably the cause of "anti-NMDAR encephalitis," a disease characterized by subacute onset (days to weeks) of psychiatric features, cognitive disturbance, seizures, movement disorders and often autonomic instability and coma [2]. Techniques that use denatured antigen or peptide sequences are inadequate for NMDAR-antibody detection, probably because the antibodies bind to the antigen only when the tertiary structure is preserved. To overcome this problem the most common technique employed is the cell-based assay (CBA). This uses human embryonic kidney

Nail Burnashev and Pierre Szepetowski (eds.), *NMDA Receptors: Methods and Protocols*, Methods in Molecular Biology, vol. 1677, DOI 10.1007/978-1-4939-7321-7_4, © Springer Science+Business Media LLC 2017

(HEK 293) cells that are transiently transfected with plasmids encoding the NR1 and NR2B NMDAR subunits in order to obtain expression on the outer surface of the plasma membrane. In our laboratory, this test is usually performed on live cells, which allow antibodies to bind to extracellular regions of the fully conformational membrane protein while interpretation is not complicated by nonspecific binding to intracellular components [3]. Other laboratories obtain largely similar results when cells are fixed prior to the staining procedure [4], and there is a company that distributes fixed transfected cells. Unfortunately, no systematic multi-center studies of patients with clinically defined NMDAR-antibody encephalitis have yet been performed and the true sensitivities and specificities of the different tests are unclear.

However, a useful and more generic approach that can indicate the presence of an NMDAR or related antibody is indirect immunohistochemistry (IHC) on thin sections of rodent (usually rat) brain. This procedure has been widely performed in the past to detect antibodies against intracellular neuronal components associated with neurological paraneoplastic syndromes [4]. For those antibodies, the heavy fixation often used markedly reduces the sensitivity of the assay for surface antigens such as NMDAR. For this reason, a protocol for IHC on rat brain slices optimized for surface antigens has been proposed [5].

Here, we provide details of the identification of NMDAR-antibodies combining these two complementary techniques, as used by the laboratory that discovered the NMDAR-antibodies [4] but with some modifications as developed within our own laboratory. IHC is considered the most sensitive method, especially when CSF is used [4], but since there are other antigens (e.g., AMPAR, GABARs, LGI1, CASPR2) expressed in the rat hippocampus, use of an antigen-specific technique such as the NMDAR-CBA is essential to confirm the specific target.

2 Materials

2.1 Immunohistochemistry on Rat Brain Slices

1. Trizma-buffered saline (TBS): For 1 L of TBS: Mix 12 g of Trizma base powder and 9 g of sodium chloride, then add 800 mL of H_2O. Adjust pH to 7.4. Make up to 1 L with H_2O.
2. 4% PFA-TBS: For 1 L (to prepare under the fume hood): 40 g of paraformaldehyde in 800 mL of TBS heated under constant stirring to 60 °C. Add a few drops of 1 N NaOH until the solution becomes clear, then let it cool down and filter. If necessary, correct the pH to 7.4. Store at 4° for several weeks or in aliquots at −20 °C for months.

3. Hematoxylin solution: hematoxylin solution 1:5 or 1:10 with H_2O, according to the desired level of counter staining. Store protected from light at room temperature.
4. Other reagents: goat anti-Human IgG biotinylated antibody. Avidin-Biotin kit (ABC). Vectastain 3,3′-diamenobenzydine (DAB) kit (all from Vector labs, Burlingame, California, US) (*see* **Note 1**).

2.2 Live Cell CBA

2.2.1 Cells and Cell Culture Reagents

1. Poly-L-Lysine (PLL) 0.01%: 1% PLL with sterilized H_2O. Store in 5 mL aliquots at −20 °C for later use. Defrost and dilute 5 mL aliquots in 500 mL of sterilized H_2O. Filter sterilize before use. The unfrozen aliquots can be stored at 4° for several weeks.
2. Phosphate-Buffered Saline (PBS): 10 PBS tablets in 1 L of sterile H_2O.
3. Trypsin: prepare aliquots of 1–5 mL of trypsin and store at −20 °C for later use. When needed, defrost and prepare a 1:10 dilution of trypsin-PBS. This can be stored at 4 °C for several weeks.
4. Dulbecco's Modified Eagle's Medium with high glucose (DMEM containing 4500 mg/L of glucose, L-glutamine, sodium pyruvate and sodium carbonate).
5. Fetal bovine serum (FBS).
6. Antibiotic/antimycotic solution (A/A, penicillin, streptomycin and amphotericin, Life technologies, Carlsbad, California, US).
7. Complete Culture medium: DMEM with 10% FBS and 1:100 A/A.
8. Polyethylenimine (PEI): low molecular weight (25,000) polyethylenimine (PEI) in sterile H_2O heated to 80 °C (final concentration 1 mg/mL). Stir until dissolved completely, let it cool to room temperature and adjust pH to 7. Stock in 0.2–1 mL aliquots at −80 °C for later use.

2.2.2 Staining Reagents

1. 4% paraformaldehyde (PFA)-PBS: follow the same procedure described above using PBS instead of TBS.
2. 20 mM (4-(2-hydroxyethyl)-1-piperazineethanesulfonic acid) (HEPES) in DMEM: 2.38 g of HEPES in 500 mL of DMEM.
3. Bovine serum albumin (BSA).
4. Blocking medium: DMEM-HEPES plus 1% BSA.
5. Goat anti-Human IgG (Fc specific).
6. Fluorescent donkey anti-goat (Alexafluor, 568).
7. Fluorescent mounting medium with 4′,6-Diamidino-2-Phenylindole (DAPI).

3 Methods

3.1 Immunohistochemistry on Rat Brain Slices

3.1.1 Freezing the Brain

1. Sacrifice Sprague-Dawley rat with CO_2, and extract the brain without perfusing the animal.
2. Chop the brain in half along the sagittal plane and immerse in 4% PFA-TBS for 1 h at room temperature (20–22 °C).
3. Transfer the brain to 40% Sucrose-TBS solution and store at 4 °C until the brain starts sinking (usually 48 h, *see* **Note 2**).
4. Remove the brain from the sucrose and aspirate the excess sucrose to prevent the formation of air bubbles when freezing.
5. Embed in Optimal Cutting Temperature compound (OCT) until the half brain is covered, and let rest for 15 min.
6. Place a metal cup containing isopentane in a box with dry ice and let it cool down for at least 20 min with the lid on, until the isopentane reaches −80 ° C.
7. Dip the embedded brain in the isopentane until frozen (usually 1–2 min are sufficient).
8. Store at −80 °C for later use.

3.1.2 Obtaining Rat Brain Slices

1. Prepare the cryostat chamber for slicing (chamber temperatures between −21 °C and −25 °C are ideal).
2. Remove the brain from the −80 °C freezer and put inside the cryostat. Wait for at least 30 min for the brain to reach the same temperature as the chamber.
3. Slice at 7 μm thickness (*see* **Note 3**).
4. Let the slides dry overnight at room temperature or 2 h at 37 °C.
5. Use slides immediately, or store at −20 °C (*see* **Note 4**).

3.1.3 Staining (See Note 5)

All the following steps are performed at room temperature unless stated otherwise, and using a humid chamber to prevent the slides from drying.

1. If slides are frozen, allow at least 20–30 min at room temperature for them to defrost.
2. Draw a line around each slice with a dako-pen (Dako, Glostrup, Denmark).
3. Rehydrate slices placing 150 μL of TBS on the top of each slide for 1 min, then tap off.
4. Incubate for 15 min with 0.3% H_2O_2-TBS at 4 °C to quench residual peroxidase activity.
5. Wash three times with TBS (5 min per wash, slides can be immersed in a chamber filled with TBS).

6. Block for 1 h with TBS-10% normal goat serum (NGS).
7. Without washing, tap off blocking solution and incubate with patient's IgG in TBS-5% NGS (serum 1:200 dilution, cerebrospinal fluid 1:2 dilution) overnight at 4 °C.
8. Wash three times in TBS, as previously described.
9. Incubate for 2 h with secondary biotinylated anti-human IgG 1:1000 in TBS-5% NGS at room temperature. Prepare the ABC complex and store at 4 °C (must be made at least 30 min before use in **step 11**).
10. Wash three times in TBS as previously described.
11. Incubate for 1 h with ABC complex (prepared at least 30 min before use).
12. Wash three times in TBS as previously described.
13. Develop chromogenic reaction with Vectastain kit for 6–7 min (this step should be performed under the fume hood since DAB is carcinogenic).
14. (Optional) counterstain for 10–15 s with hematoxylin solution (*see* **Note 6**).
15. Dehydrate the slides in increasing EtOH concentration and xylene (EtOH 50% for 5 min, EtOH 70% for 5 min, EtOH 100% for 5 min, EtOH 100% for 5 min, xylene I for 5 min, xylene II for 5 min) (This step should be performed under the fume hood since Xylene is carcinogenic).
16. Mount slides with DPX mounting medium, and air-dry under a fume hood (usually overnight).
17. Evaluate under a light microscope. NMDAR antibodies should provide a specific staining of the neuropil of the hippocampus (*see* Fig. 1).

3.2 Live Cell-Based Assay

To maintain sterility perform all the steps under a laminar flow hood until the incubation with patient sera.

3.2.1 Coating Coverslips

1. Put circular glass coverslips (13 mm) in the wells of a 6-well plate (3 or 4 coverslips/well can be handled easily). Alternatively, cells can be grown on coverslips placed into 24-well plates to prevent coverslips from overlapping.
2. Cover and submerge the coverslips with 3 mL of PLL solution per well (make sure no coverslips are floating on the surface of the liquid).
3. After 1 h at room temperature aspirate the PLL solution.
4. Leave the 6-well plates open under sterile conditions until completely dry (usually about 1–2 h). Then close and store at room temperature for later use (PLL-plates can be stored for several weeks).

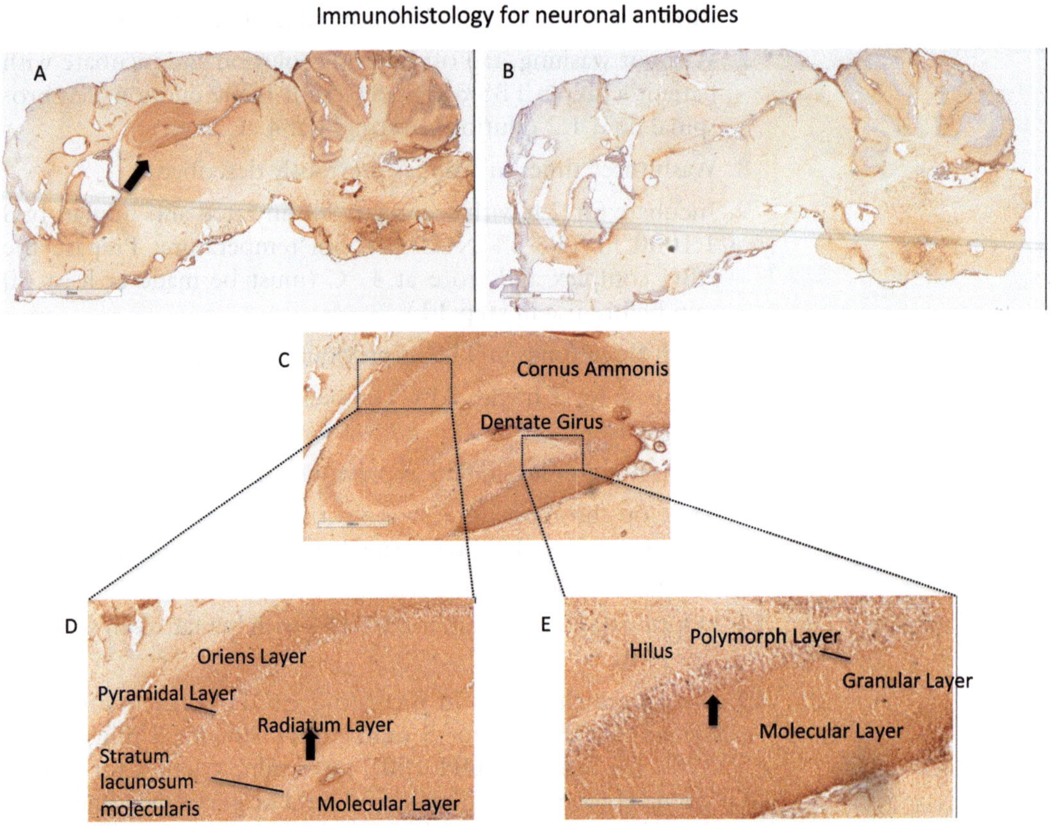

Fig. 1 Immunohistochemistry on rat brain slices. Indirect immunohistochemistry using peroxidase labeled goat anti-human IgG as the secondary antibody on sagittal rat brain slices with hematoxylin counterstaining. Hippocampal staining is visible with patient serum NMDAR antibody (**a**, *black arrow*) but not with serum from a healthy individual (**b**). Detail of the hippocampus (**c**, **d** and **e**). In the cornus ammonis the staining pattern involves the oriens layer and the radiatum layer, sparing the pyramidal layer; a thin rim of stronger staining is seen adjacent to the stratum laconosum molecularis (**d**, *black arrows*). In the dentate gyrus, the staining is visible in the molecular layer with sparing of the granular and polymorph layer and the hilus; a thin rim of stronger staining is seen adjacent to the pyramidal layer (**e**, *black arrows*). The serum was diluted 1:200 but CSF can be used undiluted if sufficient is available and gives a very clean and specific result

3.2.2 Cell Culture Preparation and Seeding

1. Grow HEK293T cells in culture medium in a T175 flask until 90–100% confluent.
2. Detach cells from the flask using trypsin. Count cells and seed approximately 450,000–550,000 cells into each well in complete culture medium (total volume 3 mL/well; seeding concentration may have to be corrected according to cell growth rate).
3. After about 32 h, make sure that cells are at least 90–95% confluent (*see* **Note** 7).
4. Before proceeding with the transfection procedure, replace complete culture medium in all wells.

3.2.3 Transfection

1. Prepare a mixture of the reagents necessary for transfection following this order:

   cDNA, total amount of 3 μg/well (NR1: 2.1 μg/well; NR2B: 0.7 μg/well; Enhanced Green Fluorescent Protein (EGFP): 0.2 μg/well) (*see* **Note 8**).
   DMEM (50 μL/well).
   PEI (1.5 μg/well).
2. Mix well, and let it rest for at least 10 min protected from light (PEI is photosensitive).
3. Add the mixture to the wells and incubate for 12–14 h (*see* **Note 9**).
4. Replace the complete culture medium from the wells with complete culture medium plus 500 μM ketamine, to prevent glutamate toxicity.
5. Let the cells incubate for 24 h for optimal surface expression.

3.2.4 Staining

1. Invert one of the coverslips with transfected cells face down onto a microscope slide onto a drop of fluorescent mounting medium. Evaluate under a fluorescence microscope the presence of an adequate number of cells and of green cells expressing EGFP.
2. Prepare samples (serum diluted 1:20 in blocking medium, CSF 1:2 or undiluted) directly in the wells of a 24-well plate (total volume 250 μL/well).
3. Transfer the coverslips with transfected cells into the 24 well-plates (with the cells facing up) and incubate for 1 h at room temperature protected from light.
4. Aspirate primary antibody (diluted patient's serum or CSF) and wash three times with DMEM-HEPES (250 μL/well).
5. Incubate the coverslips with 4% PFA in PBS for 5 min at room temperature.
6. Wash three times with DMEM-HEPES.
7. Incubate with goat anti-human IgG (Fc specific) 1:750 in blocking medium for 1 h at room temperature (*see* **Note 10**).
8. Wash three times with DMEM-HEPES (250 μL/well).
9. Incubate for 45 min with donkey anti-goat fluorescent secondary antibody 1:750 in blocking medium.
10. Wash three times with DMEM-HEPES and once with PBS. Leave the coverslips in PBS.
11. Invert the coverslips with cells face down onto one drop of antifade mounting medium per coverslip on microscope slides with (maximum six coverslips per slide).
12. Protect from the light and air-dry for 15–20 min (to avoid damaging the microscope lenses with wet mounting medium).

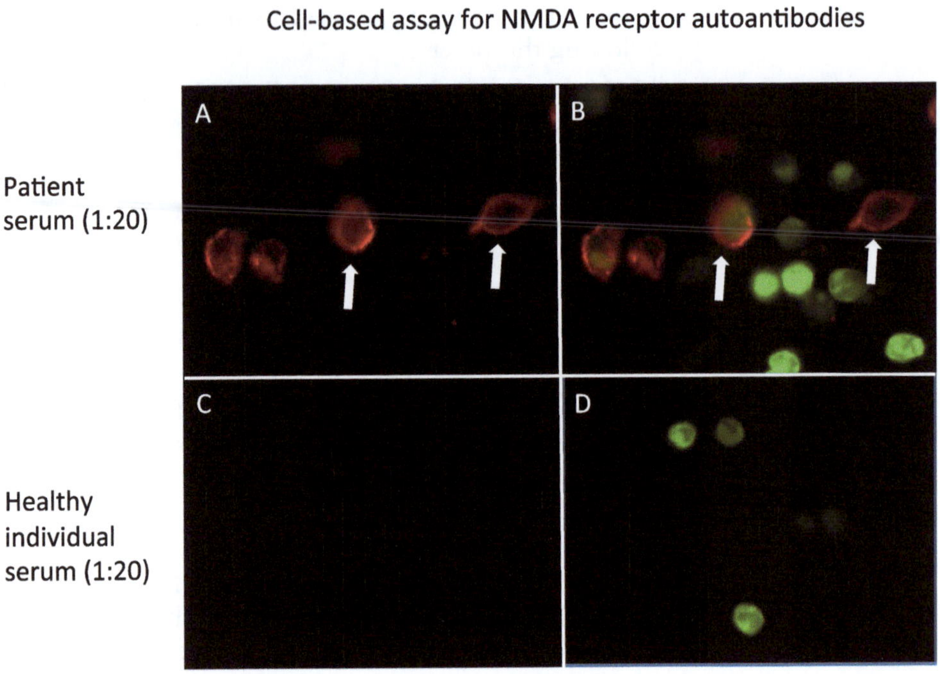

Fig. 2 Live cell-based assay. The figure shows the serum of a patient with anti-NMDAR-encephalitis (**a–b**) and the serum from a healthy individual (**c–d**) tested on the live cell-based assay for NMDAR antibodies. The rim of red staining around the cell represents the binding of the NMDAR antibodies on the cell surface revealed with a fluorescent-labeled secondary antibody (Alexa fluor 568, **a** and **b**, *white arrows*). The red staining is absent with the negative control serum (**c–d**). EGFP has been co-transfected with the NR1 and NR2B subunit plasmids to demonstrate good transfection, but does not necessarily colocalize with the NMDAR antibodies. Sera were diluted 1:20, but CSF undiluted or 1:1 can be used. Magnification 40×

13. Evaluate under a florescent microscope. The presence of NMDAR antibodies will be revealed by a thin rim of fluorescent staining on the surface of the cells (*see* Fig. 2).

4 Notes

1. The reagents indicated are used in our laboratory for reproducibility. Any chromogenic method using DAB is suitable, if freshly prepared.
2. Sinking of the brain is variable and often incomplete. In most cases after 48 h the brain has sunk almost halfway or just below the surface; this provides sufficient degree of cryoprotection and does not affect the quality of the tissue.
3. Since the blade has to cut through materials with different consistency, the slicing can be particularly challenging. Often, the slices can show several artifacts such as folds and bubbles when they are picked up. The following tips could help in the

procedure, but it must be considered that slicing with the cryostat requires experience and optimal settings must be set in each laboratory:

- Start cutting at 20 μm thickness, and progressively reduce the thickness as you become more experienced.
- Reducing the specimen temperature helps in keeping the slices flat, but a specimen that is too cold can present cutting artifacts. Some cryostats allow one to set the chamber and the specimen temperature separately. This makes the cutting easier, because the specimen temperature can be lowered more quickly. Chamber temperature has to be set at the same temperature as the specimen, or just a few degrees warmer, to prevent slides from sticking to the chamber.
- Using warm slides (37 °C) can help to pick up the rat brain slices quickly, and avoid air bubbles.

4. We usually place two slides facing each other and wrap them in aluminum foil. We then store them at −20 °C in a box containing silica gel. Best results are always obtained with freshly prepared tissue, but for practical reason long-term storage is often needed. A reduction in staining intensity after several weeks of storage may occur.
5. At every stage of the staining the slices must be fully covered by buffer. Usually, 120–150 μL/coverslip (slice) are sufficient to cover the whole brain without overfilling.
6. Hematoxylin counterstaining is optional. Conditions must be optimized in each laboratory, but we tend to perform a very light counterstain for optimal detection of antibody binding.
7. NR1/NR2B transfection is harmful to the cells because activation of the NMDAR by glutamate in the medium can lead to excessive influx of calcium and some cells will die after transfection. Healthy and confluent cells at this stage are crucial to perform a reliable staining.
8. Even though pathogenic antibodies target the NR1 subunit, in the assay optimized in our laboratory a combination of NR1 and NR2B improved the surface expression of the channel. The cells are co-transfected with EGFP to check the efficiency of the transfection before the staining procedure.
9. We usually perform this procedure overnight, even though shorter times are probably sufficient for an efficient transfection. The medium has to be replaced in any case before 12–16 h to avoid excessive glutamate toxicity.
10. Alternatively, other groups use directly an anti-gamma-specific labeled secondary antibody.

Disclosures: Angela Vincent, Patrick Waters and the University of Oxford hold patents for antibody tests (MuSK, LGI1, CASPR2). AV and PW receive a proportion of royalties.

References

1. Traynelis SF, Wollmuth LP, McBain CJ, Menniti FS, Vance KM, Ogden KK, Hansen KB, Yuan H, Myers SJ, Dingledine R (2010) Glutamate receptor ion channels: structure, regulation, and function. Pharmacol Rev 62(3):405–496. doi:10.1124/pr.109.002451
2. Dalmau J, Gleichman AJ, Hughes EG, Rossi JE, Peng X, Lai M, Dessain SK, Rosenfeld MR, Balice-Gordon R, Lynch DR (2008) Anti-NMDA-receptor encephalitis: case series and analysis of the effects of antibodies. Lancet Neurol 7(12):1091–1098. doi:10.1016/S1474-4422(08)70224-2
3. Irani SR, Bera K, Waters P, Zuliani L, Maxwell S, Zandi MS, Friese MA, Galea I, Kullmann DM, Beeson D, Lang B, Bien CG, Vincent A (2010) N-methyl-d-aspartate antibody encephalitis: temporal progression of clinical and paraclinical observations in a predominantly non-paraneoplastic disorder of both sexes. Brain 133(6):1655–1667. doi:10.1093/brain/awq113
4. Graus F, Dalmau J (2012) Paraneoplastic neurological syndromes. Curr Opin Neurol 25(6):795–801. doi:10.1097/WCO.0b013e328359da15
5. Gresa-Arribas N, Titulaer MJ, Torrents A, Aguilar E, McCracken L, Leypoldt F, Gleichman AJ, Balice-Gordon R, Rosenfeld MR, Lynch D, Graus F, Dalmau J (2014) Antibody titres at diagnosis and during follow-up of anti-NMDA receptor encephalitis: a retrospective study. Lancet Neurol 13(2):167–177

Part II

Studying Structure–Function Relations in NMDARs

Chapter 5

Magnetofection™ of NMDA Receptor Subunits GluN1 and GluN2A Expression Vectors in Non-Neuronal Host Cells

Nadine Bruneau and Pierre Szepetowski

Abstract

The functional study of reconstituted NMDA receptors (NMDARs) in host cells requires that the corresponding vectors for the expression of the NMDAR subunits are co-transfected with high efficiency. Magnetofection™ is a technology used to deliver nucleic acids to cells. It is driven and site-specifically guided by the attractive forces of magnetic fields acting on magnetic nanoparticles that are associated with nucleic acid vectors. In magnetofection™, cationic lipids form self-assembled complexes with the nucleic acid vectors of interest. Those complexes are then associated with magnetic nanoparticles that are concentrated at the surface of cultured cells by applying a permanent magnetic field. Magnetofection™ is a simple method to transfect cultured cells with high transfection rates. Satisfactory expression levels are obtained with very low amounts of nucleic acid vector. Moreover, incubation time with host cells is less than 1 h, as compared with the several hours needed with standard transfection assays.

Key words NMDA receptors, GluN1, GluN2A, Magnetofection™, Mammalian cells

1 Introduction

NMDARs are cation channels that are gated by the major excitatory neurotransmitter glutamate. NMDAR-mediated signaling is involved in development, plasticity, learning, memory, and high cognitive functions [1–3]. NMDARs play an important role in temporal integration of neuronal network activity and long-term alterations in synaptic structure and function. Typically, NMDARs are composed of two obligatory GluN1 subunits and two modulatory subunits GluN2(A-D) or GluN3(A, B) to form either di- or tri-heterotetrameric channels [3–5]. The GluN2 (GluN3) subunits have different temporal and spatial expression patterns in the brain and NMDAR subtypes also vary according to the cell types and subcellular localization. The identity of the GluN2 subunit is cru-

Nail Burnashev and Pierre Szepetowski (eds.), *NMDA Receptors: Methods and Protocols*, Methods in Molecular Biology, vol. 1677, DOI 10.1007/978-1-4939-7321-7_5, © Springer Science+Business Media LLC 2017

cial for determining several gating properties, including maximal channel open probability, agonist sensitivity, and deactivation kinetics; hence, GluN1/GluN2A have the fastest excitatory postsynaptic current decay, a key parameter in the control of synaptic integration.

NMDARs subunits are assembled early in the endoplasmic reticulum, and the association of both GluN1 and GluN2 subunits is necessary for successful cell surface targeting [6]. Because functional NMDARs are heterotetrameric channels assembled from four subunits [7], the successful co-expression in a single cell of all target subunits obviously is a key prerequisite. Hence, an efficient method to achieve the co-expression of various combinations of NMDAR subunit-encoding cDNAs in cultured cells is needed. Magnetofection™, a method invented by Christian Plank and his collaborators [8], is a process in which magnetic lipocomplexes are formed from magnetic nanoparticles associated with nucleic acids previously combined with cationic lipids [9].

2 Materials

2.1 Cell Cultures

1. Human embryonic kidney cell line HEK293T: American Type Culture Collection, ATCC CRL-3216.
2. Modified DMEM (Dulbecco's Modified Eagle's Medium) containing: 4.5 g/L glucose supplemented with 10% fetal bovine serum, 2 mM L-glutamine, 100 U/mL penicillin, 100 μg/mL streptomycin, and 0,25 μg/mL amphotericin B.
3. OptiMEM: reduced serum Medium with GlutaMax supplement.
4. 0,01% Poly-D-Lysine solution.
5. LabTek separable, on glass slide, 4 or 8 culture chambers.
6. Six-well plates.
7. L-glutamine stock solution (200 mM solution).
8. Stock solution of Antibiotic/Antimycotic (100×) contains: 10,000 U/mL penicillin, 10 mg/mL streptomycin, and 25 μg/mL amphotericin B.
9. Ketamine stock solution (100 mg/mL).
10. Dulbecco's Phosphate-Buffered Salines (DPBS, no calcium, no magnesium): 2.6 mM KCl, 1.4 mM, KH_2PO_4, 138 mM NaCl, and 8 mM Na_2HPO_4-$7H_2O$, pH = 7.2.
11. 0.05% Trypsin-EDTA phenol red.
12. Sterile DNase/RNase-free water.

2.2 Transfection

1. MagnetoFectamine™ Transfection reagent. Magnetofectamine™ is the association of Lipofectamine™ 2000 with the CombiMag Transfection Reagent.
2. Mega Magnetic Plate.

2.3 Constructs

GluN1 and GluN2A mammalian expression vectors.

3 Methods

3.1 Cell Cultures

All the procedures are performed under a cell culture hood.

Routinely, human embryonic kidney HEK293T cells are grown in a modified DMEM medium in 75 cm^2 culture flasks in a sterile incubator at 37°C with 5% CO_2. Cells are subcultured every 2–3 days with subcultivation ratio at 1:3–1:8 into the new flasks, as follows.

1. Rinse cell layer with Ca^{2+}/Mg^{2+} free D-PBS to remove any trace of trypsin inhibitor-containing serum.
2. Add 1 mL of Trypsin-EDTA solution and incubate for 2 min.
3. Add 10 mL of complete growth medium.
4. Aspirate cells by gentle pipetting and centrifuge at 200 × *g* for 2 min at room temperature (20–22 °C).
5. Resuspend the cell pellet in the appropriate volume.
6. Add cells to new culture flasks and incubate at 37 °C with 5% of CO_2.

3.2 Transfection (Fig. 1)

3.2.1 Coating Cell Culture Plates

1. Incubate cell culture plates for 30–60 min at room temperature with 0.01% poly-D-Lysine [10].
2. Remove the poly-D-Lysine solution, wash three times with sterile water, and dry the plates under the hood. Coated cell culture plates can be stored at 4 °C for up to 3 weeks.

3.2.2 Cell Preparation

One day before transfection, HEK293T cellsCell are seeded in modified DMEM in six-well plates at a concentration of 2 × 10^5 cells/well to reach 70% confluence (*see* **Notes 1** and **2**).

3.2.3 Transfection Assay

1. The day after (day of transfection), replace modified DMEM with OptiMEM (*see* **Notes 3** and **4**).
2. Prepare vector solution: in a 250 μL final volume of OptiMEM medium, mix 1 μg of GluN1 expression construct and either 3 μg of wild-type or mutant GluN2A expression construct (to test the wild-type and homozygous mutant conditions, respectively), or 1.5 μg of each wild-type and mutant GluN2A expression constructs (to test the heterozygous condition) (*see* **Note 5**).

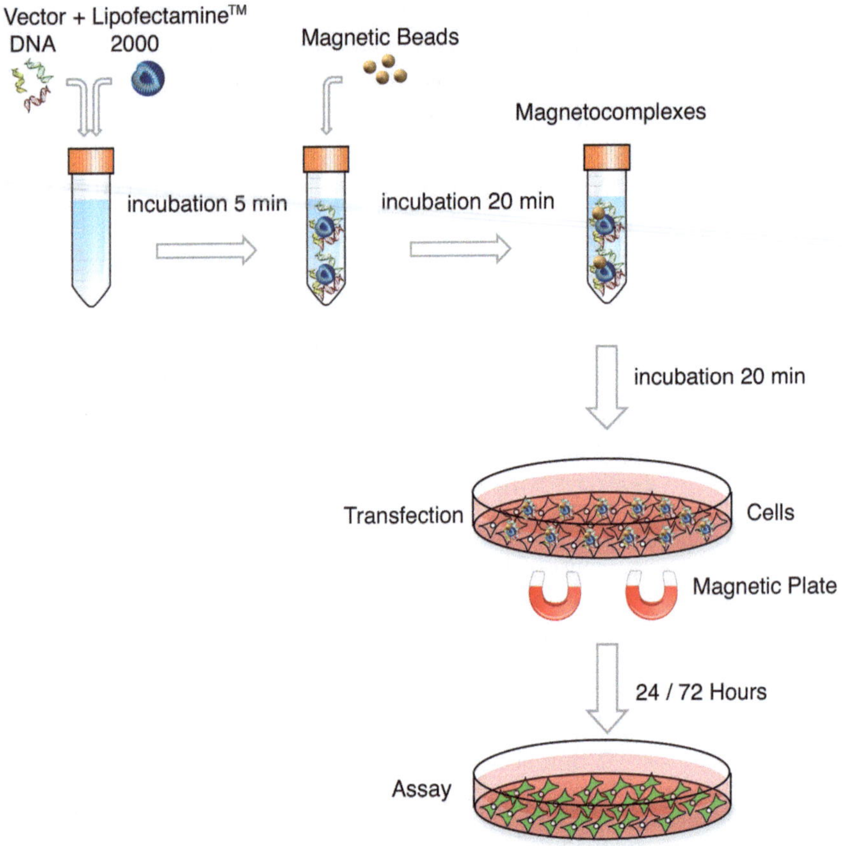

Fig. 1 Schematic view of Magnetofection™ protocol

3. Prepare Lipofectamine solution: dilute 12 μL Lipofectamine 2000 in OptiMEM medium (250 μL final volume).
4. Mix the vector and lipofectamine solutions and incubate for 5 min at room temperature.
5. Add 6 μL of magnetic particle preparation (CombiMag reagent), vortex gently before use, to the 500 μL vector + lipofectamine mix.
6. Mix gently and incubate for 20 min at room temperature.
7. Add magnetocomplexes into the wells containing HEK-293T cells growing in OptiMEM medium and homogenize gently on a rocking plate for 2 min to ensure uniform distribution.
8. Incubate the cells for 20 min on a magnetic plate at 37 °C in the CO_2 incubator.
9. Remove the medium and replace with fresh DMEM supplemented with 100 μM of ketamine solution (*see* **Note 6**).
10. Incubated cells at 37 °C in a CO_2 incubator for 24 h.

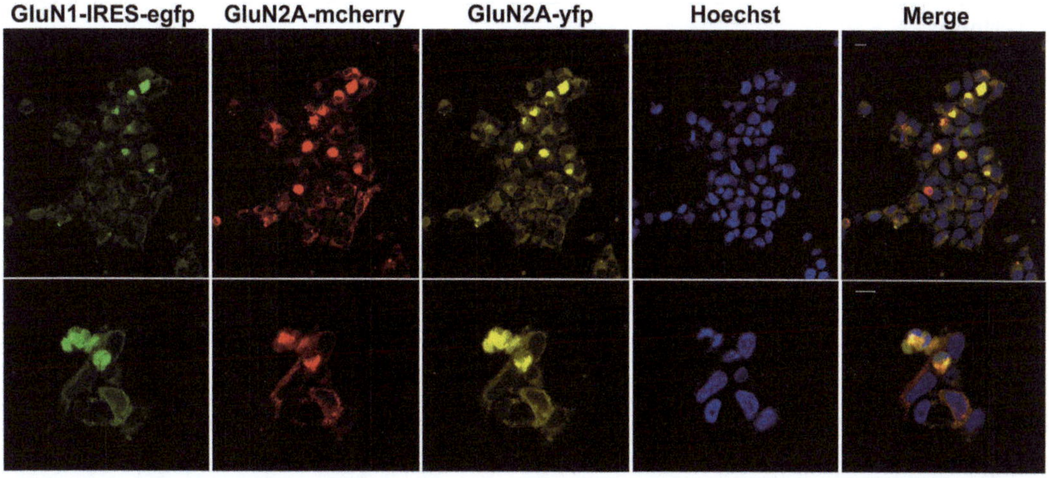

Fig. 2 Triple magnetofection™ of GluN1 expression vector and two wild-type GluN2A expression vectors in HEK-293T cells. HEK293T cells were co-transfected with GluN1 construct and with two different constructs for the expression of wild-type (wt) GluN2A subunit fused to either yfp-tag (*yellow*) or mcherry tag (*red*). GluN1 construct allowed for IRES-driven independent expression of egfp fluorescent protein (*green*). Nuclei were counterstained with Hoechst (*blue*). Bar: 10 μm. Images were captured with confocal laser scanning microscope. *IRES* internal ribosomal entry site, *yfp* yellow fluorescent protein, *egfp* enhanced green fluorescent proteineGFP

3.3 Analysis of Transfected Cells

1. In the pre-warmed modified DMEM medium, plate transfected cells onto coverslips (e.g., for electrophysiological recordings) coated with poly-D-Lysine.
2. In parallel and to check for NMDA receptor subunit expression, plate transfected cells onto Lab-Teck chambers coated with poly-D-Lysine to assess transfection efficiency.
3. Wash cells twice in DPBS 48 h later (i.e., 3 days after transfection).
4. Fix cells with 4% paraformaldehyde in PBS buffer pH 7.4 for 30 min at room temperature.
5. Wash cells three times in DPBS.
6. Incubate cells with Hoechst 33258 for nuclei staining for 10 min at room temperature.
7. Rinse cells three times in DPBS.
8. Remove buffer, then add one drop of Fluoromount-GT onto the cells. Mount cover slip and press gently. Air-dry before examination.
9. Check NMDA receptor subunit expression (fluorescence-based microscopy) (Fig. 2).

4 Notes

1. For optimal transfection efficiency, cell density is a crucial parameter. On the day of transfection, cells should be 60–90% confluent. Therefore, cells should be at a concentration of 2×10^5 cells/well in a 6-well plate 1 day before transfection.
2. Cells should be transfected during their exponential growing phase. Transfection efficiency can be affected by the presence of contaminants (mycoplasma, fungi).
3. During transfection assay, the use of antibiotics in the culture medium is not recommended.
4. Immediately before transfection, it is mandatory to replace modified DMEM with OptiMEM.
5. For double magnetofections, optimal expression was achieved with 4 μg total DNA (2:2 ratio) for 2.10^5 cells. For triple magnetofections, optimal expression was achieved with 4 μg total DNA (2:1:1 ratio) for 2.10^5 cells.
6. Ketamine (a noncompetitive antagonist of NMDA receptors) rapidly dissociates from the receptor and is added to reduce the excitotoxicity mediated by activation of NMDA receptors.

Acknowledgments

This work was supported by the European Union Seventh Framework Programme FP7/2007-2013 under the project DESIRE (grant agreement n°602531), and by INSERM (Institut National de la Santé et de la Recherche Médicale).

References

1. Paoletti P, Bellone C, Zhou Q (2013) NMDA receptor subunit diversity: impact on receptor properties, synaptic plasticity and disease. Nat Rev Neurosci 14(6):383–400. doi:10.1038/nrn3504
2. Paoletti P, Neyton J (2007) NMDA receptor subunits: function and pharmacology. Curr Opin Pharmacol 7(1):39–47. doi:10.1016/j.coph.2006.08.011
3. Traynelis SF, Wollmuth LP, McBain CJ, Menniti FS, Vance KM, Ogden KK, Hansen KB, Yuan H, Myers SJ, Dingledine R (2010) Glutamate receptor ion channels: structure, regulation, and function. Pharmacol Rev 62(3):405–496. doi:10.1124/pr.109.002451
4. Hansen KB, Tajima N, Risgaard R, Perszyk RE, Jorgensen L, Vance KM, Ogden KK, Clausen RP, Furukawa H, Traynelis SF (2013) Structural determinants of agonist efficacy at the glutamate binding site of N-methyl-D-aspartate receptors. Mol Pharmacol 84(1):114–127. doi:10.1124/mol.113.085803
5. Sheng M, Cummings J, Roldan LA, Jan YN, Jan LY (1994) Changing subunit composition of heteromeric NMDA receptors during development of rat cortex. Nature 368(6467):144–147. doi:10.1038/368144a0
6. Wenthold RJ, Prybylowski K, Standley S, Sans N, Petralia RS (2003) Trafficking of NMDA receptors. Annu Rev Pharmacol Toxicol 43:335–358. doi:10.1146/annurev.pharmtox.43.100901.135803
7. Rosenmund C, Stern-Bach Y, Stevens CF (1998) The tetrameric structure of a glutamate

receptor channel. Science 280 (5369):1596–1599

8. Scherer F, Anton M, Schillinger U, Henke J, Bergemann C, Kruger A, Gansbacher B, Plank C (2002) Magnetofection: enhancing and targeting gene delivery by magnetic force in vitro and in vivo. Gene Ther 9(2):102–109. doi:10.1038/sj.gt.3301624
9. Plank C, Zelphati O, Mykhaylyk O (2011) Magnetically enhanced nucleic acid delivery. Ten years of magnetofection-progress and prospects. Adv Drug Deliv Rev 63 (14–15):1300–1331. doi:10.1016/j.addr.2011.08.002
10. Varney MA, Lin FF, Jachec C, Rao SP, Hess SD, Johnson EC, Velicelebi G (1999) Stable expression of human NMDA receptors in cultured mammalian cells. Methods Mol Biol 128:43–59. doi:10.1385/1-59259-683-5:43

Chapter 6

Transfection in Primary Cultured Neuronal Cells

Katie F.M. Marwick and Giles E. Hardingham

Abstract

Transfection allows the introduction of foreign nucleic acid into eukaryotic cells. It is an important tool in understanding the roles of NMDARs in neurons. Here, we describe using lipofection-mediated transfection to introduce cDNA encoding NMDAR subunits into postmitotic rodent primary cortical neurons maintained in culture.

Key words Neuron, Transfection, Lipofectamine®, Lipofection, NMDA, Culture

1 Introduction

Primary cultured neuronal cells offer many advantages over other host cells for expression of N-methyl-D-aspartate receptors (NMDARs). Because they express NMDARs endogenously, they allow investigation of NMDAR trafficking and regulation in a physiological context. They also allow the impact of perturbing NMDARs to be assessed on key physiological outcomes, such as synapse formation, synaptic activity, and cell survival. Often, it is of interest to experimentally vary the NMDAR composition of primary neurons. For example, introducing a subunit before it is normally expressed, knocking down a subunit, or introducing mutant forms of NMDARs such as those carrying point mutations, tags, or larger structural rearrangements. The introduction of exogenous nucleic acids into a eukaryotic cell's nucleus by non-viral means is termed transfection. An approach to achieving the transfection of NMDAR subunits into primary neuronal cultured cells is described in this chapter.

There are a wide variety of reagent and instrument-based methods of transfection available. However, primary cultured neurons are in general more difficult to transfect than other cultured cells, because they are postmitotic and more vulnerable to cell death [1], so not all transfection methods are usable. The three transfection methods for cultured neurons that we have experienced success

Nail Burnashev and Pierre Szepetowski (eds.), *NMDA Receptors: Methods and Protocols*, Methods in Molecular Biology, vol. 1677, DOI 10.1007/978-1-4939-7321-7_6, © Springer Science+Business Media LLC 2017

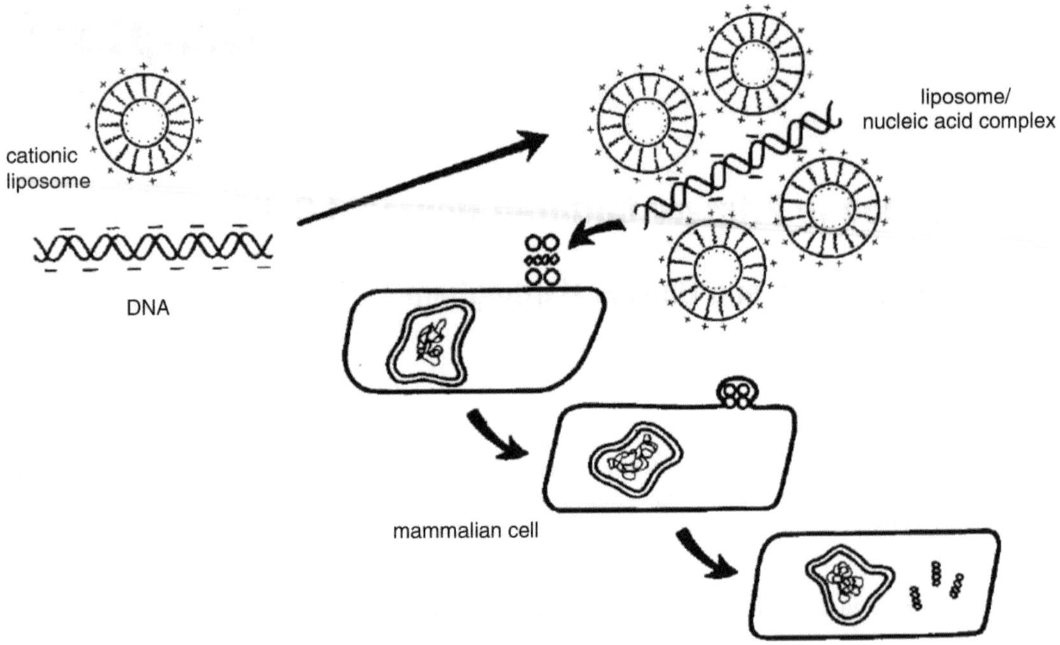

Fig. 1 Mechanism by which lipofection facilitates nucleic acid entry into cells (reproduced with permission from Felgner and Ringold, 1989)

with in our lab are nucleofection (a form of electroporation), calcium phosphate, and lipofection. The main advantage of nucleofection is that it generally achieves high transfection efficiency, with around 75% of cells transfected [2]. However, it can only be performed on the day of culturing, on neurons in suspension which are yet to develop neurites. The main advantage of calcium phosphate transfection is that it is inexpensive. However, it is very reagent and pH dependent, making it less reproducible than the other methods [1]. The technique we use most often is lipofection, and this is the subject of this chapter. Our protocol typically achieves transfection efficiency of 1–5%. This means it is most suitable for experiments involving analysis of single cells, rather than wells of cells.

Lipofection introduces DNA or RNA into cells by forming liposomes of positively charged lipids surrounding negatively charged nucleic acids. These liposomes are able to fuse with cell membranes, allowing the entry of nucleic acids into the cell [3, 4] (*see* Fig. 1) A wide range of cationic lipid reagents for lipofection are now manufactured, marketed as optimized for different cell types and different types of nucleic acids (e.g., RNA vs DNA, co-transfection of both).

This chapter will cover the steps required to transfect primary cultured rodent neurons (*see* **Note 1**) with DNA encoding NMDAR subunits by lipofection. This chapter will not cover the creation of neuronal cultures or the preparation of DNA.

Lipofection with RNA (e.g., siRNA) is very similar except additional precautions for avoiding exposure to RNAse are required.

2 Materials

Prepare all solutions using ultrapure water (prepared by purifying deionized water, to attain a sensitivity of 14 MΩ-cm at 25 °C), which has then been sterilized by autoclaving. Use analytical grade reagents. Store at 4 °C except for plasmid DNA.

2.1 Transfection Medium

Serum-free non-trophic medium. Premade versions are available (*see* **Note 2**). Below is what is used in our laboratory.

1. Salt-glucose-glycine [5]: 114 mM NaCl, 26.1 mM $NaHCO_3$, 5.3 mM KCl, 1 mM $MgCl_2$, 2 mM $CaCl_2$, 10 mM HEPES, 1 mM glycine, 30 mM D-glucose, 0.5 mM sodium pyruvate, with phenol red 0.001%. Use within a week.
2. Minimum essential media (MEM) (+Earles, −L-glutamine).
3. 1% insulin-transferrin-selenium supplement. This supplement enhances cell survival at serum concentrations below 4%.
4. 1% antibiotic/antimycotic. The choice of antibiotic and/or antimycotic required depends on what microorganisms are prevalent in your laboratory.

2.2 Lipofection Reagent

Lipofectamine® 2000 (Invitrogen) (*see* **Note 3**). Take care to keep cool during use, e.g., by use of a cool box or rack. Once opened, effectiveness appears to decline gradually: recommend using within 2 months of opening.

2.3 Plasmid DNA

This can be prepared using standard plasmid-preparation techniques such as mini-preps and recovered in standard elution buffers such as 10 mM Tris-Cl, with or without EDTA, or sterile water (*see* **Note 4**). The final concentration is easiest to work with at around 200–1000 ng/μL, but higher or lower concentrations are usable. Store at −20 °C. We find cDNAs placed behind CMV promoters are expressed well in primary neurons, other promoters are also likely to work.

3 Methods

There are a number of differences between what is recommended here and the manufacturer's guidance for Lipofectamine® 2000. This is because transfection needs to be optimized for the given cell type, culture conditions, and DNA quality. The protocol below works reliably and well for our neural cultures with our DNA preparation techniques, you may need to systematically vary factors such as DNA:lipofectamine ratio or total DNA administered per well to optimize results in your systems.

We have successfully transfected primary neurons at DIV 3–15 using this protocol. Older may well be possible. From DIV 3 onward the cells transfected are overwhelmingly neuronal [6]. If astrocytic transfection is desired transfection at DIV 0–2 is recommended. Wait 48 h after transfection before performing experiments on the cells. Expression levels of NMDAR subunits driven by CMV promoters remain high for at least 96 h post-transfection. As neurons are postmitotic transfected plasmids are not diluted as in dividing cell lines.

All volumes and masses specified are for one well of a 24-well plate. Carry out all procedures at room temperature unless otherwise specified.

3.1 Prepare the Transfection Medium

1. Prepare salt-glucose-glycine solution in a measuring cylinder at the concentrations indicated above, made up to almost the required volume with ultrapure, sterile, water.
2. Adjust to pH 7.3 by adding 0.2 M NaOH dropwise until the media's color (provided by phenol red) is a bright, pure, red, not orange-red and not purple-red (*see* **Note 5**). Bring to the required volume with ultrapure, sterile, water.
3. Mix with 10% MEM, 1% antibiotic/antimycotic and 1% insulin-transferrin-selenium. For example, for 100 mL of transfection medium, use 88 mL of salt-glucose-glycine solution, 10 mL of MEM and 1 mL each of antibiotic/antimycotic and insulin-transferrin-selenium solution.
4. Sterile filter the medium using a filter with 0.22 μm porelon channel size.

3.2 Prepare the Transfection Mixes (See Fig. 2)

We recommend preparing enough transfection mix for an extra half or whole well to overcome losses during pipetting.

1. Calculate DNA volumes required. We recommend a maximum DNA mass/well of 650 ng (*see* **Note 6**). This includes both the NMDAR subunits of interest, e.g., GluN2A, and any marker plasmid included to allow identification of transfected cells, e.g., GFP (*see* **Note** 7). We find using half as much marker DNA as DNA of interest is sufficient. Remember neurons express GluN1 so transfection of this is not required.
2. Add the DNA to transfection medium (33 μL/well) ("DNA mix"). Mix by pipetting. (e.g., for 4 wells with 500 ng DNA/well at a concentration of 500 ng/μL, add 4 ul DNA to 132 μL medium).
3. Calculate the volume of lipofectamine® 2000 required. We use 2.33 μL/well.

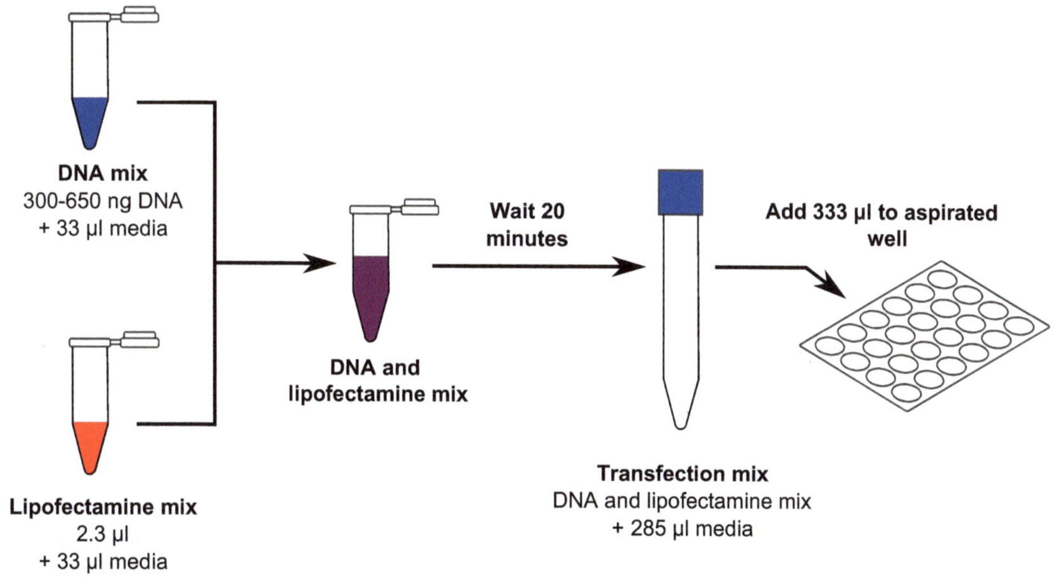

Fig. 2 Overview of preparation of transfection mix

4. Add the lipofectamine to transfection medium (33 μL/well) ("lipofectamine mix"). Mix by pipetting (*see* **Note 8**). (e.g., for 4 wells, add 9.3 μL lipofectamine to 132 μL medium).
5. Combine the DNA mix and the lipofectamine mix (*see* **Note 9**). Mix by pipetting.
6. Leave for 20 min at room temperature (*see* **Note 10**). Longer is acceptable (e.g., 90 min). Liposomes form during this time.
7. Dilute the DNA/lipofectamine mixture by adding further medium: 285 μL medium per well. Mix by pipetting. This media should be pre-warmed to 37 °C as it will shortly be applied to cells. (e.g., for 4 wells, add 1140 μL media to the 277 μL mix).

3.3 Perform the Transfection

1. Remove media from the wells (*see* **Note 11**). Keep the media. A 5 mL pipette works well for this. It is not necessary to completely remove media.
2. Add 333 μL of the final transfection mix to each well.
3. Leave for 2–5 h in a 37 °C incubator.
4. Return the conditioned media to the wells (*see* **Note 12**). Do not aspirate out the transfection mix (*see* **Note 13**).

4 Notes

1. This protocol has been optimized for cortical neurons cultured from E17.5 mouse embryos and E20.5 rat embryos. The

protocol works equally well for both species. It has not been trialled in other species. It works equally well for cells grown on plastic and on glass cover slips.

2. Opti-MEM® Reduced Serum Medium with GlutaMAX™ Supplement (Invitrogen) is an alternative serum-free non-trophic medium that works well for lipofection in HEK293 cells. We have not trialled it in neurons.
3. This chapter describes the use of one particular lipofection reagent (Lipfectamine® 2000, Invitrogen). Lipofectamine® 2000 is a commonly used and often cited reagent; however, many others are available and are likely to also work well. However, the use of a different reagent may require some variations to the protocol described here.
4. Endotoxin (lipopolysaccharide) contaminates DNA recovered from bacterial pellets to varying extents depending on the purification method used. At high levels it inhibits transfection [7]. If your transfection efficiency is low it is worth trying an endotoxin-free recovery method.
5. If you are unsure, you can use pH strips or a pH meter. This will be more accurate and may be worth doing the first time you make the media.
6. The DNA:lipofectamine ratio is important in transfection efficiency. If substantially less plasmid than this is required, e.g., <300 ng/well, consider adding a plasmid with cDNA encoding an inert protein (e.g., β globin) or an empty vector. Alternatively, reduce the volume of lipofectamine® 2000 used.
7. It is advisable when first commencing transfection to assess transfection efficiency. Fluorescent proteins offer a convenient way to do this. In our experience co-transfection rates are >96%, so it would be rare for a cell to be transfected with GFP but not the NMDAR subunit. If recordings from individual cells are planned (e.g., electrophysiological whole-cell recordings, or imaging) then a fluorescent marker is also advisable to allow identification of transfected cells. It is important to select cells that are well transfected and healthy, i.e., show extension of fluorescent protein into dendrites but are not blebbed. *See* Fig. 3.
8. Vortexing may reduce the efficiency of lipofectamine® 2000.
9. Do not combine the DNA and lipofectamine together without first diluting each in media separately. Although this is tempting, it does not work.
10. This procedure also works with media which has been heated to 37 °C, then left at room temperature.
11. Avoid removing media from more than six wells at a time, to minimize time without media.

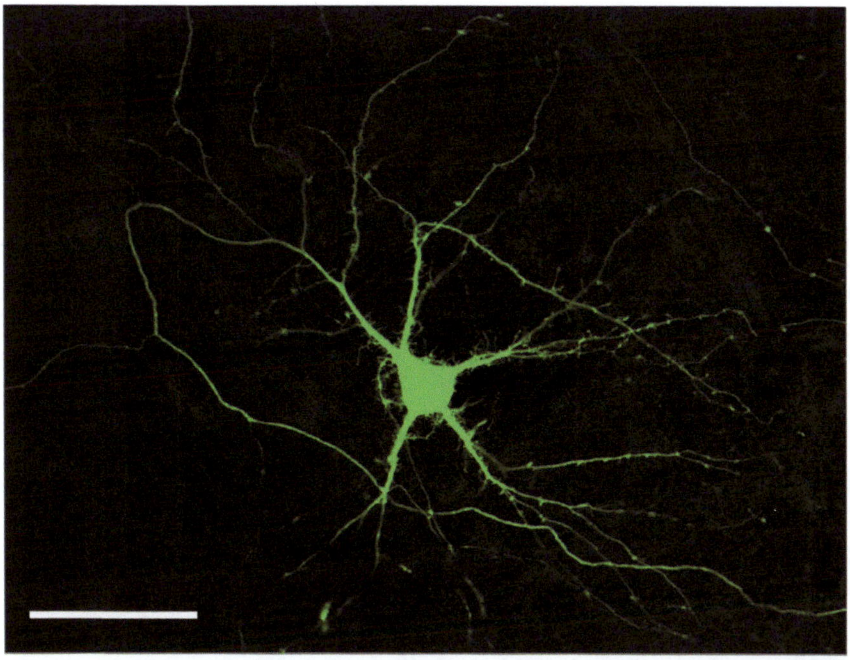

Fig. 3 A healthy and well-transfected neuron at 40× magnification. Note the extension of fluorescent protein into the dendrites. Cells with show pale fluorescence in the soma only are unlikely to be expressing transfected NMDAR subunits at high levels. Scale bar 60 μL

12. It is not necessary to add NMDAR antagonists to the media post-transfection as neurons express NMDARs endogenously.
13. It is also fine to remove the transfection mix, and/or to add fresh medium, but we have found minimizing aspiration is good and that the cells benefit from the return of conditioned media.

Acknowledgments

This work was supported by a Wellcome Trust Clinical PhD studentship grant to KFMM. Much of the work put into optimizing this protocol was done by others in GEH's lab over the last decade: particularly Sofia Papadia and Sean McKay. Michael Greenberg's lab developed the principles of transfection in neurons.

References

1. Karra D, Dahm R (2010) Transfection techniques for neuronal cells. J Neurosci 30:6171–6177. doi:10.1523/JNEUROSCI.0183-10.2010
2. Zeitelhofer M, Vessey JP, Xie Y, Tübing F, Thomas S, Kiebler M et al (2007) High-efficiency transfection of mammalian neurons

via nucleofection. Nat Protoc 2:1692–1704. doi:10.1038/nprot.2007.226

3. Felgner PL, Gadek TR, Holm M, Roman R, Chan HW, Wenz M et al (1987) Lipofection: a highly efficient, lipid-mediated DNA-transfection procedure. Proc Natl Acad Sci U S A 84:7413–7417. doi:10.1073/pnas.84.21.7413
4. Felgner PL, Ringold GM (1989) Cationic liposome-mediated transfection. Nature 337 (6205):387–388. doi:10.1038/337387a0
5. Bading H, Ginty DD, Greenberg ME (1993) Regulation of gene expression in hippocampal neurons by distinct calcium signaling pathways. Science 260:181–186. doi:10.1126/science.8097060
6. Papadia S, Soriano FX, Léveillé F, Martel M-A, Dakin KA, Hansen HH et al (2008) Synaptic NMDA receptor activity boosts intrinsic antioxidant defenses. Nat Neurosci 11:476–487. doi:10.1038/nn2071
7. Butash KA, Natarajan P, Young A, Fox DK (2000) Reexamination of the effect of endotoxin on cell proliferation and transfection efficiency. BioTechniques 29:610–619

Chapter 7

Selective Cell-Surface Expression of Triheteromeric NMDA Receptors

Feng Yi, Stephen F. Traynelis, and Kasper B. Hansen

Abstract

The NMDA-type ionotropic glutamate receptors play pivotal roles in many brain functions, but are also involved in numerous brain disorders. Seven NMDA receptor subunits exist (GluN1, GluN2A-D, and GluN3A-B) that assemble into a diverse array of tetrameric receptor subtypes with distinct functional properties and physiological roles. Most NMDA receptors are composed of two GluN1 and two GluN2 subunits, which can assemble into four diheteromeric receptor subtypes composed of GluN1 and one type of GluN2 subunit (e.g., GluN1/2A), and presumably also six triheteromeric receptor subtypes composed of GluN1 and two different GluN2 subunits (e.g., GluN1/2A/2B). Despite accumulating evidence that a large proportion of native NMDA receptors are triheteromers, little is known about their function and pharmacology due to the lack of methods to faithfully express triheteromeric NMDA receptors in heterologous expression systems. The problem is that co-expression of GluN1 with two different GluN2 subunits generates two distinct diheteromeric receptor subtypes as well as one triheteromeric receptor subtype, thereby confounding studies on a homogenous population of triheteromeric NMDA receptors. Here, we will describe a method to selectively express recombinant triheteromeric GluN1/2A/2B receptors without interfering co-expression of diheteromeric GluN1/2A and GluN1/2B receptors. This method enables quantitative evaluation of functional and pharmacological properties of triheteromeric GluN1/2A/2B receptors, which are presumably the most abundant NMDA receptors in the adult cortex and hippocampus.

Key words Ionotropic glutamate receptor, Ligand-gated ion channel, Assembly, Retention signals, Coiled-coil, Endoplasmic reticulum, Trafficking, *Xenopus* oocytes

1 Introduction

NMDA receptors have been extensively studied for more than three decades as principle mediators of excitatory synaptic transmission. Molecular cloning in the early 1990s identified cDNAs for individual NMDA receptor subunits and ensuing studies in heterologous expression systems revealed pronounced variation in pharmacological and functional properties, depending on the subunit composition of the different NMDA receptor subtypes [1]. Initially, five NMDA receptor subunits were identified, GluN1 and four different GluN2 subunits (GluN2A-D), and it was quickly discovered

Nail Burnashev and Pierre Szepetowski (eds.), *NMDA Receptors: Methods and Protocols*, Methods in Molecular Biology, vol. 1677, DOI 10.1007/978-1-4939-7321-7_7, © Springer Science+Business Media LLC 2017

that co-expression of GluN1 and GluN2 subunits is necessary for the formation of functional NMDA receptors [1]. By now, it is firmly established that two GluN1 and two GluN2 subunits assemble into functional receptors with an alternating subunit arrangement around a central ion channel pore (i.e., 1-2-1-2) [2–4] (Fig. 1a). Later, two GluN3 subunits (GluN3A-B) were identified, but the stoichiometry of GluN3-containing receptors is still unknown [5]. Recombinant studies almost exclusively describe diheteromeric NMDA receptors that are assembled from GluN1 and only one type of GluN2 (i.e., GluN1/2A, GluN1/2B, GluN1/2C, and GluN1/2D) (Fig. 1b). However, a compelling body of biochemical and functional evidence, dating back to 1994, shows that a large proportion of native NMDA receptors are triheteromers assembled from two GluN1 and two different GluN2 subunits (Fig. 1b). For example, GluN1/2A/2B is the major subtype in hippocampus and cortex of the adult rodent brain [6–10], GluN1/2B/2D is found in cerebellar Golgi cells [11], hippocampal granule cells [12], substantia nigra [13], and the subthalamic nucleus [14], and GluN1/2A/2C is identified in adult cerebellar granule cells [15, 16]. Despite the prevalence of triheteromeric NMDA receptors in vivo, the biophysical and pharmacological properties for most triheteromeric NMDA receptor subtypes are poorly understood, primarily due to our inability to study a homogenous population of triheteromeric receptors in heterologous expression systems without accompanying diheteromeric receptors [8, 15, 17–20].

In order to enable functional studies on triheteromeric NMDA receptors, we developed a method to selectively express NMDA receptors composed of two GluN1 and two different GluN2 subunits at the cell surface in heterologous expression systems (e.g., HEK cells and *Xenopus* oocytes) [18]. This method was designed to prevent surface expression of diheteromeric NMDA receptors without affecting functional and pharmacological receptor properties. Inspired by previous studies that selectively expressed heterodimeric metabotropic glutamate receptors in heterologous systems [21], we adapted the trafficking control system of the G protein-coupled $GABA_B$ receptors to control the expression of triheteromeric NMDA receptors. Functional $GABA_B$ receptors are heterodimers composed of $GABA_{B1}$ and $GABA_{B2}$ subunits [22]. The intracellular C-terminal domain (CTD) of the $GABA_{B1}$ subunit contains a retention/retrieval signal (RXR motif) that retains the subunit in the endoplasmic reticulum (ER), thereby preventing expression of homomeric $GABA_{B1}$ at the cell surface [23]. The RXR retention signal is masked through a heterodimeric coiled-coil interaction between leucine zipper motifs (LZ1 and LZ2) in the CTDs of $GABA_{B1}$ and $GABA_{B2}$ subunits, resulting in trafficking of the heterodimer to the cell surface [23–25]. The intracellular CTD of the $GABA_{B2}$ subunit is lacking ER retention/retrieval

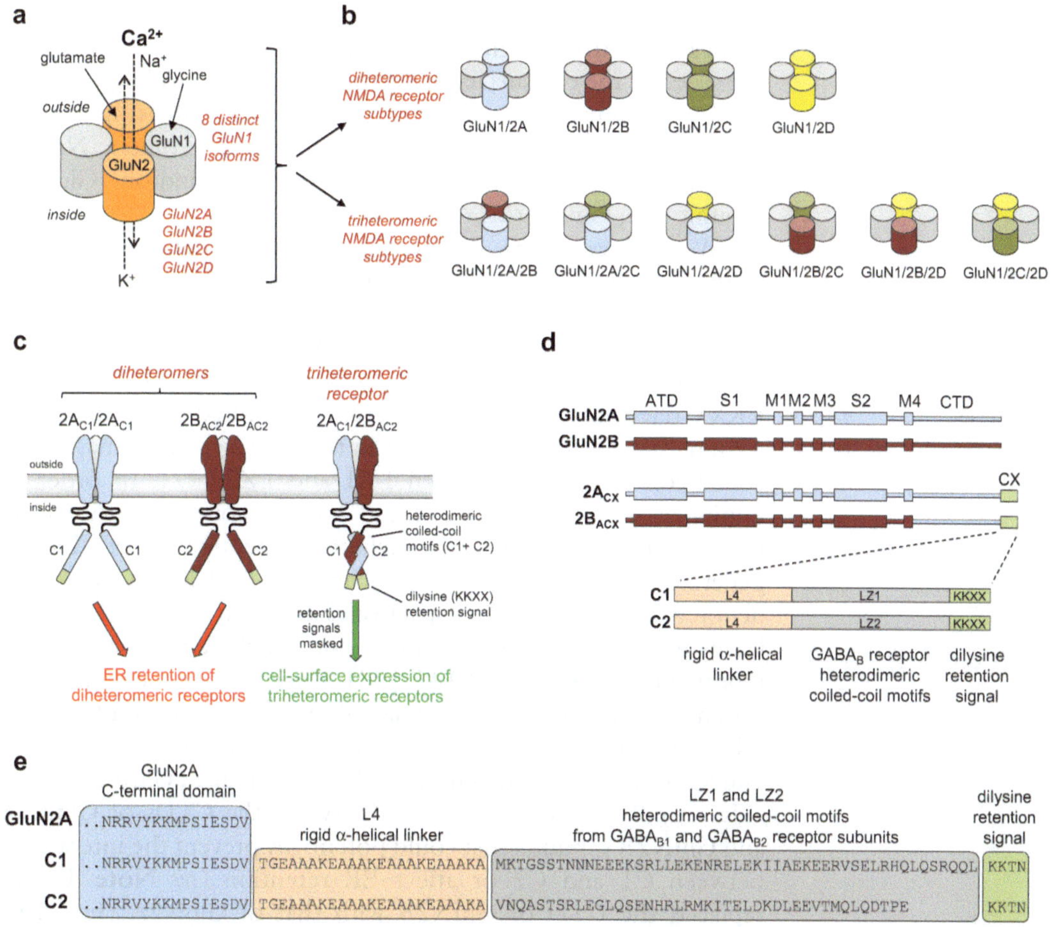

Fig. 1 Control of NMDA receptor composition using engineered peptide tags. (**a**) Schematic representation of the subunit arrangement in NMDA receptors composed of glycine-binding GluN1 and glutamate-binding GluN2 subunits. The GluN1 subunit exists in eight different isoforms (i.e., splice variants) and four different genes encoding GluN2 subunits have been identified (GluN2A-D). (**b**) Schematic representation of possible subunit compositions in NMDA receptors assembled from GluN1 and GluN2 subunits. (**c**) Co-expression of GluN1, GluN2A, and GluN2B subunits in heterologous systems normally generates three populations of functional NMDA receptors (i.e., GluN1/2A, GluN1/GluN2B, and GluN1/2A/2B). However, fusing the engineered C1 and C2 tags to the intracellular C-termini of GluN2 subunits prevents cell-surface expression of receptors that contain two C1-tagged or two C2-tagged GluN2 subunits. GluN1 is omitted for clarity. (**d**) Linear representations of the polypeptide chains show the amino-terminal domain (ATD), S1 and S2 segments that form the agonist binding domain, and the transmembrane domain formed by M1, M2, M3, and M4 of GluN2A (*blue*), GluN2B (*red*), GluN2A with C1- or C2-tags fused to the intracellular C-terminus ($2A_{CX}$), and chimeric GluN2B subunits with the C-terminal domain (CTD) replaced by that of C1- or C2-tagged GluN2A ($2B_{ACX}$). The C1 and C2 tags are composed of a rigid α-helical linker (L4), the coiled-coil motifs of $GABA_{B1}$ (LZ1) and $GABA_{B2}$ (LZ2) receptors, and a di-lysine endoplasmic reticulum (ER) retention/retrieval signal. (**e**) Sequences for the C-termini of wild-type GluN2A and subunits with C1- or C2-tags fused to the C-terminus. Adapted with permission from Hansen et al. [18]

signals, but homomeric $GABA_{B2}$ are nonfunctional [22]. Thus, this trafficking control system ensures that functional $GABA_B$ receptors expressed at the cell surface are composed of one $GABA_{B1}$ and one $GABA_{B2}$ subunit. We modified this $GABA_B$ trafficking control system to only allow surface expression of triheteromeric NMDA receptors containing GluN1 with one GluN2A, and one GluN2B subunit (i.e., GluN1/2A/2B) [18] (Fig. 1c).

We engineered two different peptide tags intended for fusion to the C-terminus of GluN2 subunits. The peptide tags are composed of the leucine zipper motifs from $GABA_{B1}$ and $GABA_{B2}$ (LZ1 and LZ2, respectively) immediately followed by C-terminal di-lysine (KKXX) ER retention/retrieval signals (Fig. 1d). The KKXX retention signals, which must be located at the distal C-termini, will localize the subunits in the ER [26] unless masked by coiled-coil formation between LZ1 and LZ2 [27]. In order to minimize steric effects that could prevent coiled-coil formation between LZ1 and LZ2, we included a peptide linker (L4) to generate the tags L4-LZ1-KKXX and L4-LZ2-KKXX (hereafter denoted as C1 and C2, respectively) (Fig. 1d). The L4 peptide linker was designed to generate a rigid α-helix composed of four repeats of amino acids EAAAK [28]. Placement of the engineered C1 and C2 tags at the intracellular GluN2 C-termini should enable selective surface expression of NMDA receptors composed of two GluN1, one C1-tagged GluN2, and one C2-tagged GluN2 subunit (Fig. 1c). To remove differences between the CTDs of GluN2A and GluN2B subunits that could potentially prevent the interaction between C1 and C2 or affect ER retention (*see* **Note 1**), we replaced the CTD of GluN2B with that of GluN2A (Fig. 1d). The intracellular CTDs of C1- and C2-tagged GluN2A and GluN2B subunits (also denoted as $2A_{C1}$, $2A_{C2}$, $2B_{AC1}$, and $2B_{AC2}$) therefore only differ by the leucine zipper motifs (LZ1 and LZ2) (Fig. 1e).

To assess whether co-expression of GluN1 with $2A_{C1}$ and $2B_{AC2}$ generates triheteromeric GluN1/$2A_{C1}$/$2B_{AC2}$ receptors and whether the method prevents expression of diheteromeric GluN1/$2A_{C1}$/$2A_{C1}$ and GluN1/$2B_{AC2}$/$2B_{AC2}$ receptors, it is critical to carefully develop a set of control experiments (*see* **Note 2**). In addition, controls can be designed to determine if fusion of the C1 and C2 tags to the GluN2 subunits affects receptor function (*see* **Note 3**). According to the experimental design, the current responses activated by glutamate + glycine from cells co-expressing GluN1 with $2A_{C1}$ and $2B_{AC2}$ should, in theory, be mediated by triheteromeric GluN1/$2A_{C1}$/$2B_{AC2}$ receptors with masked retention signals. However, some diheteromeric GluN1/$2A_{C1}$/$2A_{C1}$ and GluN1/$2B_{AC2}$/$2B_{AC2}$ receptors may also escape ER retention and could contribute to the measured current responses. To evaluate the contribution of escaped receptors (escape current) to the total current response, we generated $2A_{C1}$ and $2B_{AC2}$ subunits with

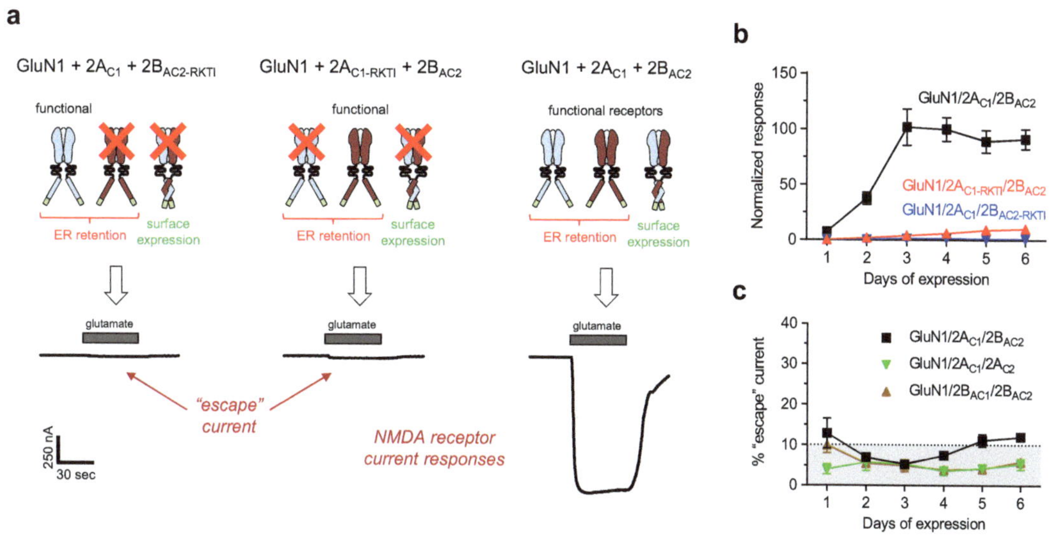

Fig. 2 Evaluation of selective triheteromeric NMDA receptor expression. (**a**) Representative two-electrode voltage-clamp recordings of responses from recombinant NMDA receptors expressed in *Xenopus* oocytes activated by 100 μM glutamate in the continuous presence of 50 μM glycine. GluN1 was co-expressed with $2A_{C1}$ + mutated $2B_{AC2}$ (R519K + T691I in $2B_{AC2\text{-}RKTI}$) (*left*), mutated $2A_{C1}$ (R518K + T690I in $2A_{C1\text{-}RKTI}$) + $2B_{AC2}$ (*middle*), or $2A_{C1}$ + $2B_{AC2}$ (*right*). The RKTI mutations abolish binding of glutamate, thereby causing NMDA receptors with this subunit to be nonfunctional (indicated by *red* X). (**b**) Co-expression of GluN1 with $2A_{C1}$ and $2B_{AC2}$ (GluN1/$2A_{C1}$/$2B_{AC2}$) produces robust current responses that increase in the days following cRNA injection (*black*), whereas current responses from diheteromeric GluN1/$2A_{C1}$/$2A_{C1}$ and GluN1/$2B_{AC2}$/$2B_{AC2}$ receptors that may have escaped ER retention remain small (i.e., escape currents). Data are normalized to the averaged response from GluN1/$2A_{C1}$/$2B_{AC2}$ on day 4, and are mean ± SEM. Each data point is from 5 batches of oocytes with 6 oocytes for each batch ($N = 30$). (**c**) The sum of the fractional currents assessed using the RKTI mutations (i.e., shown in panel **b**) provides an estimate of the percent "escape" current in oocytes co-expressing GluN1 with $2A_{C1}$ and $2B_{AC2}$ (GluN1/$2A_{C1}$/$2B_{AC2}$). The method can also be used for the expression of GluN1/$2A_{C1}$/$2A_{C2}$ and GluN1/$2B_{AC1}$/$2B_{AC2}$ receptors as indicated by the percent "escape" current included in the graph, which are similarly assessed using RKTI mutations. Adapted with permission from Hansen et al. [18]

mutations in the agonist binding pocket that abolish glutamate binding (R518K + T690I in $2A_{C1\text{-}RKTI}$ and R519K + T691I in $2B_{AC2\text{-}RKTI}$) (*see* **Note 4**). Thus, the RKTI mutations effectively render the subunit nonfunctional. NMDA receptors containing one copy of a nonfunctional subunit are nonresponsive to glutamate (i.e. in the presence of glycine), since agonist occupancy at all four subunits is required for NMDA receptor activation [29, 30] (Fig. 2a). Thus, co-expression of GluN1 + $2A_{C1}$ + $2B_{AC2\text{-}RKTI}$ produces only one population of functional receptors, namely escaped GluN1/$2A_{C1}$/$2A_{C1}$. Comparison of current responses from cells expressing GluN1 + $2A_{C1}$ + $2B_{AC2}$ with responses from cells co-expressing GluN1 + $2A_{C1}$ + $2B_{AC2\text{-}RKTI}$ allows estimation of the fractional current mediated by escaped GluN1/$2A_{C1}$/$2A_{C1}$. Similarly, comparison to current responses from cells expressing GluN1 + $2A_{C1\text{-}RKTI}$ + $2B_{AC2}$ allows estimation of the fractional current mediated by escaped GluN1/$2B_{AC2}$/$2B_{AC2}$ (Fig. 2b). The sum of the fractional currents from escaped GluN1/$2A_{C1}$/$2A_{C1}$ and GluN1/$2B_{AC2}$/$2B_{AC2}$ will provide an estimate of the percent "escape" current in cells co-expressing GluN1 + $2A_{C1}$ + $2B_{AC2}$. Alternative control experiments can be designed to evaluate the efficiency by which the engineered peptide tags express a homogenous population of triheteromeric receptors. However, the RKTI controls have the distinct advantage that they would also detect if C1 and C2 tags interact between receptors, rather than within one receptor as intended (Fig. 2c). For example, GluN1/$2A_{C1}$/$2A_{C1}$ and GluN1/$2B_{AC2}$/$2B_{AC2}$ could in theory interact via the C1 and C2 tags, thereby enabling surface expression. This scenario would manifest in current responses for cells co-expressing GluN1 + $2A_{C1}$ + $2B_{AC2\text{-}RKTI}$ or GluN1 + $2A_{C1\text{-}RKTI}$ + $2B_{AC2}$. Alternative control experiments could evaluate current responses from cells co-expressing GluN1 + $2A_{C1}$ or GluN1 + $2B_{AC2}$, which should not allow surface expression, but these controls would not assess whether inter-receptor interactions occur between C1 and C2 tags of GluN1/$2A_{C1}$/$2A_{C1}$ and GluN1/$2B_{AC2}$/$2B_{AC2}$. This problem emphasizes the need to carefully design control experiments to assess the effectiveness of the method, especially when the experimental conditions are changed.

The method, which was described in 2014, has been applied in multiple studies and revealed surprising pharmacological and functional properties of triheteromeric GluN1/2A/2B receptors that are distinct from the properties of the respective diheteromeric GluN1/2A and GluN1/2B receptors [18, 20, 31–35]. In addition, the method is facilitating new opportunities to develop therapeutic agents that target triheteromeric NMDA receptors with one copy of a disease-causing mutation [32–37]. The method also paves the way for new experiments into the relationship between NMDA receptor structure and function by enabling unprecedented control of subunit composition. Future work will show if the method can be optimized for the expression of triheteromeric NMDA receptors with other subunit compositions (e.g., GluN1/2B/2D or GluN1/2A/2C) (*see* **Note 5**).

2 Materials

The protocol described here uses *Xenopus laevis* oocytes as heterologous expression system for triheteromeric GluN1/2A/2B receptors, and recordings of receptor function are performed using two-electrode voltage-clamp electrophysiology. DNA constructs for GluN1 as well as C1- and C2-tagged GluN2A and GluN2B subunits can be designed as described in Fig. 1.

2.1 Linearization of Plasmid DNA and In Vitro cRNA Transcription

1. Purified plasmid DNA (5–10 μg) and suitable restriction enzymes.
2. QIAquick PCR Purification Kit (Qiagen).
3. RNase-free 3 M sodium acetate, pH 5.2 (Sigma-Aldrich).
4. 96–100% ethanol and 70% ethanol prepared using RNase-free water.
5. TE buffer (DNA/RNA storage buffer): 1 mM EDTA, 10 mM Tris·Cl, pH 8.0 prepared using RNase-free water.
6. mMESSAGE mMACHINE High Yield Capped RNA Transcription Kit (Ambion) with a suitable RNA polymerase (SP6, T3, or T7).
7. RNase-free 1.5 mL microcentrifuge tubes.
8. Refrigerated centrifuge for 1.5 mL microcentrifuge tubes.

2.2 Injection and Maintenance of Xenopus Oocytes

1. Stereomicroscope with at least ten-fold magnification and a working distance of at least 10 cm.
2. Microinjector and micromanipulator: Cytoplasmic injections in oocytes can be performed in volumes between ~2 and 70 nL using either an automated injection device (e.g., Nanoject III, Drummond Scientific) or using a manual injection device. Precise positioning of the microinjector requires a micromanipulator.
3. Micropipette puller (e.g., PC-10 Puller, Narishige).
4. Glass capillaries suitable for the injection device (Drummond Scientific or World Precision Instruments) and mineral oil (e.g., light oil for molecular biology, Sigma-Aldrich) for back filling the glass capillaries using a flexible needle (e.g., MicroFil, 34 ga., World Precision Instruments).
5. Incubator: This method requires incubation of the oocytes at 19 °C (*see* **Note 6**).
6. Injection chamber: Can be custom-made with different designs (*see* **Note** 7).
7. Pipettes for handling *Xenopus* oocytes: Can be prepared by breaking the tip of glass Pasteur pipettes (9″, Fischer Scientific) to create a tip with an internal diameter of 1.5–2.5 mm. It is important to fire polish the tip to prevent damage to the oocytes.
8. Modified Barth's solution (MBS): 88 mM NaCl, 1.0 mM KCl, 2.4 mM $NaHCO_3$, mM 0.41 $CaCl_2$, 0.82 mM $MgSO_4$, 0.33 mM $Ca(NO_3)_2$, 10 mM HEPES, pH 7.5 supplemented with 1 IU/mL penicillin, 1 μg/mL streptomycin (i.e., add 100 μL from Penicillin/Streptomycin solution (10,000 IU/mL and 10,000 μg/mL, respectively; Invitrogen)), and 50 μg/mL gentamycin.
9. Defolliculated *Xenopus* oocytes (*see* **Note 8**).

2.3 Electrophysiological Recordings from Xenopus Oocytes

1. Two-electrode voltage-clamp setup (*see* **Note 9**).
2. Extracellular recording solution: 90 mM NaCl, 1 mM KCl, 0.5 mM $BaCl_2$, 0.01 mM EDTA, 10 mM HEPES, pH 7.4 (*see* **Note 10**).
3. NMDA receptor agonists: Glycine and L-glutamate.

3 Methods

3.1 Linearization of Plasmid DNA

1. To produce template DNA for cRNA synthesis, dilute 5–10 μg plasmid DNA in 200 μL of the suitable restriction enzyme buffer, add 20–40 units of restriction enzyme, and incubate for 2–3 h at 37 °C. Restriction enzyme that leaves a 5′ protruding overhang is preferred (*see* **Note 11**). Proceed to purify the linearized DNA or store the reaction at −20 °C after incubation.
2. The QIAquick PCR Purification Kit will be used for purification of template DNA: Following incubation at 37 °C, add 1 mL (5 volumes) of Buffer PB to the reaction and mix. Transfer the DNA solution to a spin column and centrifuge at room temperature (RT) until the solution has passed the column. Apply half of the 1.2 mL DNA solution (i.e., 200 μL reaction +1 mL Buffer PB) at a time, since the spin column can only hold 800 μL. Remove the solution after it has passed the column. The DNA is now bound to the column.
3. Wash by adding 750 μL Buffer PE and centrifuge at RT until the solution has passed the column. Remove the solution and centrifuge spin column for an additional 3 min at max speed to remove residual ethanol. Transfer spin column to a new, RNase-free 1.5 mL tube.
4. To elute DNA, add 100 μL Buffer EB (RNase-free) to the spin column and centrifuge at max speed for 1 min. The eluted DNA can be stored at −20 °C.
5. Add 1/10 volume (10 μL) 3 M sodium acetate, pH 5.2 (ice cold), mix, and then add 2.5 volume (250 μL) 96–100% ethanol (ice cold). Mix well and incubate 10 min on dry ice (or 30 min to overnight at −80 °C).
6. Centrifuge at 20,000 × *g* for 30 min (4 °C) and carefully remove the supernatant without disturbing the barely visible pellet containing the DNA. Add 200 μL 70% ethanol (ice cold). Do not resuspend the DNA. Centrifuge at 20,000 × *g* for 30 min (4 °C).
7. Carefully remove the supernatant. Dry the pellet at RT or low heat (e.g., 37 °C for 3–5 min). Do not over-dry, since this will make it difficult to resuspend the DNA.

8. Resuspend pellet in 5–10 μL TE buffer (*see* **Note 12**). Run 0.5 μL on an agarose gel to assess the concentration and the quality of the purified DNA. The plasmid DNA should be completely linearized and appear as a single band on the gel (*see* **Note 13**). The concentration of DNA can be determined using a spectrophotometer or estimated by comparing band intensity with those of bands in a DNA ladder with known concentration. Store the linearized DNA template at −20 °C.

3.2 In Vitro cRNA Transcription

1. Prepare the reaction for RNA transcription using 1–2 μg linearized DNA template according to the instructions included in the mMESSAGE mMACHINE High Yield Capped RNA Transcription Kit (Ambion). Incubate the reaction at 37 °C for 2 h.
2. Add 1 μL TURBO DNase (included in the transcription kit) to the reaction, mix carefully, and incubate at 37 °C for 15 min. This step is required to digest the template DNA prior to cRNA precipitation.
3. Precipitate the cRNA by adding 30 μL RNase-free water (included in the transcription kit) and 30 μL LiCl precipitation solution (included in the transcription kit), vortex, and incubate at −20 °C for ≥30 min or overnight.
4. Transfer the frozen tubes to a refrigerated centrifuge and pellet the cRNA for 30 min at 20,000 × *g* (4 °C). Carefully remove the supernatant without disturbing the pellet, carefully add 200 μL 70% ethanol (do not resuspend), and re-centrifuge for 30 min at 20,000 × *g* (4 °C). The ethanol can be added to the side of the tube in order not to disturb the pellet.
5. Carefully remove the ethanol, briefly dry the cRNA (e.g., at 37 °C for a couple of minutes), and resuspend the cRNA in 20 μL TE buffer (*see* **Note 12**). Run 1 μL on an agarose gel to assess the quality of the RNA, which should appear as a single band on the gel (*see* **Note 13**). It is recommended that the cRNA concentration is determined using small-volume spectrophotometer. Keep RNA ice cold at all times and store at −80 °C.

3.3 Expression of Triheteromeric NMDA Receptors in Xenopus Oocytes

1. Prepare injection pipettes using a standard pipette puller. The tips of the injection pipettes are broken under the stereomicroscope to an outer diameter of about 10–20 μm. Use gloves when handling injection pipettes and protect injection pipettes from contamination by RNases.
2. Place oocytes in the injection chamber containing MBS solution. The oocytes can be manipulated using customized Pasteur pipettes.
3. Back-fill an injection pipette with mineral oil and mount it to the microinjector syringe according to the manufacturer's instructions. The plunger will force mineral oil out of the tip of the

injection pipette, resulting in a complete seal of oil between the tip and the plunger. Make sure there are no air bubbles in the injection pipette. The injection pipette can be moved and precisely positioned using a manual micromanipulator once it has been mounted to the microinjector syringe.

4. Combine cRNAs encoding GluN1 as well as C1- and C2-tagged GluN2A and GluN2B for injection. In order to evaluate escape currents from diheteromeric receptors, make the following combinations of cRNAs:

   (a) GluN1 + $2A_{C1}$ + $2B_{AC2}$ (i.e., to measure total current amplitudes)

   (b) GluN1 + $2A_{C1}$ + $2B_{AC2\text{-}RKTI}$ (i.e., escape currents from GluN1/$2A_{C1}$/$2A_{C1}$)

   (c) GluN1 + $2A_{C1\text{-}RKTI}$ + $2B_{AC2}$ (i.e., escape currents from GluN1/$2B_{AC2}$/$2B_{AC2}$)

   In order to enable comparisons of current amplitudes, it is critical that the quantity of total injected cRNA and the molar ratios of the cRNAs are kept constant among the three combinations. Optimization of total injected cRNA and molar ratios is likely required to minimize escape currents for the specific constructs used and under the given experimental conditions (*see* **Note 14**).

5. Transfer cRNA to the injection pipette by placing a small volume, usually 2–3 μL, in the cap from a 1.5 mL RNase-free microcentrifuge tube (separate the cap from the tube using a scissor) under the stereomicroscope. Load the cRNA into the injection pipette using the microinjector. Watch the cRNA solution as it is drawn into the injection pipette to ensure the tip has not been clogged and that there are no air bubbles in the injection pipette.

6. Place the injection chamber containing oocytes under the stereomicroscope and inject cRNA by positioning the injection pipette over each oocyte and gently lowering the tip until it pierces the membrane. Some oocytes may leak after the injection pipette has been pulled back out of the membrane. To allow proper diffusion of the RNA inside the oocyte, it is advisable to wait approximately 5 s after the cRNA has been injected before retracting the injection pipette. Each oocyte can be injected with anywhere between 10 nL and 100 nL RNA solution depending on the desired level of expression, which usually increase as more RNA is injected. Use a new injection pipette for each type of cRNA to prevent cross-contamination.

7. Repeat **steps 5–6** for each cRNA combination, ensuring that the volume of injected cRNA is constant for all injections.

8. Following cRNA injection, place the oocyte in a petridish containing Barth's solution and incubate at 19 °C until the day of recording (*see* **Note 14**).

3.4 Electrophysiological Recordings of NMDA Receptor Function

1. In order to evaluate expression of triheteromeric GluN1/2A/2B receptors as described in Fig. 2, make the following solutions:
   (a) Wash (i.e., extracellular recording solution with 50 μM glycine).
   (b) 100 μM L-glutamate in wash.
2. Apply solutions to the perfusion system of the two-electrode voltage clamp setup and adjust flow rates.
3. Place the injected oocyte in the recording chamber of the two-electrode voltage clamp setup, penetrate the membrane with the two electrodes, and clamp the oocyte at the desired holding potential (i.e., membrane potential). Holding potentials between −80 and −40 mV are suitable for recordings of NMDA receptor-mediated currents, albeit the holding potential must be kept constant for all oocytes to enable comparison of current amplitudes.
4. Start the recording when the baseline current has stabilized and activate the NMDA receptor current response by applying glutamate, i.e., start in wash (solution 1), switch to glutamate (solution 2), and return to wash after a steady-state response is achieved. Typically, 30–60 s applications of glutamate are sufficient to reach a steady state (i.e., equilibrium) response or to determine if no response can be observed (i.e., due to the lack of functional NMDA receptor expression). It is important though to ensure you reach equilibrium to accurately measure the response amplitude (*see* **Note 15**).
5. Determine the glutamate-activated current response (i.e., steady-state current in the presence of glutamate minus the baseline current) for an equal number of oocytes injected with cRNA from each of the three combinations. The number of oocytes should be sufficient to allow an estimate of the average current response for each cRNA combination. Typically, this can be achieved using 6–8 oocytes per cRNA combination, depending on response variability. It is important that current responses are measured using the same experimental conditions for each cRNA combination.
6. Calculate the mean current response for each cRNA combination. The mean current response from oocytes expressing GluN1 + $2A_{C1}$ + $2B_{AC2}$ divided by the sum of the mean current responses from oocytes expressing GluN1 + $2A_{C1}$ + $2B_{AC2\text{-}RKTI}$ and GluN1 + $2A_{C1\text{-}RKTI}$ + $2B_{AC2}$ provides an estimate of the contribution of escaped diheteromeric receptors to the total current response (i.e., escape current). This method should consistently yield <10% escape currents (i.e., >90% of the total current response is mediated by triheteromeric GluN1/2A/2B). However, the method may underperform for some batches

of oocytes (*see* **Note 16**) and if the cRNA quality is poor (i.e., due to degradation). It is therefore strongly advised that the estimation of escape current is performed on the day of all quantitative pharmacological experiments (e.g., determination of concentration-response relationships for triheteromeric GluN1/2A/2B receptors).

4 Notes

1. The idea behind replacing the GluN2B CTD with the GluN2A CTD in triheteromeric GluN1/2A/2B receptors was two-fold. First, the CTD replacement ensures that the structure of the two GluN2 subunit is identical, thereby resulting in a symmetrical placement of the C-termini with the coiled-coil motifs and the retention signals relative to the membrane and the GluN1 subunits. Thus, the CTD replacement would ideally optimize coiled-coil formation in triheteromeric GluN1/2A_{C1}/2B_{AC2} and minimize biased coiled-coil formation in diheteromeric GluN1/2A_{C1}/2A_{C1} and GluN1/2B_{AC2}/2B_{AC2}. Second, variation may exist among GluN2 CTDs with respect to their trafficking properties, which could interfere with the method in a subunits-specific manner.
2. Ideally, selective antagonists for GluN1/2A_{C1}/2A_{C1} and GluN1/2B_{AC2}/2B_{AC2} would enable measurements of escape currents in oocytes co-expressing GluN1 with 2A_{C1} and 2B_{AC2}, but such ligands with no effects on triheteromeric GluN1/2A/2B receptors are currently not available.
3. The functional properties of diheteromeric GluN1/2B receptors containing the GluN2A CTD are indistinguishable from those of wild-type GluN1/2B [18, 38, 39]. In addition, fusion of the C1 and C2 tags at the C-terminus of the GluN2 subunits has no detectable effects on the pharmacological and functional properties of the receptors [18]. This can be shown by comparing function and pharmacology of wild-type GluN1/2A and GluN1/2B receptors with those of GluN1/2A_{C1}/2A_{C2} and GluN1/2B_{AC1}/2B_{AC2}, respectively [18].
4. Diheteromeric GluN1/2A receptors with the R518K mutation in both GluN2A subunits (GluN1/2A_{RK}/2A_{RK}) are not activated by 100 μM glutamate (in the presence of glycine). However, small current responses are activated by 100 μM glutamate for triheteromeric GluN1/2A receptors with only one R518K mutation (GluN1/2A/2A_{RK}) (data not shown). Thus, in order to completely abolish glutamate binding, the T690I mutation was introduced to the GluN2A subunit in addition to the R518K mutation [18]. Similarly, both the

R519K and T691I mutations are required to completely abolish glutamate binding to the GluN2B subunit [18].

5. The method described here works well for the expression of GluN1/2A$_{C1}$/2A$_{C2}$ and GluN1/2B$_{AC1}$/2B$_{AC2}$ receptors (Fig. 2c), which can be used in studies on the relationship between NMDA receptor structure and function by enabling expression of receptors with mutations in only one of the two GluN2 subunits [18]. In theory, the method could also be adapted and optimized for the expression of other triheteromeric NMDA receptors, such as GluN1/2A/2C or GluN1/2B/2D. However, expression of these subunit combinations is complicated by functional differences endowed by the different GluN2 subunits. The probability that the channel will be open when all the agonist binding sites are fully occupied by agonists (i.e., the open probability) is ~0.5 for diheteromeric GluN1/2A, ~0.1 for GluN1/2B, and <0.05 for GluN1/2C and GluN1/2D [1]. For example, escaped diheteromeric GluN1/2A will produce large current responses compared to escaped diheteromeric GluN1/2C, which may be problematic if the open probability of triheteromeric GluN1/2A/2C is closer to that of GluN1/2C.
6. The success of the method to express triheteromeric receptors is highly dependent on incubation temperature during cRNA expression in oocytes. At 16 °C, the expression of triheteromeric receptors is small compared to escaped diheteromeric receptors. However, this problem was solved by raising the incubation temperature to 19 °C.
7. An injection chamber is required to prevent oocytes from moving during injection. Such a chamber can be custom-made by carving a V-shaped groove in a silicone elastomer, which has been applied to the bottom of a petridish. Alternatively, a chamber can be made by gluing a fine polypropylene mesh to the bottom of a dish.
8. Defolliculated healthy-looking stage V-VI oocytes are needed for cRNA injection and NMDA receptor expression. Oocytes can be purchased from several vendors, but can also be surgically removed from captive-bred *Xenopus laevis* and treated with collagenase to remove the surrounding follicular layer. Several publications provide detailed protocols on the maintenance and surgery of *Xenopus laevis* as well as treatment, isolation, and selection of oocytes [40–44].
9. Most electrophysiological recordings using oocytes are performed with a two-electrode voltage-clamp setup. The principles and designs of the two-electrode voltage-clamp setup are described in detail elsewhere [41].

10. The concentrations of divalent cations and the pH of the extracellular recording solution can be varied to meet special requirements of the receptor of interest and the experimental design. During two-electrode voltage-clamp recordings, the oocytes are stable when divalent cations (e.g., Ca^{2+}, Ba^{2+}, or Mg^{2+}) are included in the extracellular recording solution, but the oocytes are unstable in the complete absence of divalent cations. The central ion channel pore of NMDA receptors is blocked by extracellular Mg^{2+}, whereas the channel is highly permeable to extracellular Ca^{2+}. NMDA receptor-mediated Ca^{2+} influx into the cytoplasm of oocytes strongly activates calcium-activated chloride channels, which will impede electrophysiological recordings of NMDA receptor responses [45, 46]. Thus, Ca^{2+} and Mg^{2+} should generally be avoided, leaving Ba^{2+} as the preferred divalent cation in the extracellular recording solution for studies on NMDA receptors expressed in *Xenopus* oocytes.
11. Several factors affect the expression of NMDA receptor subunits in *Xenopus* oocytes, but one important factor is the choice of DNA template. The open reading frame of the subunits can be inserted into a vector immediately following a RNA polymerase promoter (e.g., from SP6, T7, or T3 bacteriophage) for successful transcription and translation. Expression may be increased by using a vector specifically designed for expression in *Xenopus* oocytes, such as pGEMHE [47] or pXOOM [48]. The plasmid DNA must be linearized with a restriction enzyme that leaves a 5′ protruding overhang (or blunt end), since a 3′ protruding overhang can function as a primer for unintended RNA synthesis and generate antisense RNA that strongly interferes with translation. To increase the stability of the RNA, the site of linearization should be located following an artificial poly(A) tail or 200–500 base pairs from the stop codon.
12. Patience and a great deal of pipetting are often required when resuspending DNA or RNA pellets in storage buffers (e.g., TE buffer). In addition, resuspension can be further complicated if the DNA/RNA pellets have been "over-dried." DNA/RNA that has not been properly resuspended may be absent from the agarose gel. If the pellets have been "over-dried" or are otherwise resisting resuspension, then repeated freeze/thawing of the sample can facilitate dissolution of DNA/RNA.
13. The quality of the linearized DNA template critically affects RNA transcription. It is important that the linearized DNA appear as a single band on the gel. The presence of higher order bands could indicate incomplete linearization of the DNA plasmid, which will produce higher order poorly expressing RNA products during in vitro transcription (i.e., RNA polymerase will continue transcription on the DNA plasmid instead

of producing several run-off transcripts). RNA that does not appear as a single band on the gel, but rather appear as a smear, is often caused by the presence of RNases during transcription and/or in the RNA solution. To prevent RNase contamination, it is important to always use cloves and preferentially filter tips when handling template DNA and RNA solutions.

14. It is advised that the molar ratios of cRNA for the different subunits, the total amount of cRNA injected, and the duration of expression are optimized to maximize the expression of triheteromeric receptors and to minimize the escape of diheteromeric receptors. In the previously published study, we injected a total of approximately 10 ng cRNA encoding GluN1, $2A_{C1}$, and $2BA_{C2}$ at a 1:6:6 ratio in a total volume of 50 nL, which consistently resulted in selective expression of triheteromeric receptors on days 2–4 after injection [18]. However, these experimental conditions are likely dependent on multiple factors (e.g., the plasmids used, the quality of the cRNA, or the preparation of oocytes) and will presumably vary among experimenters and laboratories.
15. Activity-dependent increases in NMDA receptor response amplitudes can be observed for *Xenopus* oocytes expressing GluN2A- or GluN2B-containing NMDA receptors (e.g., GluN1/2A and GluN1/2B) [49–51]. This "run-up" of responses can strongly interfere with the ability to perform quantitative pharmacological experiments, such as the determination of ligand concentration-response relationships. However, activity-dependent increases in response amplitude can be prevented by gentle injection with 20 nL of 50 mM BAPTA (in water and adjusted to pH 7.4 using KOH) approximately 10–30 min before the recordings are performed [49]. Injection with BAPTA will not interfere with the protocol described here [35]. However, long-term viability is low for oocytes injected with BAPTA and they are usually discarded following the subsequent recording.
16. *Xenopus laevis* oocytes have low-level expression of endogenous NMDA receptor subunits that can, under some circumstances, assemble with recombinant NMDA receptor subunits to produce functional receptors [52–54]. In most batches of oocytes, the expression of *Xenopus* NMDA receptor subunits is too low to interfere with the expression of recombinant NMDA receptors. However, there are occasional batches of oocytes with unusually high levels endogenous *Xenopus* NMDA receptor subunits, resulting in unexpected functional and pharmacological results. Such batches of oocytes are also expected to interfere with the protocol described here and should be avoided.

Acknowledgments

This work was supported by grants from National Institutes of Health (R01NS065371, P20GM103546, and R01NS097536.

References

1. Traynelis SF, Wollmuth LP, McBain CJ, Menniti FS, Vance KM, Ogden KK, Hansen KB, Yuan H, Myers SJ, Dingledine R (2010) Glutamate receptor ion channels: structure, regulation, and function. Pharmacol Rev 62 (3):405–496
2. Lee CH, Lu W, Michel JC, Goehring A, Du J, Song X, Gouaux E (2014) NMDA receptor structures reveal subunit arrangement and pore architecture. Nature 511(7508):191–197
3. Ulbrich MH, Isacoff EY (2007) Subunit counting in membrane-bound proteins. Nat Methods 4(4):319–321
4. Karakas E, Furukawa H (2014) Crystal structure of a heterotetrameric NMDA receptor ion channel. Science 344(6187):992–997
5. Perez-Otano I, Larsen RS, Wesseling JF (2016) Emerging roles of GluN3-containing NMDA receptors in the CNS. Nat Rev Neurosci 17(10):623–635
6. Al-Hallaq RA, Conrads TP, Veenstra TD, Wenthold RJ (2007) NMDA di-heteromeric receptor populations and associated proteins in rat hippocampus. J Neurosci 27 (31):8334–8343
7. Rauner C, Kohr G (2011) Triheteromeric NR1/NR2A/NR2B receptors constitute the major N-methyl-D-aspartate receptor population in adult hippocampal synapses. J Biol Chem 286(9):7558–7566
8. Tovar KR, McGinley MJ, Westbrook GL (2013) Triheteromeric NMDA receptors at hippocampal synapses. J Neurosci 33 (21):9150–9160
9. Luo JH, Wang YH, Yasuda RP, Dunah AW, Wolfe BB (1997) The majority of N-methyl-D-aspartate receptor complexes in adult rat cerebral cortex contain at least three different subunits (NR1/NR2A/NR2B). Mol Pharmacol 51(1):79–86
10. Sheng M, Cummings J, Roldan LA, Jan YN, Jan LY (1994) Changing subunit composition of heteromeric NMDA receptors during development of rat cortex. Nature 368 (6467):144–147
11. Brickley SG, Misra C, Mok MH, Mishina M, Cull-Candy SG (2003) NR2B and NR2D subunits coassemble in cerebellar Golgi cells to form a distinct NMDA receptor subtype restricted to extrasynaptic sites. J Neurosci 23 (12):4958–4966
12. Pina-Crespo JC, Gibb AJ (2002) Subtypes of NMDA receptors in new-born rat hippocampal granule cells. J Physiol 541(Pt 1):41–64
13. Jones S, Gibb AJ (2005) Functional NR2B- and NR2D-containing NMDA receptor channels in rat substantia nigra dopaminergic neurones. J Physiol 569(Pt 1):209–221
14. Swanger SA, Vance KM, Pare JF, Sotty F, Fog K, Smith Y, Traynelis SF (2015) NMDA receptors containing the GluN2D subunit control neuronal function in the subthalamic nucleus. J Neurosci 35(48):15971–15983
15. Chazot PL, Coleman SK, Cik M, Stephenson FA (1994) Molecular characterization of N-methyl-D-aspartate receptors expressed in mammalian cells yields evidence for the coexistence of three subunit types within a discrete receptor molecule. J Biol Chem 269 (39):24403–24409
16. Cathala L, Misra C, Cull-Candy S (2000) Developmental profile of the changing properties of NMDA receptors at cerebellar mossy fiber-granule cell synapses. J Neurosci 20 (16):5899–5905
17. Hatton CJ, Paoletti P (2005) Modulation of triheteromeric NMDA receptors by N-terminal domain ligands. Neuron 46(2):261–274
18. Hansen KB, Ogden KK, Yuan H, Traynelis SF (2014) Distinct functional and pharmacological properties of Triheteromeric GluN1/GluN2A/GluN2B NMDA receptors. Neuron 81(5):1084–1096
19. Vicini S, Wang JF, Li JH, Zhu WJ, Wang YH, Luo JAH, Wolfe BB, Grayson DR (1998) Functional and pharmacological differences between recombinant N-methyl-D-aspartate receptors. J Neurophysiol 79(2):555–566
20. Stroebel D, Carvalho S, Grand T, Zhu S, Paoletti P (2014) Controlling NMDA receptor subunit composition using ectopic retention signals. J Neurosci 34(50):16630–16636
21. Kniazeff J, Bessis AS, Maurel D, Ansanay H, Prezeau L, Pin JP (2004) Closed state of both binding domains of homodimeric mGlu

receptors is required for full activity. Nat Struct Mol Biol 11(8):706–713

22. Bettler B, Kaupmann K, Mosbacher J, Gassmann M (2004) Molecular structure and physiological functions of GABA(B) receptors. Physiol Rev 84(3):835–867
23. Margeta-Mitrovic M, Jan YN, Jan LY (2000) A trafficking checkpoint controls GABA(B) receptor heterodimerization. Neuron 27 (1):97–106
24. Kammerer RA, Frank S, Schulthess T, Landwehr R, Lustig A, Engel J (1999) Heterodimerization of a functional GABAB receptor is mediated by parallel coiled-coil alpha-helices. Biochemistry 38(40):13263–13269
25. Burmakina S, Geng Y, Chen Y, Fan QR (2014) Heterodimeric coiled-coil interactions of human GABAB receptor. Proc Natl Acad Sci U S A 111(19):6958–6963
26. Zerangue N, Malan MJ, Fried SR, Dazin PF, Jan YN, Jan LY, Schwappach B (2001) Analysis of endoplasmic reticulum trafficking signals by combinatorial screening in mammalian cells. Proc Natl Acad Sci U S A 98(5):2431–2436
27. Brock C, Boudier L, Maurel D, Blahos J, Pin JP (2005) Assembly-dependent surface targeting of the heterodimeric GABAB receptor is controlled by COPI but not 14-3-3. Mol Biol Cell 16(12):5572–5578
28. Arai R, Wriggers W, Nishikawa Y, Nagamune T, Fujisawa T (2004) Conformations of variably linked chimeric proteins evaluated by synchrotron X-ray small-angle scattering. Proteins 57(4):829–838
29. Benveniste M, Mayer ML (1991) Kinetic analysis of antagonist action at N-methyl-D-aspartic acid receptors. Two binding sites each for glutamate and glycine. Biophys J 59 (3):560–573
30. Clements JD, Westbrook GL (1991) Activation kinetics reveal the number of glutamate and glycine binding sites on the N-methyl-D-aspartate receptor. Neuron 7(4):605–613
31. Cheriyan J, Balsara RD, Hansen KB, Castellino FJ (2016) Pharmacology of triheteromeric N-methyl-D-aspartate receptors. Neurosci Lett 617:240–246
32. Yi F, Mou T-C, Dorsett KN, Volkmann RA, Menniti FS, Sprang SR, Hansen KB (2016) Structural basis for negative allosteric modulation of GluN2A-containing NMDA receptors. Neuron (in press)
33. Serraz B, Grand T, Paoletti P (2016) Altered zinc sensitivity of NMDA receptors harboring clinically-relevant mutations. Neuropharmacology 109:196–204
34. Hackos DH, Lupardus PJ, Grand T, Chen Y, Wang TM, Reynen P, Gustafson A, Wallweber HJ, Volgraf M, Sellers BD, Schwarz JB, Paoletti P, Sheng M, Zhou Q, Hanson JE (2016) Positive allosteric modulators of GluN2A-containing NMDARs with distinct modes of action and impacts on circuit function. Neuron 89(5):983–999
35. Yi F, Mou TC, Dorsett KN, Volkmann RA, Menniti FS, Sprang SR, Hansen KB (2016) Structural basis for negative allosteric modulation of GluN2A-containing NMDA receptors. Neuron 91(6):1316–1329
36. Yuan H, Hansen KB, Zhang J, Pierson TM, Markello TC, Fajardo KV, Holloman CM, Golas G, Adams DR, Boerkoel CF, Gahl WA, Traynelis SF (2014) Functional analysis of a de novo GRIN2A missense mutation associated with early-onset epileptic encephalopathy. Nat Commun 5:3251
37. Khatri A, Burger PB, Swanger SA, Hansen KB, Zimmerman S, Karakas E, Liotta DC, Furukawa H, Snyder JP, Traynelis SF (2014) Structural determinants and mechanism of action of a GluN2C-selective NMDA receptor positive allosteric modulator. Mol Pharmacol 86 (5):548–560
38. Maki BA, Aman TK, Amico-Ruvio SA, Kussius CL, Popescu GK (2012) C-terminal domains of N-methyl-D-aspartic acid receptor modulate unitary channel conductance and gating. J Biol Chem 287(43):36071–36080
39. Punnakkal P, Jendritza P, Kohr G (2012) Influence of the intracellular GluN2 C-terminal domain on NMDA receptor function. Neuropharmacology 62(5–6):1985–1992
40. Bossi E, Fabbrini MS, Ceriotti A (2007) Exogenous protein expression in Xenopus oocytes: basic procedures. Methods Mol Biol 375:107–131
41. Stuhmer W (1998) Electrophysiologic recordings from Xenopus oocytes. Methods Enzymol 293:280–300
42. Goldin AL (1992) Maintenance of Xenopus laevis and oocyte injection. Methods Enzymol 207:266–279
43. Goldin AL, Sumikawa K (1992) Preparation of RNA for injection into Xenopus oocytes. Methods Enzymol 207:279–297
44. Matten WT, Vande Woude GF (1995) Microinjection into Xenopus oocytes. Methods Enzymol 254:458–466
45. Leonard JP, Kelso SR (1990) Apparent desensitization of NMDA responses in Xenopus oocytes involves calcium-dependent chloride current. Neuron 4(1):53–60

46. Moriyoshi K, Masu M, Ishii T, Shigemoto R, Mizuno N, Nakanishi S (1991) Molecular cloning and characterization of the rat NMDA receptor. Nature 354(6348):31–37
47. Liman ER, Tytgat J, Hess P (1992) Subunit stoichiometry of a mammalian K+ channel determined by construction of multimeric cDNAs. Neuron 9(5):861–871
48. Jespersen T, Grunnet M, Angelo K, Klaerke DA, Olesen SP (2002) Dual-function vector for protein expression in both mammalian cells and Xenopus laevis oocytes. BioTechniques 32(3):536–538, 540
49. Williams K (1993) Ifenprodil discriminates subtypes of the N-methyl-D-aspartate receptor - selectivity and mechanisms at recombinant Heteromeric receptors. Mol Pharmacol 44 (4):851–859
50. Logan SM, Rivera FE, Leonard JP (1999) Protein kinase C modulation of recombinant NMDA receptor currents: roles for the C-terminal C1 exon and calcium ions. J Neurosci 19(3):974–986
51. Zheng X, Zhang L, Wang AP, Bennett MV, Zukin RS (1997) Ca2+ influx amplifies protein kinase C potentiation of recombinant NMDA receptors. J Neurosci 17(22):8676–8686
52. Terhag J, Cavara NA, Hollmann M (2010) Cave Canalem: how endogenous ion channels may interfere with heterologous expression in Xenopus oocytes. Methods 51(1):66–74
53. Schmidt C, Hollmann M (2009) Molecular and functional characterization of Xenopus laevis N-methyl-D-aspartate receptors. Mol Cell Neurosci 42(2):116–127
54. Schmidt C, Klein C, Hollmann M (2009) Xenopus laevis Oocytes endogenously express all subunits of the ionotropic glutamate receptor family. J Mol Biol 390(2):182–195

Chapter 8

Functional Analysis of Recombinant Channels in Host Cells Using a Fast Agonist Application System

Andrei Rozov and David Jappy

Abstract

A reduced recombinant system provides a unique opportunity to study the biophysical properties of NMDAR channels with known subunit compositions, by using a point mutation approach to analyze the structural determinants of receptor function (Wollmuth and Sobolevsky, Trends Neurosci 27:321–328, 2004). However, in addition to the well-developed repertoire of molecular biological techniques, these types of studies also require electrophysiological methods that allow a wide range of receptor activation protocols that can adequately assess desensitization, inactivation, ion permeability, and other properties of the channels. Currently, one of the most well-developed techniques suitable for addressing these issues is use of the fast agonist application system for rapid activation of ligand gated ion-channels (Colquhoun et al., J Physiol 458:261–287, 1992; Jonas and Sakmann, J Physiol 455:143–171, 1992).

Key words Recombinant NMDA Receptors, Kinetics, Calcium permeability, Magnesium block, Fast agonist application

1 Introduction

The kinetics of ligand gated ion-channel responses depends crucially on the concentration time course of agonist administration. In most of the central synapses mediator concentration rises within a few microseconds and declines in the submillisecond range [4]. Thus to mimic "physiological" activation of recombinant channels expressed in host cells lines like HEK-293, CHO, and HeLa cells, one needs to develop an application system that can provide rapid wash in and washout of the compounds to be tested. Over the last few decades there have been multiple attempts to design an experimental approach allowing channel activation with on- and offset times comparable to those measured in native synapses. Previously proposed techniques ranged from single capillary high pressure applications and U-tube perfusion to piezo-driven multicapillary applications [5–7]. Although these methods could produce consistent ligand/drug applications for 10–100 ms, all of them failed to

Nail Burnashev and Pierre Szepetowski (eds.), *NMDA Receptors: Methods and Protocols*, Methods in Molecular Biology, vol. 1677, DOI 10.1007/978-1-4939-7321-7_8, © Springer Science+Business Media LLC 2017

ensure reliable channel activation on the time scale of synaptic neurotransmitter transients. The major breakthrough was in 1992 in Bert Sakmann's laboratory by Colquhoun et al. [2, 3]. They substantially developed a technique suggested by Franke, Hatt and Dudel in 1987 [8] by using application pipettes pulled from θ-glass capillaries. In this case, the septum between two barrels and, therefore, the interface between the streams of two different solutions was a width of a few micrometers. Thus, nearly instant piezo-driven movement of the attached application pipette guaranteed not only a fast onset of drug administration, but also a rapid return to drug-fee solution. Moreover, the proposed technique allowed not only a vast range of application times, but also the ability to test numerous drugs on the same cell or outside out patch. In this chapter, we describe the technical aspects of building a piezo-driven θ-glass pipette fast agonist application system, and how this approach can be used to study the properties of the NMDAR channels.

2 Materials

2.1 Piezo-Driven θ-Glass Pipette Application System Design

The minimal requirement for an upgrade from a regular electrophysiological setup to a rig suitable for fast application experiments is the following: a piezo unit consisting of an appropriate piezo actuator attached to a 3D manipulator and piezo amplifier/controller and an application solution perfusion system (solution supply system, solution switching valves, and θ-glass pipettes (Fig. 1a).

2.1.1 The Piezo Unit

1. Currently, Physik Instrumente (PI) offers a great range of different piezo actuators with suitable piezo amplifiers ranging from the more expensive, but also more reliable linear actuators to bender multilayer actuators as a more budget option. The most important characteristic of the piezo actuator in the construction of a fast agonist application system is the linear travel distance. Taking into account that the optimal tip diameter of the θ-glass application pipette is about 200 μm, a travel distance of 180 μm is usually sufficient.
2. For fine positioning of the application pipette in the recording chamber the piezo actuator should be mounted on a 3D manipulator.

2.1.2 The Application Solution Perfusion System

1. Constant flow speed of the solutions through the application pipette during the experiment is an absolute requirement for obtaining reliable and reproducible data. Therefore, the proper choice of perfusion system is very important for building and adjusting a fast application setup. Although the use of gravity-driven perfusion in this type of experiment has been reported in some publications, the weak point of this approach is its limited

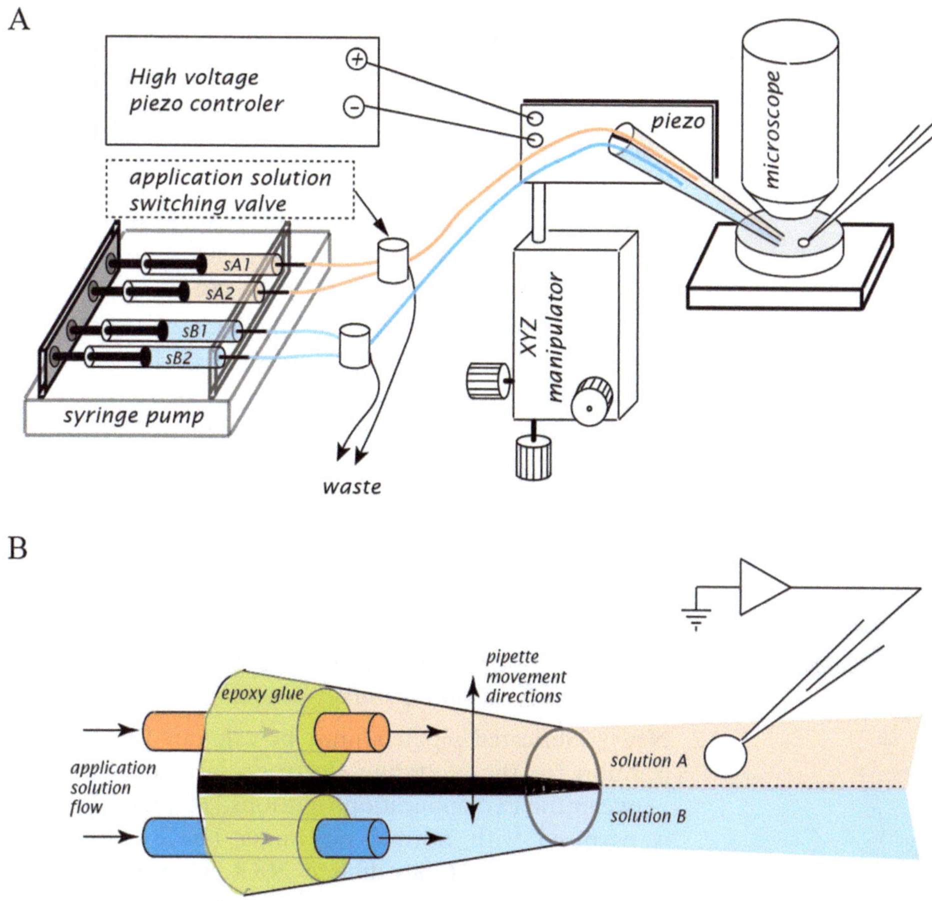

Fig. 1 (**a**) Schematic drawing of the fast agonist application system components and design. (**b**) Schematic drawing of the tip of the application pipette and the cell location during the experiment

possibility to regulate the speed of the application flow. Relatively good results can be achieved with modern multichannel peristaltic pumps. However, the best choice is a multichannel syringe pump, since they allow the best pressure control and, therefore, the best opportunity to adjust the final speed of outflow from the application pipette.

2. In some experiments there is a need to test several compounds/application solutions on the same cell/patch. Rapid switching between different solutions can be achieved by the use of either commercially available or custom made pinch-valve systems (a possible design is described in [9]).

2.1.3 Application Pipette

1. The proper fabrication and adjustment of the application pipette is one of the most important steps in the construction of a fast agonist application system. Currently, there are two types of θ-glass capillaries on the market, with outer diameters

of 2 or 1.5 mm and a wall/septum thickness of 0.25 or 0.3 mm, respectively. The choice of the particular type depends on individual preferences and existing equipment (for instance, the diameter of the pipette holders). The application pipette can be pulled from capillaries on conventional pullers. The best tip diameter is about 200 μm.

2. Often, when the tip edge is not perpendicular to the septum and the shank of the pipette, additional polishing on a fine grain spinning disk is necessary (*see* **Note 1**).
3. The next step is inserting and gluing the inflow capillaries. Thin polyethylene tubing should be inserted as close as possible to the tip of each barrel of the application pipette. Then both the barrels should be filled with water-resistant polymerizing material (for instance, epoxy glue) almost to the tips of the supplying tubes. After glue polymerization, the pipette is ready to be mounted and connected to the rest of the application system (Fig. 1b).

2.1.4 Application Pipette Positioning and Calibration

1. In a recording chamber, the application pipette should be placed at least 0.5 mm above the cultured cells or slice to avoid turbulence and maintain laminar flow.
2. Newly fabricated pipettes must be calibrated to find the best position for the recording electrode relative to the interface between the two application solutions. To this end, one barrel is perfused with normal rat ringer (NRR) solution while another with ten times diluted NRR. Due to their different optical characteristics, the interface between the two solutions should be clearly visible. If this is not the case *see* **Note 2**.
3. Next the open-tip patch pipette should be positioned in the stream of control solution (100% NRR). Usually, the best position for a recording pipette is approximately 50 μm away from the tip of the application pipette and 10–30 μm from the solutions interface (*see* **Note 3**).

2.1.5 Cells, Application, and Intracellular Solutions for Studying NMDAR Properties

1. Cultured HEK-293, CHO, or HeLa cells transfected with the appropriate constructs for the expression of recombinant NMDAR subunits (*see* Chapter 5).
2. Normal rat ringer solution: 135 mM NaCl, 5.4 mM KCl, 1.8 mM $CaCl_2$, 1 mM $MgCl_2$, 10 mM HEPES, 0.01 mM Glycine (pH 7.2 with NaOH).
3. High Ca^{2+}extracellular solution: 105 mM *N*-methyl-d-glucamine (NMDG), 30 mM $CaCl_2$, 5 mM HEPES, 0.01 mM Glycine (pH 7.2 with HCl).
4. Mg^{2+}-free extracellular solution: 135 mM NaCl, 5.4 mM KCl, 1.8 mM $CaCl_2$, 10 mM HEPES 0.2 mM EDTA, 0.01 mM Glycine (pH 7.2 with NaOH).

5. Intracellular solution: 135 mM CsCl, 4 mM Mg-ATP, 5 mM HEPES, 0.5 mM EGTA (pH 7.2 with NaOH).

3 Methods

3.1 Measurements of the Properties of NMDA Receptor Kinetics

3.1.1 Desensitization of Recombinant NMDAR Channels Expressed in Host Cells

1. After obtaining whole-cell configuration, lift the cell from the coverslip and position the cell in front of the application pipette into the flow of the control solution (normal rat ringer). The experimental solution in the second barrel should contain normal rat ringer solution with 1 mM L-glutamate.
2. For current recordings apply agonist using the fast agonist application system. Measurements are done in voltage clamp mode, and the holding potential depends on the needs of the experiment. Since the desensitization time of NMDAR channels strongly depends on subunit composition, the optimal application time is 1 s or longer. The interval between subsequent applications should allow for complete recovery from desensitization (10–15 s). For a reliable calculation of the desensitization time constant the average of at least five responses should be taken.
3. For an analysis, fit the decaying phase of the current with either a single exponential and a constant component: $I(t) = A \exp(-t/\tau) + I_{SS}$) or the sum of two exponentials and a constant component: ($I(t) = A_1 \exp(-t/\tau_1) + A_2 \exp(-t/\tau_2) + I_{SS}$), where A is the current amplitude, τ is the time constant, and I_{SS} is the amplitude at a steady-state level (time $t = 0$ at the peak of the response) (*see* **Note 4**).

3.1.2 Deactivation of Recombinant NMDAR Channel Expressed in Host Cells

1. The experimental design and protocol for deactivation measurements is similar to the one described above. However, the duration of glutamate application should be shortened to 10–20 ms. For an analysis, the decaying phase should be fitted with either a single exponential and a constant component or the sum of two exponentials and a constant component with $t = 0$ shortly after the end of the application.

3.1.3 Measurements of Current-Voltage Relations

1. For recordings of current-voltage relations of recombinant NMDAR channels evoke currents by agonist application at different holding potentials. The duration of application depends on the type of agonist. For glutamate (1 mM) 5–20 ms applications are sufficient. For NMDA (0.1 mM) application, a longer exposure to the agonist is necessary to reach the maximal response.
2. Average traces recorded at the same potential (at least three responses for each holding potential) and measure current amplitudes.

3. Plot the measured current amplitudes versus corresponding voltages to obtain the current-voltage relations for the measured voltage range.
4. Fit the averaged data points with 6th or 8th order polynomials to obtain the reversal potentialPotential.

3.2 Measurements of Ca^{2+} Permeability

1. Measure the reversal potential (V_{revNa}) for normal rat ringer as described above.
2. Switch application system to a high Ca^{2+}extracellular solution and measure the new reversal potential (V_{revCa}) value.
3. Determine the relative Ca^{2+} permeability (relative to Na^+) P_{Ca}/P_{Na} value according to the following equation:

$$\frac{P_{Ca}}{P_{Na}} = \frac{0.25\, a_{Na}}{a_{Ca}\{\exp[2V_{revCa} - V_{revNa})F/RT] + \exp[2V_{revCa} - V_{revNa})F/RT]\}},$$

where a_{Na} and a_{Ca} are the activities of Na^+ and Ca^{2+} in the extracellular solutions, respectively [10], *F*, *R*, and *T* are standard thermodynamic parameters.

3.3 Measurements of the Strength of the Mg^{2+} Block

1. To estimate the strength and voltage dependence of the Mg^{2+} block of NMDAR channels, record NMDAR currents at different potentials and plot a current voltage curve in a Mg^{2+}-free extracellular solution as described above.
2. Measure NMDAR mediated currents at the same holding potentials, but in the presence of known Mg^{2+} concentration (for instance in normal rat ringer) and plot the current-voltage curve.
3. Calculate the effect of Mg^{2+} as a ratio of the NMDAR current amplitude measured in the Mg^{2+}-free solution ($NMDA_{Mg0}$) to the value obtained in the presence of Mg^{2+} ($NMDA_{Mg1}$).

4 Notes

1. In the case where the final aim is to achieve sub-millisecond application durations, additional thinning of the walls and the septum can be achieved by briefly dipping of the tip of the pipette into hydrofluoric acid with subsequent extensive neutralization in a KOH-containing solution and washing in distilled water.
2. Perfusion speed must be increased.
3. If the position is optimal, 1 ms application of 10% NRR results in a fast rising (~100 μs) and decaying (~150 μs) liquid-junctional current of about 1 ms in duration. Moreover, longer

stimulations (100 ms) should not lead to a more than 5% enhancement of the current amplitude.

4. Depending on subunit composition NMDARs can be blocked by intracellular Ca^{2+}, to minimize the effect of the block on the kinetic properties the concentration of EGTA in the intracellular solution should be increased up to 10 mM. Alternately, 0.5 mM EGTA in intracellular solution can be substituted with 1 mM BAPTA [11, 12].

Acknowledgments

This work was supported by the program of competitive growth of Kazan Federal University and the subsidy allocated to Kazan Federal University for the state assignment in the sphere of scientific activities.

References

1. Wollmuth LP, Sobolevsky AI (2004) Structure and gating of the glutamate receptor ion channel. Trends Neurosci 27:321–328
2. Colquhoun D, Jonas P, Sakmann B (1992) Action of brief pulses of glutamate on AMPA/kainate receptors in patches from different neurons of rat hippocampal slices. J Physiol 458:261–287
3. Jonas P, Sakmann B (1992) Glutamate receptor channels in isolated patches from CA1 and CA3 pyramidal cells of rat hippocampal slices. J Physiol 455:143–171
4. Sabatini BL, Regehr WG (1999) Timing of synaptic transmission. Annu Rev Physiol 61:521–542
5. Ji D, Lape R, Dani JA (2001) Timing and location of nicotinic activity enhances or depresses hippocampal synaptic plasticity. Neuron 31:131–141
6. Klink R, de Kerchove d'Exaerde A, Zoli M, Changeux JP (2001) Molecular and physiological diversity of nicotinic acetylcholine receptors in the midbrain dopaminergic nuclei. J Neurosci 21:1452–1463
7. Nashmi R, Dickinson ME, McKinney S et al (2003) Assembly of alpha4beta2 nicotinic acetylcholine receptors assessed with functional fluorescently labeled subunits: effects of localization, trafficking, and nicotine-induced upregulation in clonal mammalian cells and in cultured midbrain neurons. J Neurosci 23:11554–11567
8. Franke C, Hatt H, Dudel J (1987) Liquid filament switch for ultra-fast exchanges of solutions at excised patches of synaptic membrane of crayfish muscle. Neurosci Lett 77:199–204
9. Komal P, Evans G, Nashmi R (2011) A rapid agonist application system for fast activation of ligand-gated ion channels. J Neurosci Methods 198:246–254
10. Brusa R, Zimmermann F, Koh DS et al (1995) Early-onset epilepsy and postnatal lethality associated with an editing-deficient GluR-B allele in mice. Science 270:1677–1680
11. Rozov A, Burnashev N (2016) Fast interaction between AMPA and NMDA receptors by intracellular calcium. Cell Calcium 60(6):407–414
12. Valiullina F, Zakharova Y, Mukhtarov M, Draguhn A, Burnashev N, Rozov A (2016) The relative contribution of NMDARs to excitatory postsynaptic currents is controlled by Ca(2+)-induced inactivation. Front Cell Neurosci 10:12. doi:10.3389/fncel.2016.00012

Chapter 9

GluN2B Subunit Labeling with Fluorescent Probes and High-Resolution Live Imaging

Cécile Perrio, Olivier Nicole, and Alain Buisson

Abstract

Laser Scanning Confocal Microscopy (LSCM) imaging using an appropriate fluorescent probe enables the visualization of a molecular target with high resolution, and represents a method of choice for studying expression, subcellular location, and trafficking of receptors in living cells. The chemical, physical, and pharmacological properties of the probe remain essential. Here, we describe (1) the preparation of a specific probe for NMDAR GluN2B receptor by conjugation of fluorescein to an ifenprodil-based ligand, (2) an in vitro functional assay by calcium imaging for GluN2B binding and inhibition evaluation of the probe, and (3) the labeling and confocal imaging of GluN2B in DS-red labeled living cortical neurons.

Key words Confocal microscopy, Calcium imaging, ifenprodil, Fluorescein, GluN2B

1 Introduction

N-Methyl-D-aspartate receptors (NMDARs) exist in different forms owing to multiple combinations of subunits (GluN1, GluN2A-D, GluN3A-B) that can assemble into a functional receptor [1–5]. Typically, endogenous NMDARs are di-heteromers comprising two GluN1 subunits and two GluN2 (or GluN3) subunits, which assemble as a dimer of dimers to form a tetrameric complex. The expression of different NMDAR subtypes is not static, but regulated dynamically in a cell- and synapse-specific manner within the central nervous system [1, 6, 7]. Location of NMDARs at either synapses or extrasynaptic sites is mainly related to the nature of GluN2 (mainly GluN2A and GluN2B) subunit [8]. The subcellular localization and physiological functions of NMDARs are closely related, and an intense speculation still remains about whether specific NMDAR subtypes and/or locations are responsible for inducing synaptic plasticity and excitotoxicity [9]. Laser Scanning Confocal Microscopy (LSCM) represents a unique opportunity to investigate the regulation/dysregulation of the segregated subcellular localization in situations of health and

Nail Burnashev and Pierre Szepetowski (eds.), *NMDA Receptors: Methods and Protocols*, Methods in Molecular Biology, vol. 1677, DOI 10.1007/978-1-4939-7321-7_9, © Springer Science+Business Media LLC 2017

disease [10]. Based on the molecular tagging using an appropriate fluorescent probe, this high-resolution live imaging technique is a powerful tool for allowing the detailed mapping of probe/receptor in cells and therefore for distinguishing location for receptors. Such an approach provides important insights into receptor dynamics and trafficking as well as into biophysical microenvironment of the binding site, and turns crucial the rethinking of the current knowledge of the pharmacological role ascribed to receptors.

Fluorescent probes are usually obtained by creating a covalent bond between a ligand specific to the molecular target of interest and a fluorescent moiety [11]. Challenges often are, first, to modify the parent ligand (especially when it is a nonpeptidic small organic molecule with poor functionality) to provide an attachment point for the fluorophore; second, to perform conjugation without significantly disturbing the natural binding function. The introduction of a spacer with appropriate length and chemical constitution linking the ligand to the fluorophore is often necessary. Among the different NMDAR subunits, GluN2B present the advantage of possessing a specific ifenprodil binding site that coincides with the *N*-terminal domain (NTD) of the subunit [12–15].

Ifenprodil remains the lead compound of the major class of antagonists of NMDARs with marked selectivity for NR2B [16–25]. To date, it is the best characterized subtype-specific NMDA antagonist both in vitro and in vivo. A major characteristic of ifenprodil at NMDARs is a high affinity, noncompetitive, and voltage-independent inhibition of receptors endowed with the GluN2B subunit. Ifenprodil has become widely used as a tool for studying the expression, properties, function, and distribution of NMDAR subtypes. The potency of ifenprodil at NR2B is due to the key structural elements that incorporate a phenolic aromatic ring connected to a second phenyl ring via a linker containing a tertiary basic amine [12–15]. The piperidine linker provides a scaffold to hold the two aryl rings in the optimum spatial orientation in the binding site which itself contains a lipophilic pocket that accommodates the phenyl ring, as well as a hydrophobic and hydrogen bond acceptor pocket that accepts the phenol group. The basic nitrogen atom contributes to the positioning by electrostatic interaction. By mutagenesis, it has been proposed that ifenprodil binds to the central, interlobe cleft of GluN2B NTD and promotes cleft closure through a hinge-bending motion (Venus flytrap mechanism) [13]. The recent characterization of binding sites by experimental and computational approaches has revealed that the phenol ring of ifenprodil points toward the entrance of the cleft, whereas the phenyl ring makes contact with the interlobe hinge [14]. Such a location makes free the aliphatic hydroxyl group as an anchoring point for the fluorophore.

Here, we report the first fluorescent probe **A**, specific to GluN2B, designed from an ifenprodil-based ligand **B** modified at

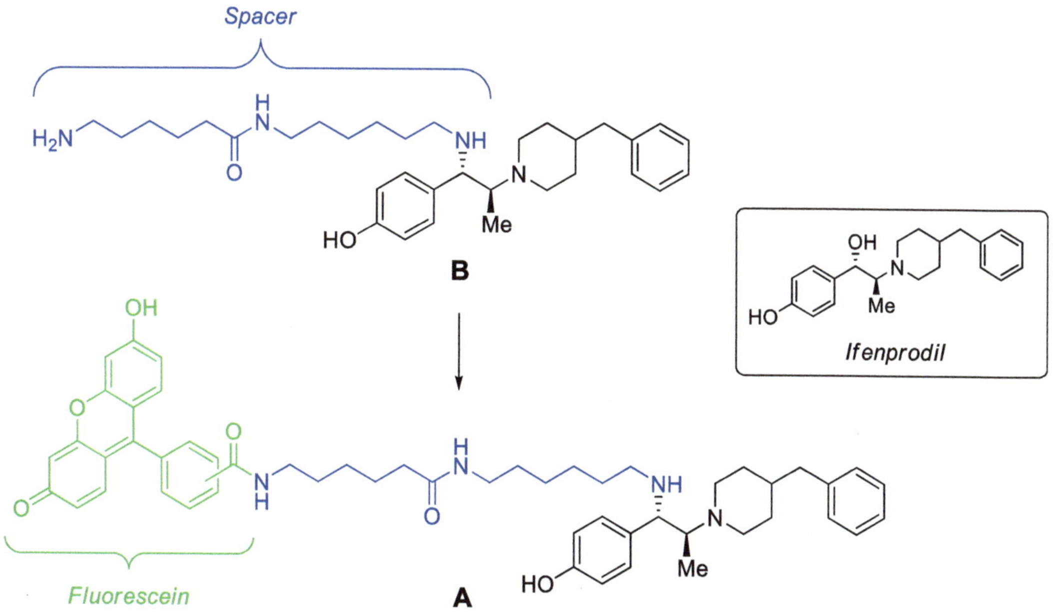

Fig. 1 GluN2B-specific fluorescent ifenprodil-based probe **A** and its synthesis by the conjugation of ifenprodil-modified ligand **B** to fluorescein

the aliphatic hydroxyl position (Fig. 1) [26]. In **B**, the aliphatic hydroxyl group was replaced by a 1,6-hexamethylenediamine chain that plays the role of a spacer. The structure of **B** resulted from structure-activity relationship optimization studies [27]. We describe the preparation of the fluorescent GluN2B specific probe **A** by the conjugation of **B** to fluorescein ($\lambda_{ex}/\lambda_{em} = 490/525$ nm). We also detail the in vitro functional assay by calcium imaging for the evaluation of GluN2B binding and inhibition potency of the probe **A** by comparison with parent ifenrpodil. For this purpose, we monitored intracellular calcium concentrations by videomicroscopy of HEK-293 cells transfected with the cDNA encoding either for GluN1-1a and GluN2B, or for GluN1-1a and GluN2A subunits (Fig. 2a). As NMDARs are calcium-permeable ion channels [28], we measured calcium ([Ca^{2+}]) influx induced by the application of NMDA (at 100 μM for 30 s) in the presence of glycine (10 μM) and of increasing concentrations of potential antagonists. The application of NMDA induced a rapid and sustained increase in [Ca^{2+}] (determined as the 100% value), whereas the simultaneous application of NMDA and the probe led to a lesser increase in [Ca^{2+}] (Fig. 2b). Dose–response curves were thus obtained (Fig. 2c, d). We finally performed the labeling and confocal imaging of GluN2B in DS-red labeled cortical neurons using the specific probe **A**. Primary neuronal cultures (at DIV 10) were transfected with the DS-red fluorescent protein ($\lambda_{ex}/\lambda_{em} = 540/570$ nm). Imaging experiments were performed at DIV 14 when cortical neurons expressed a subset of NMDARs similar to those observed in adults.

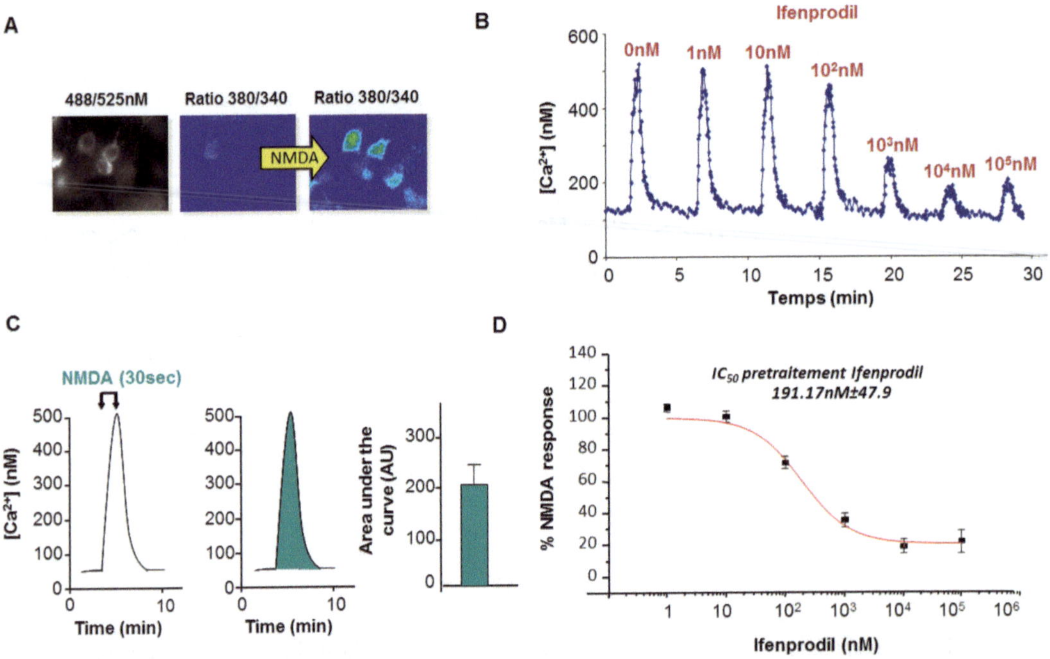

Fig. 2 Evaluation of the antagonistic properties of ifenprodil of GluN2B by measurement of inhibition of NMDA-induced Ca^{2+} influx in HEK 293. (**a**) Illustration of GFP fluorescence observed in HEK 293 transfected with cDNA encoding for GluN2B-GFP and GluN1 subunit. The Fura-2 fluorescence is evaluated by measuring the ratio of emission fluorescence at 380 and 340. Bath application of 100 μM of NMDA induces an increase of 380/340 ratio only in GluN2B/GluN1 transfected cells. (**b**) Representative experiment of the NMDA-induced calcium influx observed in HEK 293 transfected with cDNA encoding for GluN2B and GluN1 subunit with or without increasing concentration of ifenprodil. (**c**) The [Ca^{2+}] variation is measured by integrating the area under the curve (AUC) on a 2 min period after the beginning of the NMDA application. (**d**) Dose-response curve of ifenprodil for the inhibition of NMDA-induced calcium influx through NR1-1a/NR2B receptors

When transfected with cDNA encoding for DS-red, neurons displayed a cytosolic expression of the fluorescent protein that allowed the visualization of the dendritic morphology (Fig. 3a). After the exposure to fluorescent probe **A**, a punctate green fluorescence was demonstrable (Fig. 3b). The merged image showed the yellow coloration that characterizes the exact colocalization of red and green fluorescence, especially on small protrusions along the dendrites that could correspond to postsynaptic sites (Fig. 3c). A high magnification of a spine labeled with probe **A** proved the localization of the staining on the protrusions (Fig. 3g). To assess the specificity of the labeling, exposure with probe **A** was preceded by a pre-incubation with ifenprodil (10 μM). Under these conditions, image acquisition with the same amplification setting failed to reveal any green fluorescence (Fig. 3d–f). Inhibition of probe **A** staining by ifenprodil strongly revealed that the fluorescent image

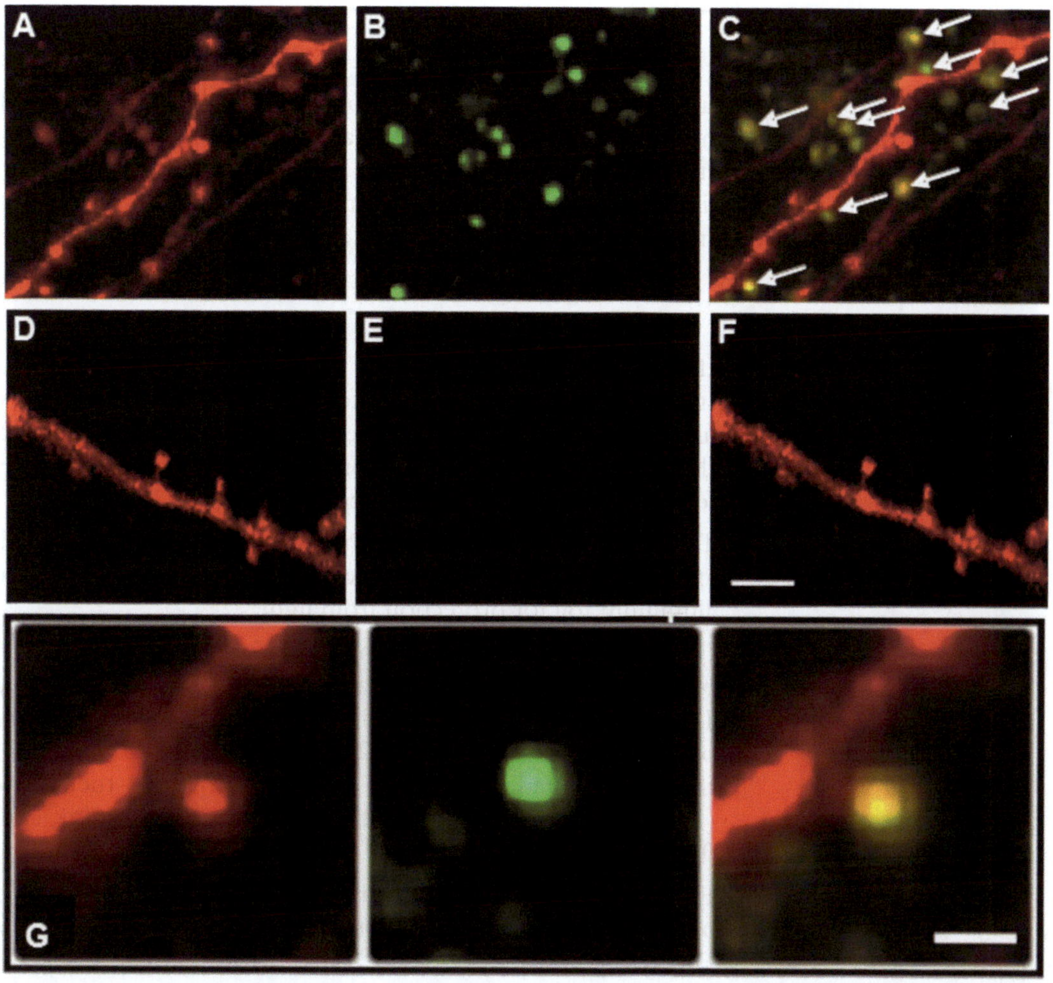

Fig. 3 Confocal microscopy imaging. (**a**) Dendrite of Ds-red transfected living cortical neurons (*red*, emission band 550–590 nm). (**b**) After 15 min exposition to 10 μM fluorescent ifenprodil-based probe **A** (*green*, emission band 513–530 nm). (**c**) Merge image showing colocalization of the fluorescence in *yellow* with dendritic spines (*arrows*). (**d–f**) Pre-incubation with non-fluorescent ifenprodil (15 min, 10 μM) fully blocked the fluorescent ifenprodil-based probe **A** labeling. Scale bar 5 μm. (**g**) High magnification of a dendritic spine after 15 min exposition to 10 μM probe **A** as in (**a–c**). Scale bar 1 μm

in Fig. 3b was due to a specific binding of probe on GluN2B-containing NMDARs. Thus, we demonstrate the ability of the probe **A** to tag and visualize specifically GluN2B receptor. Subsequently, investigators successfully exploited the ifenprodil-modified ligand **B** for the preparation of a near-IR imaging probe for NMDA GluN2B visualization by Multi-Spectral Optoacoustic Tomography (MSOT) [29].

2 Materials

Prepare all the solutions using ultrapure water (obtained as purified deionized water with sensitivity reaching 18 MΩ-cm at 25 °C), analytic grade reagents, and HPLC grade solvents. Store the fluorescein reagent and the fluorescent probe in freezer at −20 °C. Prepare and store all other reagents and solutions at room temperature unless indicated otherwise.

2.1 Fluorescent Probe

1. Ifenprodil-modified precursor was prepared from racemic (threo)-*O*-benzyl ifenprodil according to a two-step synthesis including a stereospecific mesylation/amination sequence through the use of *N*-Boc-monoprotected α,ω-hexyldiamine followed by an acidic double deprotection [26, 27].
2. The fluorescein reagent and the fluorescent probe obtained after chemical conjugation were kept in bottles wrapped with aluminum foil for protection from light.
3. HPLC analysis and purification was performed using a reverse-phase MS C18 Waters Bondapak column (7.8 × 300 mm, 10 μm) eluting with acetonitrile (ACN) and aqueous trifluoroacetic acid (TFA, 0.1%, pH 2) as follows: ACN 0% (5 min), then linear gradient from 0 to 90% (45 min), then 90% (10 min), then linear gradient from 90 to 0% (10 min), and 0% (10 min)—at a flow rate of 3 mL/min.
4. Fluorescence emission measurements were performed using probe at 0.003 mM in solution in water containing 0.3% DMSO.

2.2 HEK293 Cell Culture and Plasmid Transfection

1. Human Embryonic Kidney 293 also often referred to HEK 293 is a specific cell line originally derived from human embryonic kidney cells grown in tissue culture. HEK 293 cells (ATCC® CRL-1573™) are purchased from ATCC®. HEK 293 cells are maintained in Dulbecco's Modified Eagle Medium (DMEM) medium supplemented with 2 mM L-Glutamine, 10% fetal bovine serum. Antibiotics are not recommended; however, 5 mL/L of Penicillin-Streptomycin may be used when required. Cells should be incubated at 37.0 °C in 5% CO_2 and expected to double approximately every 24–48 h. The medium should be renewed every 2 or 3 days. Subculturing is usually performed in T-25 flasks.
2. For calcium imaging, HEK293 should grow on glass support (35 mm Dish, No. 1.5 Coverslip, 12 mm Glass Diameter). To improve cell adhesion, dishes are pretreated overnight at 37 °C with a solution of Matrigel diluted 300 times in DMEM.
3. Trypsin-EDTA.

4. Opti-MEM® (GIBCO®).
5. Plasmids encoding the full-length sequences of GluN1-1a and GluN2B-GFP were a generous gift from Dr. Stefano Vicini, (Georgetown University Medical Center, Washington). The DNA is purified by using an Endotoxin-free Maxi-prep kit and the DNA solution is finally prepared for transfection to desired concentrations in water. A concentration of 1 μg/μL is generally used.
6. D-2-amino-5-phosphonovalerate or D-APV is a widely used competitive NMDA antagonist. AP-V powder is solubilized in ultrapure water at 100 mM. Aliquots were stored at −20 °C for up to 6 months.

2.3 Calcium Imaging

1. Fura-2 AM: 1 mg of Fura-2 AM is dissolved in 1,1,1,3,3,3-hexafluoro-2-propanol (HFIP) to 1 mM and the solution is quickly spared in 50 μL aliquots. The HFIP is then removed by evaporation and the dried HFIP film was stored at −20 °C for up to two months. The peptide film is resuspended to make a 1 mM solution in 50 μL of anhydrous dimethyl sulfoxide (DMSO) supplemented with 0.2% (w/v) pluronic acid.
2. HEPES and Bicarbonate Buffered Saline Solution or HBBSS: 116 mM NaCl, 5.4 mM KCl, 1.8 mM $CaCl_2$, 0.8 mM $MgSO_4$, 1.3 mM NaH_2PO_4, 12 mM HEPES, 5.5 mM Glucose, 25 mM bicarbonate, 10 μM glycine, pH = 7.45. This solution is prepared in ultrapure water obtained by purifying deionized water, to attain a sensitivity of 18 MΩ-cm at 25 °C. The solution can be stored at 4 °C during 24–48 h.
3. Experiments are performed on an inverted microscope with the use of a quartz light pathway, a xenon arc lamp (100 W), and an oil immersion objective (40× NA. 1.3 optimized for fluorescence imaging). An intensified charge-coupled device camera recorded the fluorescence emissions (>510 nm) evoked by excitation through narrow band-pass filters (340 ± 5 nm/380 ± 6.5 nm) housed in a computer-controlled filter wheel. Fluorescence images were acquired with a PC-based computer at a maximal time resolution of 2 s per successive measurement with the use of Metafluor 6.3 (Universal Imaging Corporation).

2.4 Primary Cortical Neuron Culture, Transfection, and Confocal Imaging

1. Cerebral cortices were dissected, dissociated, and cultured in Dulbecco's modified eagle medium (DMEM) containing: 5% fetal bovine serum, 5% horse serum, and 2 mM glutamine.
2. Neurons were seeded on 35 mm glass-bottom dishes. All dishes were beforehand coated with 0.1 mg/mL poly-D-lysine and 0.02 mg/mL laminin.
3. DMKY solution: 1 mM kynurenic acid, 0.05% phenol red, 0.9 mM NaOH, 0.5 mM HEPES, 10 mM $MgCl_2$, pH: 7.4.

4. HBS solution: 25 mM HEPES, 140 mM NaCl, 0.750 mM Na_2HPO_4, pH = 7.06.
5. Neurons were imaged on a microscope equipped with a confocal system. The images were acquired with the appropriate software for a confocal imaging system, using a 60× oil immersion objective with a numerical aperture of 1.4.

3 Methods

3.1 Synthesis of Fluorescent Probe

1. To a solution of 6-[fluorescein-5(6)-carboxamido]hexanoic acid *N*-hydroxysuccinimidyl ester (6.1 mg, 10.4 μM) in NMP (*N*-methyl-2-pyrrolidinone, 150 μL), add a solution of ifenprodil-modified precursor (5.3 mg, 9.8 μM) and DIEA (*N*,*N*-diisopropylethylamine, 3.5 μL, 20.2 μM) in NMP (200 μL).
2. Wrap the reaction vessel with aluminum foil for protection from light.
3. Stir the reaction mixture at rt. for 18 h and then dilute with water (2.3 mL).
4. Perform purification by semi-preparative HPLC by successive injections of volumes ranging from 600 to 800 μL.
5. Collect fractions corresponding to the UV peak at the retention time (t_R) of 24.3 min.
6. Check the following quality parameters:
   - Lyophilization of the collected fractions should yield to the fluorescent probe as a yellow powder (5 mg, 45%).
   - HPLC analysis should give a purity >95%.
   - MS and HRMS analyses should confirm the obtention of the fluorescent probe: ESI+/MS/MS m/z (%) 895.8 ([M ($C_{54}H_{62}N_4O_8$) + $H]^+$). ESI+/HMRS calculated for $C_{54}H_{63}N_4O_8$ (MH^+) 895.4646, found 895.4666.
   - Fluorescence emission measurements should correspond to $\lambda_{ex}/\lambda_{em} = 490/525$ nm.

3.2 HEK293 Cell Culture and Plasmid Transfection

1. The culture medium is quickly removed out of the flask and the cell layer is briefly rinsed with Trypsin-EDTA solution to remove all traces of serum that contains a trypsin inhibitor.
2. Add 1 mL Trypsin-EDTA solution to flask and observe cells under an inverted microscope until the cell layer is dispersed (usually within 5–10 min).
3. Add 5 mL of complete growth medium to inhibit the trypsin and aspirate cells by gently pipetting.

4. Aliquot of the cell suspension is added to 35 mm dish pre-coated with matrigel.
5. HEK293 cells are incubated at 37.0 °C in 5% CO_2 until the cell density reaches 50–80% confluent.
6. The day of transfection, remove growth medium from cells and replace with 1.5 mL of complete growth medium.
7. For each dish of HEK293 cells to be transfected, dilute 2 μg of GluN1 and 2 μg of GluN2A or GluN2B in 500 μL of Opti--MEM without serum. cDNAs encoding the full-length sequences of GluN1-1a and GluN2B-GFP were a generous gift from Dr. Stefano Vicini (Georgetown University Medical Center, Washington) (*see* **Note 1**).
8. For each dish, add 8 μL of Lipofectamine 2000® Reagent into the above diluted Opti-MEM®/DNA solution, mix gently, and incubate 30 min at room temperature to form DNA-Lipofectamine 2000® reagent complexes.
9. After 30 min incubation, add 500 μL of the DNA-Lipofectamine 2000® reagent complexes directly to each dish containing HEK 293 cells and mix gently by rocking the plate back and forth.
10. 2 h after transfection, add directly in the medium, 2 μL of AP-v (final concentration 100 μM) to prevent the toxic effect of Ca^{2+} overflow-related to NMDA receptor activation.
11. One or two days after transfection, the GFP-GluN2B subunit can be visualized to evaluate the efficiency of transfection (Fig. 2a) before the use for calcium imaging.

3.3 Calcium Imaging

1. The day of calcium imaging, 500 mL of HBBSS is warmed to 37 °C in a water bath.
2. A dye solution is prepared in a 1.5 mL eppendroff tube by adding 494 μL of HBBSS, 5 μL of Fura2-AM (final concentration 10 μM) with pluronic acid and 1 μL of D-APV (final concentration 100 μM). The solution is vigorously vortexed.
3. The 35 mm dish containing HEK 293 cells transfected with GluN1 and GluN2B is washed twice with 2 mL of warmed HBBSS.
4. The HBBSS is next removed and the dye solution is transferred on the HEK 293.
5. The cells are incubated at 37 °C for 45 min in a dark incubator.
6. The dye solution is next washed out twice with 2 mL of warmed HBBSS.
7. Place a drop of oil on the 40× objective and the 35 mm dish is placed and secured on the microscope stage.

8. Two glass pipettes homemade are placed closed to the recorded area and connected through two plastic tubes to a peristaltic pump that permanently controls the perfusion of the HEK293 cells. The input line is connected to a HBBSS and the output line is aspirated to the solution in the trash. Turn on the perfusion at 2 mL/min as quickly as possible. Take care to prevent the formation of bubbles in the 35 mm dish.
9. Examine the cells with the GFP filter (*Ex.* 488 nm, *Em.* 525 nm) to localize the transfected cells and adjust the analyzed area.
10. Examine the HEK 293 using the camera and set the gain and the exposure time of the camera to generate an image at 380 nm closed to the saturation and largely below the saturation for the 340 nm image. Once set, do not change the camera settings.
11. A frame is taken at each excitation wavelength from an area of the coverslip devoid of cells to obtain a background level. The analyzed images are all background-subtracted.
12. The regions of interest tool (ROI) are next drawn and should delimitate the cell body of transfected cells in which the calcium measurement will be done. At the end of the experiment, it is useful to play the movie of the images to control that the cells or the focus does not move during the course of the experiment.
13. Set up the time-lapse interval to collect images (one image every 2 s).
14. Start the experiment.
15. After 2 min of basal recording, apply the NMDA solution (at 100 μM) for 30 s, each application being separated by a minimum of 5 min to allow the full recovery to the basal level of $[Ca^{2+}]_I$ (*see* **Note 2**).
16. Because the slow rate of inhibition of ifenprodil and derived ligands, apply drugs 60 s prior to and during NMDA application.
17. Stop the experiment and start to analyze.
18. Use the software to collect time-lapse ratio measurements for each ROI in each image.
19. Convert the ratio measurements to intracellular calcium values using the following formula:

$$[Ca^{2+}] = Kd[(R - R_{min})/(R_{max} - R)]F_0/F_s$$

where R is the ratio for the observed 340/380 fluorescence ratio, R_{min} is a ratio for a Ca^{2+}-free solution, R_{max} is the ratio for a saturated Ca^{2+} solution, Kd = 135 nM (the dissociation constant for Fura-2), F_0 is the intensity of a Ca^{2+}-free solution at 380 nm, and F_s is the intensity of a saturated Ca^{2+} solution at 380 nm. One way of measuring Fura-2 ratio values as a function of calcium concentrations is to use a calibration kit (Cat. N°: F6774, Molecular Probes®) which contain precise instructions for use in vitro (Fig. 2b).

20. Plot the obtained calcium values against time to generate graphics of the intracellular Ca^{2+} trend in response to stimulation by NMDA in the presence of dose response increase of ligand.
21. Measure the $[Ca^{2+}]$ variation for each drug application by integrating the area under the curve (AUC) on a 2 min period after the beginning of the NMDA application (Fig. 2c).
22. Analyze dose-response curves on the standard slope with GraphPad Prism 5.0 (Fig. 2d).

3.4 Primary Cultures of Cortical Neuron Culture and Transfection

1. Culture mouse cortical neurons from 14- to 15-days embryos [30].
2. Dissect and dissociate cerebral cortices and culture in Dulbecco's modified eagle medium (DMEM) containing 5% fetal bovine serum, 5% horse serum, and 2 mM glutamine.
3. Seed neurons on 35 mm glass-bottom dishes at a final concentration of two cortical hemispheres per dish for confocal experiments.
4. Maintain cultures at 37 °C in a humidified atmosphere containing 5% CO_2–95% air. The day before the experiments, bathing media from the primary cortical culture dishes were changed for DMEM without serum. Perform treatments on neuronal cultures at 14–15 DIV.
5. Perform transfections on cortical neuron cultures after 10 DIV with the calcium phosphate precipitation technique.
6. Remove growth medium (DMEM and serums) and keep until the last step of transfection.
7. Wash cells into DMKY for 1–1.5 h. Mix 3 μg of plasmid encoding for DS-red ($\lambda_{ex}/\lambda_{em}$ = 540/570 nm) (*see* **Note 3**) with 120 mM $CaCl_2$ in HBS and leave for 20 min to precipitate the DNA.
8. Apply plasmids to cells for 120 min.
9. Replace transfection media with conditioned growth medium and return cultures to the incubator until use at DIV 14–15.

3.5 Confocal Imaging

1. Place transfected neurons in the HBBSS solution 1.5–2 h before experiments.
2. Image neurons on a microscope equipped with a confocal system.

3. Acquire images with the appropriate software for the confocal imaging system, using a 60× oil immersion objective. Z-stacks were performed with 0.3 μm per step. Collect images immediately before and after the treatment with the fluorescent probe ($\lambda_{ex}/\lambda_{em}$ = 490/525 nm) dissolved in DMSO at the final concentration of 10 μM.
4. Monitor the evolution of the fluorescence intensity at the two wavelengths of interest. Perform the analysis of the fluorescent signal on a maximum projection image and analyze with software.

4 Notes

1. In order to evaluate the selectivity of modified ligands, fluorescent ligands should be also tested on a NMDA-induced calcium influx in HEK 293 transfected with GluN2A/GluN1-1a.
2. The NMDA-induced calcium influx should be stable at least for three consecutive responses before starting the NMDA receptor antagonist application.
3. Here DsRed is used as a marker of neuronal morphology.

Acknowledgment

This work was supported by the CNRS, CEA and UNICAEN for financial support.

References

1. Sanz-Clemente A, Nicoll RA, Roche KW (2013) Diversity in NMDA receptor composition: many regulators, many consequences. Neuroscientist 19:62–75
2. Traynelis SF, Wollmuth LP, McBain CJ, Menniti FS, Vance KM, Ogden KK, Hansen KB, Yuan H, Myers SJ, Dingledine R (2010) Glutamate receptor ion channels: structure, regulation, and function. Pharmacol Rev 62:405–496
3. Laube B, Kuhse J, Betz H (1998) Evidence for a tetrameric structure of recombinant NMDA receptors. J Neurosci 18:2954–2961
4. Loftis JM, Janowsky A (2003) The N-methyl-D-aspartate receptor subunit NR2B: localization, functional properties, regulation, and clinical implications. Pharmacol Ther 97:55–85
5. Paoletti P (2011) Molecular basis of NMDA receptor functional diversity. Eur J Neurosci 33:1351–1365
6. Papouin T, Oliet SHR (2014) Organization, control and function of extrasynaptic NMDA receptors. Philos Trans R Soc Lond B Biol Sci 369(1654):20130601
7. Cull-Candy SG, Leszkiewicz DN (2004) Role of distinct NMDA receptor subtypes at central synapses. Sci STKE 2004(255):re16
8. Hardingham GE, Bading H (2010) Synaptic versus extrasynaptic NMDA receptor signalling: implications for neurodegenerative disorders. Nat Rev Neurosci 11:682–696
9. Parsons MP, Raymond LA (2014) Extrasynaptic NMDA receptor involvement in central nervous system disorders. Neuron 82:279–293
10. Sousa JB, Frescol P, Diniz C (2014) Imaging receptors with Laser Scanning Confocal Microscopy: qualitative and quantitative analysis. In: Méndes-Vilas A (ed) Microscopy: advances in scientific research and education, Microscopy book series #6, vol 1. Formatex

Research Center Publisher, Badajoz, Spain, pp 201–208
11. McGrath JC, Arribas S, Daly CJ (1996) Fluorescent ligands for the study of receptors. Trends Pharmacol Sci 17:393–399
12. Masuko T, Kashiwagi K, Kuno T, Nguyen ND, Pahk AJ, Fukuchi J, Igarashi K, Williams K (1999) A regulatory domain (R1-R2) in the amino terminus of the N-methyl-D-aspartate receptor: effects of spermine, protons, and ifenprodil, and structural similarity to bacterial leucine/isoleucine/valine binding protein. Mol Pharmacol 55:957–969
13. Perin-Dureau F, Rachline J, Neyton J, Paoletti P (2002) Mapping the binding site of the neuroprotectant ifenprodil on NMDA receptors. J Neurosci 22:5955–5965
14. Mony L, Krzaczkowski L, Leonetti M, Le Goff A, Alarcon K, Neyton J, Bertrand H-O, Acher F, Paoletti P (2009) Structural basis of NR2B-selective antagonist recognition by N-methyl-D-aspartate receptors. Mol Pharmacol 75:60–74
15. Ng F-M, Geballe MT, Snyder JP, Traynelis SF, Low C-M (2008) Structural insights into phenylethanolamines highaffinity binding site in NR2B from binding and molecular modeling studies. Mol Brain 1:1–11
16. Reynold IJ, Miller R (1989) Ifenprodil is a novel type of N-methyl-D-aspartate receptor antagonist: interaction with polyamines. Mol Pharmacol 36:758–765
17. Williams K (2001) Ifenprodil, a novel NMDA receptor antagonist: site and mechanism of action. Curr Drug Targets 2:285–298
18. Borza I, Domany G (2006) NR2B selective NMDA antagonists: the evolution of the ifenprodil-type pharmacophore. Curr Top Med Chem 6:687–695
19. Layton ME, Kelly MJ, Rodzinak KJ (2006) Recent advances in the development of NR2B subtype-selective NMDA receptor antagonists. Curr Top Med Chem 6:697–709
20. MacCauley JA (2006) Amide-containing NR2B/NMDA receptor antagonists. Expert Opin Ther Patents 16:863–870
21. MacCauley JA (2005) NR2B subtype-selective NMDA receptor antagonists: 2001–2004. Expert Opin Ther Patents 15:389–407
22. Nikam SS, Meltzer LT (2002) NR2B selective NMDA receptor antagonists. Curr Pharm Des 8:845–855
23. Rosahl TW, Wingrove PB, Hunt V, Fradley RL, Lawrence JMK, Heavens RP, Treacey P, Usala M, Macauley A, Bonnert TP, Whiting PJ, Wafford KA (2006) A genetically modified mouse model probing the selective action of ifenprodil at the N-methyl-D-aspartate type 2B receptor. Mol Cell Neurosci 33:47–56
24. Grimwood S, Richards P, Murray F, Harrison N, Wingrove PB, Hutson PH (2000) Characterisation of N-methyl-Daspartate receptor-specific [3H]ifenprodil binding to recombinant human NR1a/NR2B receptors compared with native receptors in rodent brain membranes. J Neurochem 75:2455–2463
25. Avenet P, Leonardon J, Besnard F, Graham D, Frost J, Depoortere H, Langer SZ, Scatton B (1996) Antagonist properties of the stereoisomers of ifenprodil at NR1A/NR2A and NR1A/NR2B subtypes of the NMDA receptor expressed in Xenopus oocytes. Eur J Pharmacol 296:209–213
26. Marchand P, Becerril-Ortega J, Mony L, Bouteiller C, Paoletti P, Nicole O, Barré L, Buisson A, Perrio C (2012) Confocal microscopy imaging of NR2B containing NMDA receptor using fluorescent ifenprodil-based conjugates. Bioconjug Chem 23:2126
27. Dhilly M, Becerril-Ortega J, Colloc'h N, MacKenzie ET, Barré L, Buisson A, Nicole O, Perrio C (2013) Synthesis and in vitro characterization of ifenprodil-based fluorescein conjugates as GluN1/GluN2B N-Methyl-D-Aspartate receptor antagonists. Chembiochem 14:759769
28. Dingledine R, Borges K, Bowie D, Traynelis SF (1999) The glutamate receptor ion channels. Pharmacol Rev 51:7–61
29. Sim N, Gottschalk S, Pal R, Delbianco M, Degtyaruk O, Razansky D, Westmeyer GG, Ntziachristos V, Parker D, Mishra A (2015) Wavelength-dependent optoacoustic imaging probes for NMDA receptor visualization. Chem Commun 51:15149–15152
30. Léveillé F, El Gaamouch F, Gouix E, Lecocq M, Lobner D, Nicole O, Buisson A (2008) Neuronal viability is controlled by a functional relation between synaptic and extrasynaptic NMDA receptors. FASEB J 22:4258–4271

Chapter 10

Design of Light-Sensitive NMDARs by Genetically Encoded Photo-Cross-Linkers

Meilin Tian and Shixin Ye

Abstract

Genetic code expansion exploiting unnatural amino acids (Uaas) is a powerful technique to create novel protein function in vivo. Here, we provide a protocol for the incorporation of two UV-sensitive crosslinking Uaas into NMDA receptors (NMDARs), a type of glutamate-gated ion channels mediating fast synaptic transmission. Through heterologous expression in *Xenopus laevis* oocytes, we have identified light-sensitive NMDARs of GluN2B subtype by using the two-electrode voltage electrophysiology measurement in combination with online-UV application. Immunoblotting analysis has been used to confirm inter-subunit crosslinking.

Key words Genetic code expansion, Photo-cross-linkers, *p*-Azido-L-phenylalanine, *Xenopus laevis* oocytes

1 Introduction

The design of protein receptors responsive to light stimuli has wide-ranging applications from structure-function studies to pharmacology. An attractive strategy to engineer light responsiveness is the site-specific incorporation of photoreactive unnatural amino acids (Uaas)Amino acids into a targeted protein [5–9]. The methodology relies on the read-through of an unassigned codon (commonly the amber stop codon) in an mRNA by a suppressor tRNA aminoacylated with a desired UAA (Fig. 1a). The development of genetically engineered orthogonal suppressor tRNA/aminoacyl-tRNA synthetase pairs with altered amino acid specificity allowed for aminoacylation in the expression system in situ [1]. This method provided a major step forward by advancing the Uaa technology to the all-genetic-based level, also known as "the genetic code expansion" [2, 3, 4].

We have applied the genetic code expansion technique to engineer light-sensitive NMDA receptors (NMDARs), which are obligatory heterotetramers commonly composed of two glycine-binding

Nail Burnashev and Pierre Szepetowski (eds.), *NMDA Receptors: Methods and Protocols*, Methods in Molecular Biology, vol. 1677, DOI 10.1007/978-1-4939-7321-7_10, © Springer Science+Business Media LLC 2017

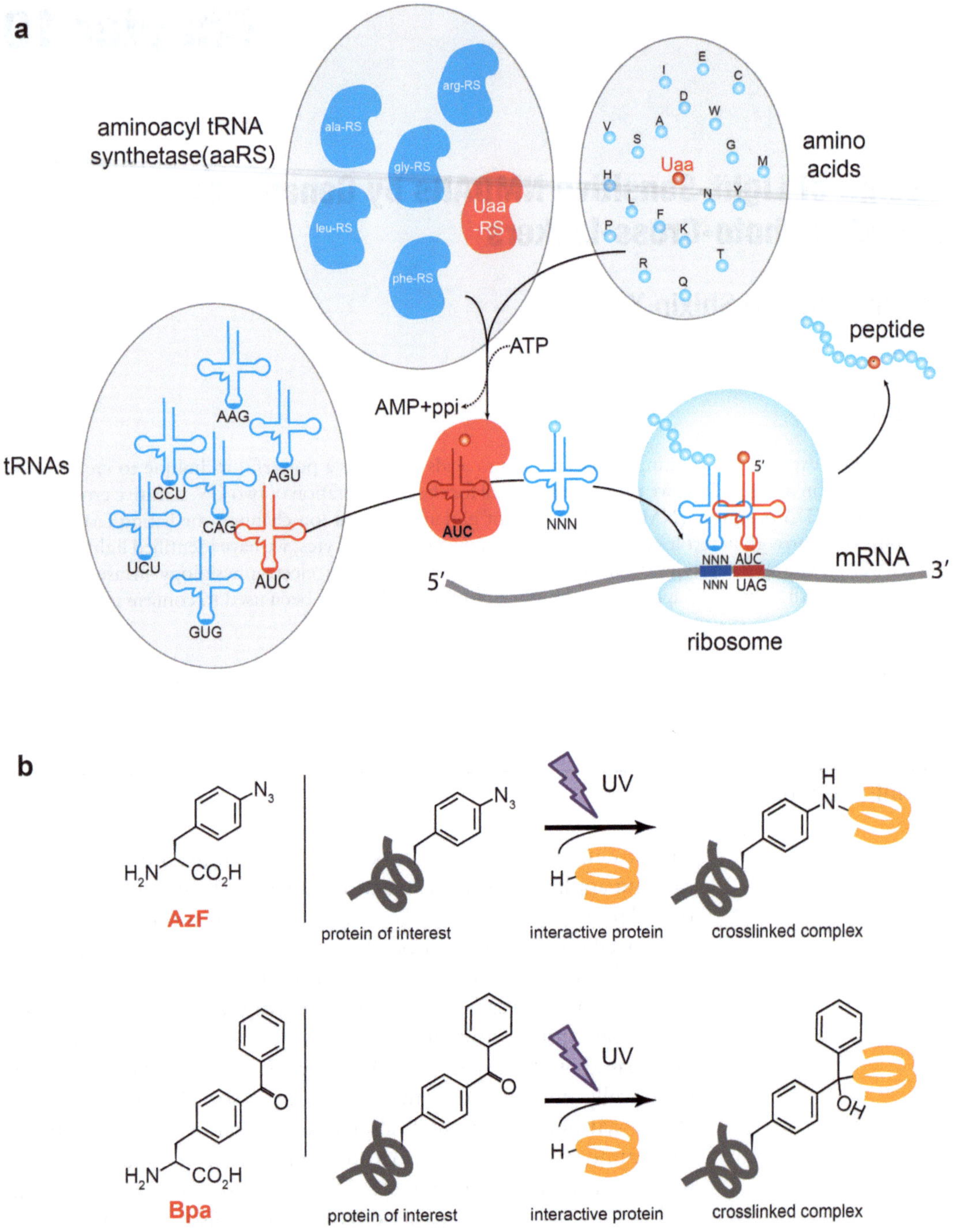

Fig. 1 Schematic illustration of the genetic code expansion. (**a**) The exogenously added orthogonal aa-RS, suppressor tRNA and the UAA are colored in *red*. The endogenous aa-RS, tRNA and amino acids are colored in *blue*. The orthogonal aa-RS can aminoacylate the suppressor tRNA with the desired Uaa in situ. The resulted aminoacylated suppressor tRNA participates in the protein translation and read-through an amber stop codon in the mRNA. The translated full-length protein polypeptide chain with the UAA incorporates at the location where the amber stop codon is introduced. (**b**) Bpa and AzF serve as photo-crosslinkers. When exposed to UV light, Bpa and AzF generate bi-radicals and form covalent linkages with nearby side-chains

GluN1 subunits and two glutamate-binding GluN2 subunits [10, 11]. Although GluN1 is encoded by a single gene, there are four subtypes of GluN2 (GluN2A, 2B, 2C, and 2D) encoded by four different genes, which endow NMDARs with different properties including channel open probabilities (P_o) and sensitivities to allosteric modulators [11].

This work presents the incorporation of two photo-crosslinker Uaas into NMDARs of GluN2B subtype (Fig. 2a). By analyzing

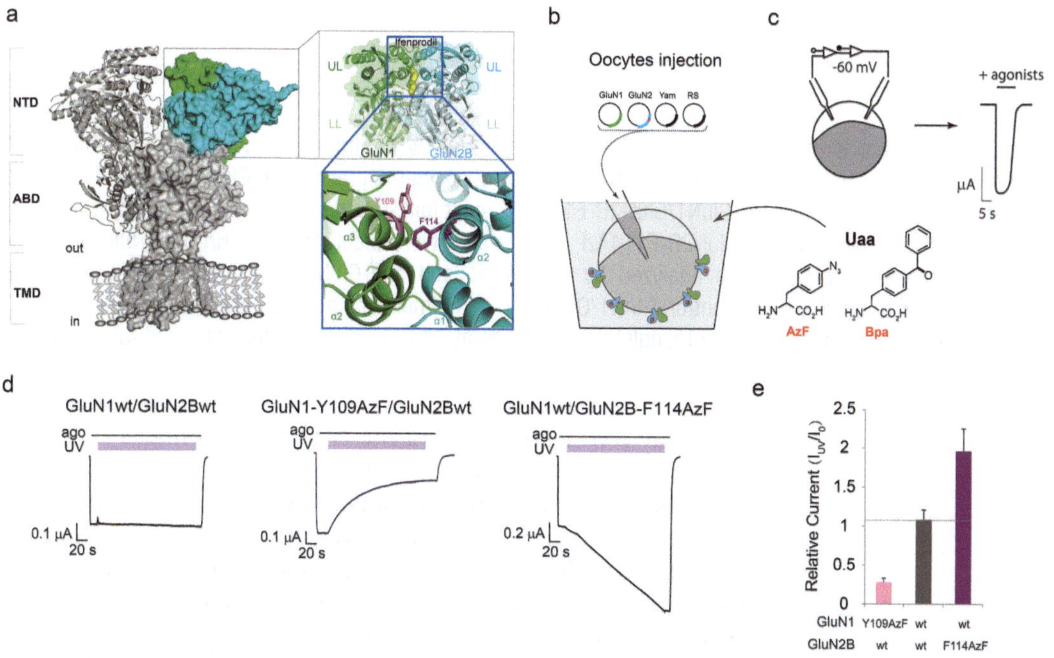

Fig. 2 Incorporation of a photo-cross-linking Uaa at the NTD dimer interface. (**a**) Full-length GluN1/GluN2B receptor crystal structure reveals a tetrameric complex assembled as a dimer of dimers [21, 22]. Side view with ifenprodil molecule omitted from clarity (PDB 4PE5). There are three major domains: N-terminal domain (NTD) that harbors several allosteric modulators; agonist binding domain (ABD) that binds glycine (or D-serine) in GluN1 and glutamate in GluN2 subunits; and transmembrane domain (TMD) that comprises the ion-channel pore. These three domains NTD, ABD, and TMD are arranged in layers. One NTD dimer in complex with ifenprodil (*yellow sphere*) is highlighted. Helices from the upper-lobe upper-lobe (UL-UL) interface (helix α2 & α3 of GluN1, and helix α1 & α2 of GluN2B) are enlarged. Residues in the GluN1-Y109 (*pink*) and GluN2B-F114 (*purple*) subjected to amber mutation are represented as colored sticks. (**b**) Oocytes co-injected with four plasmids: wt GluN1, GluN2 amber mutant, Yam and AzF-RS. For each injection, oocytes were spitted into two batches: one batch was incubated in the medium with 1 mM Uaa supplemented into the medium. For each condition, 6–52 oocytes were tested, currents >10 nA were plotted; the other batch was incubated in the absence of AzF. At least five oocytes were tested. (**c**) The current responses to co-agonists glutamate and glycine (100 μM Glu/Gly) at −60 mV were measured on the second to fourth day after co-injection. (**d**) Representative current traces measured from oocytes expressing wt and AzF mutant receptors during UV illumination. Two mutants (GluN1-Y109AzF/GluN2B and GluN1/GluN2B-F114AzF) have UV sensitivity, with GluN1-Y109 mutant showing UV-induced inhibition whereas GluN2B-F114AzF showing UV-induced potentiation. (**e**) Relative currents (I_{UV}/I_0) measured on oocytes expressing GluN2B receptors of AzF mutants: GluN2Bwt (1.08 ± 0.12, $n = 26$), GluN1-Y10AzF (0.28 ± 0.05, $n = 18$), GluN2B-F114AzF (1.97 ± 0.29, $n = 21$)

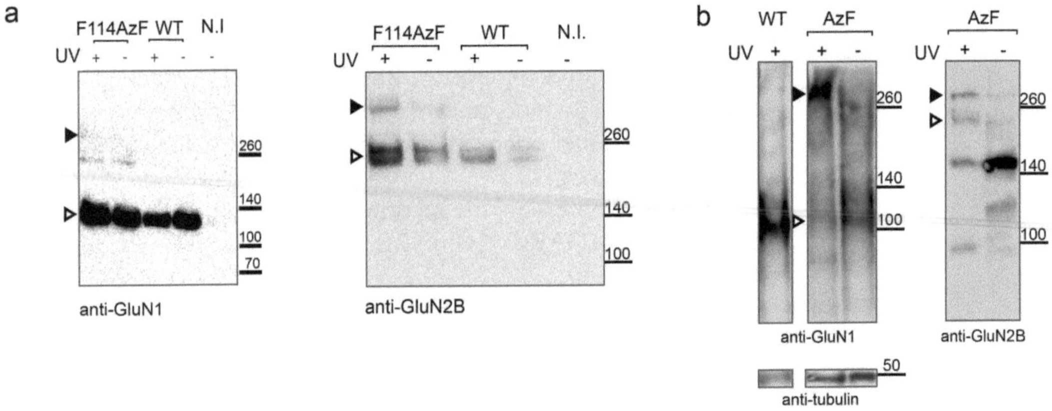

Fig. 3 Heterodimer formation of GluN1/GluN2B-F114AzF after UV treatment detected by immunoblots from *Xenopus* oocytes. Immunoblots detected from cell lysates of oocytes expressing (**a**) GluN1/GluN2B-F114AzF or wild-type receptors and (**b**) HA-GluN1/GluN2B-F114AzF receptors. Oocytes were either treated with (+) or without (−) UV. For HA-GluN1/GluN2B-F114AzF, cell lysates were incubated with HA-resin and eluted for western analysis. Samples were analyzed by anti-GluN1 and anti-GluN2B antibodies. Anti-tubulin was detected as an internal loading control. GluN1/GluN2B heterodimer runs at ~290 kDa (solid triangle), GluN1 monomer runs at ~180 kDa (empty triangle), GluN2B runs at ~110 kDa (empty triangle), and tubulin runs at ~50 kDa

static x-ray crystal structures of the N-terminal domains (NTD) of the GluN1/GluN2B receptors (Fig. 2a), we introduced photo-cross-linking *p*-azido-L-phenylalanine (AzF) or *p*-benzoyl-L-phenylalanine (Bpa) site-specifically into at the NTD dimer interface (Figs. 1b, 2b). This protocol takes advantage of the recent development of the genetic code expansion in *Xenopus laevis* oocytes [10], which is a classical vehicle for heterologous expression and functional characterization of ligand-gated ion channels. We have applied functional characterizations with two-electrode voltage clamp electrophysiology (Fig. 2c) and online UV illumination (Fig. 2d). We have successfully identified two UV-sensitive GluN2B mutants (Fig. 2d, e): GluN1-Y109AzF/GluN2B (UV inhibition) [7] and GluN1/GluN2B-F114AzF (UV potentiation) [9]. For the GluN1/GluN2B-F114AzF mutant, inter-subunit crosslinking has been detected by the immunoblotting (Fig. 3).

2 Materials

All water purity should be 18 MΩ-cm grade. Filtration of solutions through 0.22 μm filter is required for room temperature storage.

2.1 Solutions and Instrumentation

1. Oocytes incubation Barth's solution: 88 mM NaCl (17.6 mL of 5 M stock), 1 mM KCl (1 mL of 1 M stock), 0.33 mM Ca $(NO_3)_2$, (0.33 mL of 1 M stock), 0.41 mM $CaCl_2$ (0.41 mL of 1 M stock), 0.82 mM $MgSO_4$, (0.82 mL of 1 M stock), 2.4 mM $NaHCO_3$, (2.4 mL of 1 M stock), 10 mM 4-(2-hydroxyethyl)-

1-piperazineethanesulfonic acid (HEPES) (10 mL of 1 M stock), fill ddH_2O to ~0.95 L. Adjust pH to 7.6 with 1 M NaOH, add ddH_2O to 1 L, store in a 18 °C fridge that oocytes are kept.

2. 1000× gentamicin stock: 500 mg gentamicin sulfate dissolve in 10 mL ddH_2O. Aliquot to store in −20 °C (*see* **Note 1**).
3. 1000× D-(−)-2-Amino-5-phosphonopentanoic acid (D-APV) stock: 98.6 mg D-APV in 10 mL ddH_2O. Aliquot to store in −20 °C (*see* **Note 1**).
4. 1000× diethylenetriaminepentaacetic acid (DTPA) stock: 1.11 g DTPA dissolve in 20 mL ddH_2O. Add 35% NaOH dropwise until DTPA completely dissolved. Adjust pH to 7.4 with NaOH. Dilute 10 fold to obtain 10 mM stock and aliquot to store in −20 °C (*see* **Note 1**).
5. Barth solution with AzF: 14.5 mL Barth, 15 μL 1000× gentamicin, 15 μL 1000× D-APV, 1.5 mL 10× AzF stock (*see* **Note 2**).
6. Electrophysiology Ringer solution: 100 mM NaCl, 5 mM HEPES, 0.3 mM $BaCl_2$, pH 7.3. Weigh 5.84 g of NaCl, 1.2 g of HEPES, 300 μL of 1 M $BaCl_2$ stock, 2.5 mL of 1 M KOH. Fill ddH_2O to ~0.95 L. Adjust pH to 7.3 with HCl, add ddH_2O to 1 L. Add 1 mL 1000× DTPA (*see* **Note 1**).
7. 1000× agonists stock solutions: 0.169 g of L-Glutamic acid monosodium salt (Glu) in 10 mL of ddH_2O, 0.075 g of Glycine (Gly) in 10 mL of ddH_2O to make 100 mM stock. Aliquot to store in −20 °C.
8. Ringer solution with 100 μM agonists: add 100 μL of 100 mM Glu stock and 100 μL of 100 mM Gly stock to 100 mL Ringer solution.
9. Two electrode voltage clamp (TEVC) apparatus major components: (1) Two micro-manipulators, each one includes a motorized linear stage and a manual XY stage. (2) One stereomicroscope. (3) One perfusion chamber and platform. (4) One PCI data acquisition board with BNC interface. (5) LED light source 105 mW (42 mW/cm^2) for white light and UV illumination of oocytes (*see* **Note 3**).

2.2 Plasmids Stocks and Uaas

1. pcDNA3.1 RS plasmid: contains the RS gene orthogonal for the Uaa of interest and ampicillin resistance gene (*see* **Note 4**).
2. pSVB.Yam plasmid: contains the orthogonal amber suppressor tRNA in response to the Uaa of interest and RS, and ampicillin resistance gene (*see* **Note 5**).
3. Overexpression pcDNA3.1 plasmids containing individual NMDAR GluN1 and GluN2B subunit, and ampicillin resistance gene (*see* **Note 6**).

4. QuikChange site-directed mutagenesis kit (*see* **Note 7**).
5. Nanoject auto-nanoliter injector.
6. 10× AzF stock, 10 mM. AzF M.W. = 206.1 g/mol. 20.6 mg dissolve in 10 mL Barth solution. Aliquot to store in −20 °C (*see* **Note 8**).
7. 100× Bpa stock, 100 mM. Bpa M.W. = 269.3 g/mol. 26.7 mg dissolve in 1 mL Barth solution. Aliquot to store in −20 °C (*see* **Note 9**).

2.3 SDS Polyacrylamide Gel

1. Precast or home-made protein gels: NuPAGE 3–8% Tri-Acetate Midi gel, 1.0 mm, 15-well.
2. SDS-PAGE running buffer (5×): 0.5 M Tris–HCl, pH 8.3. Weigh 60 g of Tris–HCl, 288 g of Glycine, 20 g of SDS. Fill ddH_2O to 4 L.
3. SDS lysis buffer: 0.02 mM Tris–HCl (pH 8.0), 50 mM NaCl, 1% n-Dodecyl ß-D-maltoside. In 10 mL of lysis buffer, add one tablet of proteases inhibitors cocktail.
4. Loading buffer (5×): 0.2 M Tris–HCl (pH 6.8), 10% SDS, 10% ß-mercaptoethanol, 0.1% bromophenol blue, 30% glycerol. Leave one aliquot at 4 °C for current use and store the remaining aliquots at −20 °C.

2.4 Immunoblotting

1. Polyvinylidine difluoride (PVDF) membranes.
2. Western blot transfer buffer: 0.025 M Tris–HCl, 0.192 M glycine, 0.025% SDS, 10% methanol.
3. Extra-thick filter papers for protein blot transfer.
4. 10× Tris-buffered saline (TBS): 1.5 M NaCl, 0.1 M Tris–HCl, pH 7.4.
5. TBS containing 0.15% Tween-20 (TBST).
6. Blocking solution: 5% milk in TBS. Store at 4 °C.
7. Plastic container.

2.5 Antibodies

1. Primary antibodies: Anti-GluN1, GluN2B and tubulin antibodies (*see* **Note 10**).
2. Goat peroxidase-conjugated anti-mouse secondary antibody.
3. BenchMark prestained molecular weight standards.

3 Methods

3.1 Mutant Plasmid Preparation

1. To the gene of interest (GluN1, and GluN2B in a pcDNA3.1 vector), introduce an amber codon (TAG) in the location that Uaa is desired through QuikChange.

2. Midi or Maxiprep plasmids and prepare DNA stocks at ~0.5–1 μg/mL concentration. Dilute DNAs with standard Tris-EDTA (TE) pH 8.0 buffer: GluN1-wt or GluN1-amb (80 ng/μL), GluN2B-wt or GluN2B-amb (80 ng/μL), Yam (10 ng/μL), AzF-RS (2.5 ng/μL), and Bpa-RS (5 ng/μL).

3.2 Expressing Mutants

1. Obtain freshly prepared *Xenopus laevis* oocytes (*see* **Note 11**).
2. To express recombinant wt NMDARs in oocytes, perform nuclear injection of 36 nL of a mixture cDNAs encoding GluN1 and GluN2 at 15 ng/μL for each subunit. For Uaa containing mutants (Fig. 2b), a 36 nL of a mixture of cDNAs containing: (1) GluN1 (20 ng/μL), (2) GluN2B-amb mutant (20 ng/μL), (3) Yam (2.5 ng/μL), and (4) AzF-RS (0.6 ng/μL) or Bpa-RS (1.25 ng/μL) was used for injection.
3. After injection, incubate oocytes at 19 °C in the Barth solution supplemented with 50 ng/mL gentamicin sulfate, 50 μM D-APV, and 1 mM AzF or 1 mM Bpa (*see* **Note 12**).

3.3 Functional Assay

1. Measure the NMDAR-mediated currents with the wild-type GluN1/GluN2B control (*see* **Note 13**). To activate the receptors at the maximum level, perfuse co-agonist solution (Ringer solution with 100 μM glutamate and 100 μM glycine, *see* **Note 14**) to the oocytes (*see* **Note 15**). To deactivate the receptors, perfuse regular Ringer solution. Record and measure currents at a holding potential of −60 mV at room temperature (*see* **Note 16**).
2. Apply UV illumination online during the TEVC measurement. If mutants are sensitive to UV light, currents can be reduced or enhanced due to the UV light treatment. Such functional assays enable the identification of light-sensitive mutants due to the incorporation of Uaa (*see* **Note 17**).

3.4 UV Photo-Cross-Linking Treatment

1. Transfer oocytes expressing wt NMDA receptors or AzF mutant receptors to a 96-well plate containing Barth solution free of AzF and agonists (one oocyte per well, animal pole facing up) on ice.
2. Irradiate cells for 30 min with a hand-held UV lamp placed on the top of the plate. After UV treatment, cells are subjected to immunoblotting assays.

3.5 Immunoblotting

Carry out all the procedures on ice unless otherwise specified.

1. Lyse oocytes. Transfer oocytes into 1.5 mL Eppendorf tubes and remove Barth solution as much as possible. Immediately add lysis buffer into the tubes (10 μL/oocyte). Smash and homogenize the oocytes with a handheld homogenizer.

2. Incubate lysates for an hour on ice. Centrifuge lysates at 4 °C and 20,000 × *g* for 10 min. Resuspend the pellet with the pipette and centrifuge at 4 °C and 20,000 × *g* for 10 min. The supernatant was saved and can be stored at −80 °C.
3. Perform immunoprecipitation of Hemagglutinin (HA)-tagged GluN1 by incubation of the lysates with anti-HA-bound beads (anti-HA-Agarose) and elute following the manufacturer's manual (*see* **Note 18**).
4. Add loading buffer to the lysate at the volume ratio of 1:2 (loading buffer: lysate). Heat samples at 65 °C for 5 min. Short spin on a benchtop centrifuge to bring down the condensate.
5. SDS-PAGE gel electrophoresis. Load samples (40 μL/well) onto 12-well NuPAGE 3–8% Tri-Acetate Midi gels with the same order repeated three times to be detected later for anti-GluN1, anti-GluN2B, and anti-tubulin antibody detection. Load the protein ladder for each gel (2 μg). Electrophorese at 90–110 V for 10–15 min until samples have entered the gel and concentrated, then continue at 130 V till the dye front (from the BPB dye in the samples) reached the bottom of the gel.
6. Following electrophoresis, pry the gel plates open with the use of a spatula. The gel remains on one of the plastic plates. Remove the stacking gel. Rinse the gel with water and transfer gently into a container with transfer buffer.
7. Cut a PVDF membrane to fit the size of the gel and immerse in methanol. Rinse twice with water and once in transfer buffer.
8. Immerse thick filter papers with transfer buffer. Lay one thick filter paper on the semi-dry transfer device. Place the membrane on the filter paper, gently place the gel onto the membrane and then followed by another filter paper. Pour transfer buffer onto the sandwich and drive away bubbles. Transfer at 0.18 A for 1 h. (0.36 A for two gels and follow the instruction of the device).
9. Immediately following transfer, gently dissemble the sandwich and transfer the membranes to 5% milk in TBST. Block the membranes for 1 h at room temperature.
10. Incubate membranes with primary antibody overnight at 4 °C. The following antibodies were diluted in 5% milk in TBST: anti-GluN1 (1:750 dilution), anti-GluN2B (1:500 dilution), and anti-α-tubulin antibody (1:1000 dilution).
11. Wash membranes 5× with TBST, 10 min each time.
12. Add anti-goat peroxidase-conjugated secondary antibody (1:20,000 dilution) and incubate for 1 h at room temperature.
13. Wash as in **step 11**.
14. Visualize protein bands with chemiluminescent substrate. Detect the chemiluminescent signal with a gel imager.

4 Notes

1. D-APV, gentamicin, and DTPA are prepared in stock solution and stored in aliquot at −20 °C. DTPA is added to Ringer solution to chelate trace amounts of contaminant Zn^{2+} [12, 13].
2. Although a potential problem associated with the reduction of AzF has been reported [14], our oocytes experiments involved in expression, functional characterization, and UV-crosslinking have not been involved with any reducing reagents (such as dithiotheritol, glutathione, and 2-mercapthanol). Therefore, no significant degradation of aryl-azide to aryl-amine has been noticed.
3. Online light treatment the pE-300^{white} (CoolLED) is a compact and simple-to-use LED illumination unit. It provides a broad spectrum of illumination.
4. The orthogonal RS was derived from *E. coli* Tyr-RS, which was originally screened in yeast [2]. These RS mutants have high aminoacyl activity toward tyrosine analogues. Their activities were later validated in mammalian HEK293 cells [15] and *Xenpous* oocytes [10]. For the incorporation of AzF, AzF-RS (Y37L/D182S/F183 M/L186A) was used [16, 17], and for the incorporation of Bpa, Bpa-RS (Y37G/D182G/L186A) was used [15].
5. The expression vector carrying an amber suppressor tRNA was described previously [15]. In brief, the tRNA was derived from the *B. stearothermophilus* tRNA^{Tyr} (*Bst*-Yam) was designed with the following sequence:

   5′AGCGCTCCGGTTTTTCTGTGCTGAACCTCAGGGG-ACGCCGACACACGTACACGTC**GGAGGGGTAGC-GAAGTGGCTAAACGCGGCGGACTC-TAAATCCGCTCCCTTTGGGTTCGGCGGTTC-GAATCCGTCCCCCTCCA***GTCCTTTTTTTG*-3′. It consists of a (1) 5′ leader sequence taken from the human tRNA^{Tyr} gene (underlined); (2) the *B. stearothermophilus* tRNA^{Tyr} gene sequence (bold) in which the anticodon was changed to CUA (bold underlined) and the 3′-terminal CCA was omitted; and (3) a 3′-terminal region containing a transcription termination sequence (italics).
6. The pcDNA3-based expression plasmids for rat GluN1–1a and mouse GluN2B have been described previously [18].
7. Any method of site-directed mutagenesis is acceptable, but the Stratgene QuikChange kit has been successfully used.

8. Sonification is required for complete dissolution of AzF. It is best to prepare 1 mM AzF solution fresh each time. AzF powder can be stored at −20 °C for a few years.
9. By adding equal molarity of NaOH to achieve complete dissolution, the stock solution remains dissolved without obvious precipitation.
10. We used anti-GluN1 (mouse monoclonal MAB1586 clone R1JHL; Millipore), anti-GluN2B antibody (mouse monoclonal 75–101 clone N59/36; NeuroMab) and anti-α-tubulin antibody (mouse monoclonal DM1A clone, Upstate).
11. Freshly prepared oocytes with reliable quality can be obtained with commercial sources. They can be also prepared with surgery on *Xenopus* frogs based on previous reports [10, 18].
12. To be confident that Uaa is incorporated at a specific site where the amber stop codon is introduced, a negative control for oocytes incubated in the absence of Uaa is often performed.
13. We typically screen all oocytes incubated in the presence of AzF. Ideally, we should have at least 50% oocytes expressing mutants 2 ~ 4 days postinjection. However, we do experience variability due to the manual injection and different batches of oocytes. All mutants we have tested have current values between 1 and 10 μA, demonstrating that all sites allow the expression of full-length proteins by the incorporation of AzF in response to the amber stop codon. We also test at least ten oocytes incubated in the absence of AzF and record the current values. We have observed that majority of oocytes have 0 μA currents 2–4 days postinjection, although few oocytes have current larger than 0.1 μA which we consider a leakage due to the mis-incorporation of natural amino acid. We have noticed that such mis-incorporation has a lower chance using Bpa-RS than the AzF-RS. By carefully testing the gene dosages of the suppressor tRNA and RS, the leakage problem can be minimized [10].
14. Ba^{2+}, rather than Ca^{2+}, was used as the divalent cation in the Ringer solution [19]. It has been well documented that long exposure of agonists could still cause secondary activation of Ca^{2+} activated Cl^- current.
15. Homodimerization of GluN1 causing "homomeric GluN1 currents" can be a specific issue with Xenopus oocytes expression system. In this situation, when injecting only GluN1 into oocytes, it may dimerize with endogenously expressed Xenopus GluN2 to generate functional receptor channels (something that is not seen in mammalian HEK293 cells). However, this is a very rare event, likely because endogenous Xenopus

GluN2 subunits are expressed at very low levels. When it does happen, the currents generated by "homomeric GluN1 receptors" are tiny (at least 50-fold smaller current amplitude than overexpressed GluN1/GluN2 currents). We perform functional test on different batches of oocytes to minimize this "homomeric GluN1 currents."

16. Unlike overexpression of receptors using mRNA injection which generate stable current, the amount of functional proteins at the membrane from DNA injection increases over time and does not saturate to generate stable current. We have noticed that the current measured on the same oocyte may change in the range of 1–5 μA, depending on the batches of oocytes.
17. Our functional assays cannot distinguish nonspecific effects due to the incorporation of Uaa into the whole proteome. However, introducing an amber suppressor tRNA into oocytes and its proteomic consequence has been verified by H. Grosjean and coworkers [20]. Using 2-dimentional electrophoresis, they have observed 57 proteins (out of 600 proteins being characterized) changed their size due to the suppression of naturally occurring amber codons when providing an amber suppressor tRNA in oocytes. This shows that only a minority of proteins is affected by the read-through of amber codons. We note that such a whole proteomic change due to the genetic code expansion approach may apply to any kind of cellular expression systems.
18. Convincing evidence should be demonstrated with western blots showing crosslinked bands with both anti GluN2B and GluN1 antibodies. When using cell lysate directly, which limited the loading capacity to four oocytes/lane, it was difficult to observe a dimer band with the anti-GluN1 antibody due to the low yield of photo-cross-linking on receptors expressed in oocytes. The UV light does not penetrate the opaque cell bodies of oocytes; therefore, only those receptors on the plasma membranes were fully exposed to the UV radiation. We managed to observe a heterodimer band using both GluN1 and GluN2B antibodies (Fig. 3a), although for the anti-GluN1 antibody (Fig. 3a, left panel), the intensity of the heterodimer band is much weaker than the GluN2B (Fig. 3a, right panel). The detection is much improved when implementing the HA- immunoprecipitation. Heterodimer band at the expected molecular weight (solid triangle) was detected using both GluN1 and GluN2B antibodies (Fig. 3b). Since the HA-tag is on the GluN1, we expected that only the GluN2B subunit that is crosslinked to the GluN1 could be

pulled down and detected. We reasoned that some GluN2B monomer and lower molecular weight proteins were present in the IP samples and can be detected by the available anti-GluN2B antibody.

Acknowledgments

We thank P. Paoletti, S. Zhu, S. Carvahol, and M. Riou for pilot experimental works. We also thank Professor Thomas P. Sakmar for the generous gift of Yam and AzF-RS construct, and Dr. Philippe Rondard at the University of Montpellier for the generous gift of the HA-GluN1 construct. We are grateful to the Proteomic Resource Center at the Curie Institute for providing technical resources. Financial support was provided by the Chinese Scholars Council (CSC fellowship to M.T.) and the Agence Nationale de la Recherche of France (ANR-JCJC grant to S.Y.). Other financial support was provided by National Natural Science Foundation of China (31528007 to S.Y.).

References

1. Wang L, Brock A, Herberich B, Schultz PG (2001) Expanding the genetic code in *E. coli*. Science 292:498–500
2. Chin JW et al (2003) An expanded eukaryotic genetic code. Science 301:964–967
3. Parrish AR, Wang W, Wang L (2006) Manipulating proteins for neuroscience. Curr Opin Neurobiol 16:585–592
4. Chin JW (2014) Expanding and reprogramming the genetic code of cells and animals. Annu Rev Biochem 83:379–408
5. Miller JC, Silverman SK, England PM, Dougherty DA, Lester HA (1998) Flash decaging of tyrosine sidechains in an ion channel. Neuron 20:619–624
6. Kang JY et al (2013) In vivo expression of a light-activatable potassium channel using unnatural amino acids. Neuron 80:358–370
7. Zhu S et al (2014) Genetically encoding a light switch in an ionotropic glutamate receptor reveals subunit-specific interfaces. Proc Natl Acad Sci U S A 111:6081–6086
8. Klippenstein V, Ghisi V, Wietstruk M, Plested AJ (2014) Photoinactivation of glutamate receptors by genetically encoded unnatural amino acids. J Neurosci 34:980–991
9. Tian M, Ye S (2016) Allosteric regulation in NMDA receptors revealed by the genetically encoded photo-cross-linkers. Sci Rep 6:34751
10. Ye S, Riou M, Carvalho S, Paoletti P (2013) Expanding the genetic code in *Xenopus laevis* oocytes. Chembiochem 14:230–235
11. Paoletti P, Bellone C, Zhou Q (2013) NMDA receptor subunit diversity: impact on receptor properties, synaptic plasticity and disease. Nat Rev Neurosci 14:383–400
12. Paoletti P, Ascher P, Neyton J (1997) High-affinity zinc inhibition of NMDA NR1-NR2A receptors. J Neurosci 17:5711–5725
13. Rachline J et al (2005) The micromolar zinc-binding domain on the NMDA receptor subunit NR2B. J Neurosci 25:308–317
14. Staros JV, Bayley H, Standring DN, Knowles JR (1978) Duction of aryl azides by thiols: implications for the use of photoaffinity reagents. Biochem Biophys Res Commun 80:568–572
15. Ye S et al (2008) Site-specific incorporation of keto amino acids into functional G protein-coupled receptors using unnatural amino acid mutagenesis. J Biol Chem 283:1525–1533
16. Ye S, Huber T, Vogel R, Sakmar TP (2009) FTIR analysis of GPCR activation using azido probes. Nat Chem Biol 5:397–399
17. Coin I et al (2013) Genetically encoded chemical probes in cells reveal the binding path of urocortin-I to CRF class B GPCR. Cell 155:1258–1269

18. Gielen M, Retchless BS, Mony L, Johnson JW, Paoletti P (2009) Mechanism of differential control of NMDA receptor activity by NR2 subunits. Nature 459:703–707
19. Leonard JP, Kelso SR (1990) Apparent desensitization of NMDA responses in Xenopus oocytes involves calcium-dependent chloride current. Neuron 4:53–60
20. Bienz M, Kubli E, Kohli J, deHenau S, Huez G, Marbaix G, Grosjean H (1981) Usage of the three termination codons in a single eukaryotic cell, the *Xenopus laevis* oocyte. Nucleic Acid Res 9:3835–3850
21. Karakas E, Furukawa H (2014) Crystal structure of a heterotetrameric NMDA receptor ion channel. Science 344:992–997
22. Lee CH et al (2014) NMDA receptor structures reveal subunit arrangement and pore architecture. Nature 511:191–197

Part III

Analysis of NMDAR Function/Activity

Chapter 11

Gene Targeted Mice with Conditional Knock-In (-Out) of NMDAR Mutations

Rolf Sprengel, Ahmed Eltokhi, and Frank N. Single

Abstract

For the genetic alterations of NMDA receptor (NMDAR) properties like Ca^{2+}-permeability or voltage-dependent gating in mice and for the experimental analysis of nonsense or missense mutations that were identified in human patients, single nucleotide mutations have to be introduced into the germ line of mice (Burnashev and Szepetowski, Curr Opin Pharmacol 20:73–82, 2015; Endele et al., Nat Genet 42:1021–1026, 2010). This can be done with very high precision by the well-established method of gene replacement, which makes use of homologous recombination in pluripotent embryonic stem (ES) cells of mice. The homologous recombination at NMDAR subunit genes (*Grin*; for glutamate receptor ionotropic NMDAR subtype) has to be performed by targeting vectors, also called replacement vectors. The targeting vector should encode part of the gene for the NMDAR subunit, the NMDAR mutation, and a removable selection maker. In these days, the targeting vector can be precisely designed using DNA sequences from public databases. The assembly of the vector is then done from isogenic NMDAR gene fragments cloned in bacterial artificial chromosomes (BACs) using "high fidelity" long-range PCR reactions. During these PCR reactions, the NMDAR mutations are introduced into the cloned NMDAR gene fragments of the targeting vector. Finally, the targeting vector is used for homologous recombination in mouse ES cells. Positive ES cell clones which have the correct mutation have to be selected and are then used for blastocyst injection to generate chimeric mice that hopefully transmit the *Grin* gene targeted ES cells to their offspring. In the first offspring generation of the founder (F1), some animals will be heterozygous for the targeted NMDAR gene mutation. In order to regulate the expression of NMDAR mutations, it is important to keep the targeted NMDAR mutation under conditional control. Here, we describe a general method how those conditionally controlled NMDAR mutations can be engraved into the germ line of mice as hypomorphic *Grin* alleles. By breeding these hypomorphic *Grin* gene targeted mice with Cre recombinase expressing mice, the hypomorphic *Grin* allele can be activated at specific time points in specific cell types, and the function of the mutated NMDAR can be analyzed in these - so called - conditional mouse models. In this method chapter, we describe in detail the different methodical steps for successful gene targeting and generation of conditional NMDAR mutant mouse lines. Within the last 20 years, several students in our Department of Molecular Neurobiology in Heidelberg used these techniques several times to generate different mouse lines with mutated NMDARs.

Key words Gene targeting, *Grin1*, GluN1, Hypomorphic allele, Southern blot, Genotyping

Nail Burnashev and Pierre Szepetowski (eds.), *NMDA Receptors: Methods and Protocols*, Methods in Molecular Biology, vol. 1677, DOI 10.1007/978-1-4939-7321-7_11, © Springer Science+Business Media LLC 2017

1 Introduction

Recently, a reasonable number of mutations in genes for NMDAR subunits have been identified in human patients with severe neurological disorders [1, 2]. Correlative studies suggested a direct contribution of mutated NMDARs in the cognitive dysfunction of patients. However, the underlying developmental or neuronal deficiency cannot be studied in patients. For a detailed morphological, molecular, and cell network analysis, mutated NMDARs have to be studied in animal models. Thus, the mutation of interest has to be experimentally introduced into the genome of experimentally accessible animals. Currently, there are two technologies available that can be used to introduce a specific mutation into the genome of animals; gene targeting by homologous recombination and the CRISPR/Cas9 system. Both methods are based on the specific genetic manipulations of single embryonic cells. The most recent method is called the CRISPR/Cas9 system [3]. CRISPR stands for "Clustered Regularly Interspaced Short Palindromic Repeats" and Cas9 for "Endonuclease Cas9." In the CRISPR/Cas9 system, plasmids encoding the essential components of the CRISPR/Cas9 system are injected into the zygote. In the nucleus of the zygote the plasmid encoded small CRISPR-RNAs and Cas9 induce a sequence-specific dsDNA break with a subsequent DNA repair of the targeted gene on the chromosome. The sequence-specific cutting at the target site is determined by the sequence of the small guide CRISPR-RNA. In addition a single stranded oligonucleotide template used can be to repair the CRISPR/Cas9 targeting site by homologous recombination [4]. In this way, specific nucleotide exchanges are introduced into the target gene. Although the CRISPR/Cas9 system seems to be a very promising and powerful tool for the rapid generation of mouse models with single point mutations [5], we do not recommend it for the generation of NMDAR mutant mice for two major reasons. First, frequently, the most interesting NMDAR mutations are dominant or co-dominant [6]. Carriers of those dominant mutations might not be viable or might not be reproductive. Therefore, it can be very difficult, if not impossible, to establish a stable mouse line after successful genomic editing of NMDARs in mouse zygotes using the CRSIPR/Cas9 system. Second, to date, the introduction of conditional NMDAR mutations, as well as conditional mutations in other genes by CRISPR/Cas9, has not yet been described in neuroscience research. In neuroscience, conditional gene regulation in selected neuronal populations is obligatory for dissecting the very diverse functions of the central nervous system. Thus, conditional regulation of NMDAR mutations permitted a comprehensive and productive analysis for the role of NMDARs in different neuronal circuits [6–10]. Moreover, the conditional gene regulation system is very advanced in the mouse [11, 12]. In particular, the Cre/loxP system in which Cre recombinase is used to destroy or to activate a loxP flanked targeted gene in selective cell types at any time point in development by Cre recombinase mediated removal or inversion of a loxP site flanked gene segment is well-established [13–15]. Many Cre expressing mouse lines have been developed and characterized in the last 20 years [16–18] and different Cre lines can be used to establish sub-lines that express the very same targeted NMDAR mutation in selected neuronal cell populations, such as dentate gyrus granule (DG) cells or CA3 pyramidal cells (e.g., line *Grin1*^{dDG} [7, 9]; *Grin1*^{dCA3} [8]). But, more importantly, we and others established Cre systems that can be activated in the brains of adult mice. Many studies use Cre recombinase fused to a modified hormone-binding site of the estrogen receptor (ER) (e.g., CreER^{T2} or ERiCreER), that can be activated by 4-hydroxy-tamoxifen but not by endogenous estrogen [19, 20]. The Cre-ER fusion proteins are trapped in the cytoplasm of neurons and remain inactive. After intraperitoneal (i.p.) injection of tamoxifen, the Cre is released from the trap in the cytoplasm and transferred to the nucleus where the loxP flanked gene fragment will be removed or inverted. Another system, the doxycycline controlled Cre expression system, is more complex since two transgenes are necessary for the deletion of loxP flanked gene segments [21]. One transgene produces the doxycycline regulated synthetic transcription factor tTA in selected cell types. The second transgene encodes a tTA controlled Cre recombinase. We used this system for successful removal of NMDARs in the DG granule and CA1 pyramidal cells of adult mice to demonstrate that hippocampal NMDARs are important for decision making but not for spatial reference memory [22, 23].

In this chapter, we will give an outline of our approach for generating mice, which express mutated NMDARs in selective populations under temporal and spatial control (conditional mutant mice). Our approach is based on the following findings: In 1996 we found that our GluA2 gene targeted mutation leads to an early death of heterozygous mutant mice during puberty. We used the classical gene targeting approach to introduce the constitutive GluA2 gene mutation into the germ line of mice. This GluA2 gene (*Gria2*) mutation turned out to be a dominant negative mutation and all heterozygous mice died from epileptic seizures within the first three postnatal weeks [24]. At the same time, we figured out that the presence of a selection marker (necessary for selective targeting, see below) silenced the targeted, dominant negative *Gria2* allele. The carriers with the silenced, dominant negative *Gria2* allele survived and produced offspring. Removal of the selection marker in principal neurons of the forebrain using the Cre/loxP technology allowed us to study the mutated GluA2 in adult animals [25, 26].

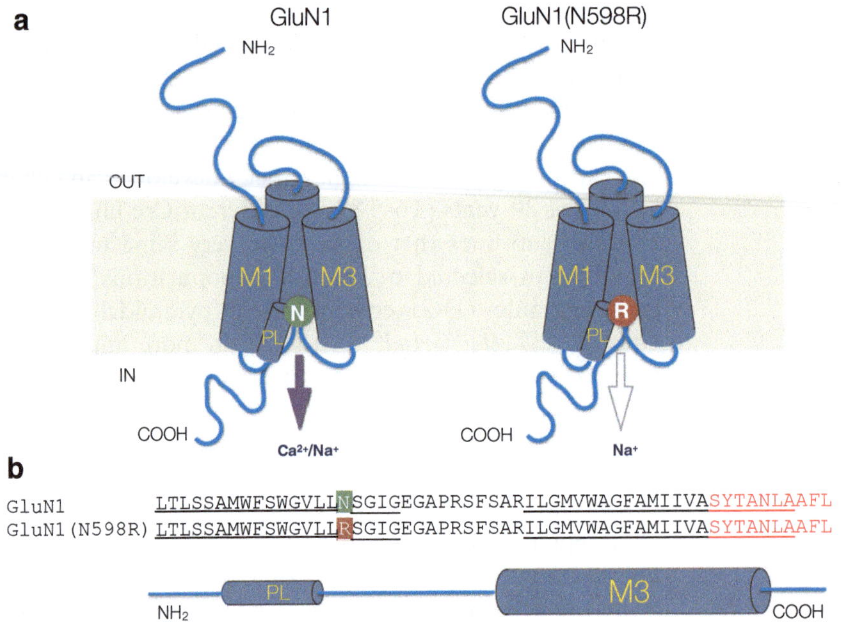

Fig. 1 Position of the N598R amino acid exchange that should be introduced by gene targeting into the GluN1 subunit. (**a**) Secondary structure of the channel structure of NMDAR subunit GluN1 and position of the N598R point mutation at the channel pore. (**b**) Amino acid sequence of pore loop (PL) and transmembrane region M3 of NMDAR subunit GluN1 and position of the N598R point mutation.

The presented method chapter will guide through all the different steps that are necessary to establish such a conditional mouse model. As an example, we describe the generation of a targeting vector for the *Grin1* gene, which encodes the principle subunit GluN1 (Fig. 1a) of all NMDARs. The GluN1 subunit of the NMDARs determines the Na^{+} and Ca^{2+} ion selectivity of the channel pore [27, 28]. For this ion selectivity, the asparagine N598 on the tip of the channel pore domain of the GluN1 subunit [29] is of central importance [30, 31]. The mutation of N598 into the positively charged arginine (N598R) makes the mutated NMDARs impermeable for Ca^{2+} (Fig. 1a, b). Thus, the GluN1(N598R) mutation destroys the coincidence detection of all NMDARs [32]. Heterozygous mice die before puberty when the GluN1 (N598R) mutation is activated in embryos but survive when the GluN1(N598R) mutation is encoded by a hypomorphic *Grin1* allele [6, 33, 34]. Activation of the expression of GluN1(N598R) mutation by Cre/loxP-mediated elimination of the silencing PGKneoPA selection marker from the hypomorphic *Grin1* allele allowed the analysis in developed mice [6, 33, 34].

The design and generation of the *Grin1* targeting vector, which will be presented in the following, is based on a vector that we used to generate conditional GluN1(N598R) expressing mice (mouse line $Grin1^{N598R}$, previously called $NR1^{+/R}$; [6]) (*see* **Note 1**).

However, the described vector design and mouse genotyping are updated using the most recent nucleotide sequence resources. Standard methods like ES cell manipulation and blastocyst injection have already been described in great detail in several textbooks [35, 36] and will not be described extensively in this chapter. Most investigators use commercial services or core facilities at their institutions for the biological parts of gene targeting experiments. This chapter will cover the genotyping of mice by PCR and nonradioactive Southern blotting but not the basic techniques of Molecular Biology described in detail in the laboratory manual: Molecular Cloning [37]. At the end of the chapter, breeding strategies for generating conditional *Grin1* mutant mice with complex genotypes are given.

2 Materials

2.1 Public Sequence Databases and PC/Mac Programs

1. Genomic sequence and genomic information: Public National Center for Biotechnology Information, NCBI: www.ncbi.nlm.nih.gov/genbank; University of California Santa Cruz Genome Browser, UCSC: https://genome.ucsc.edu; Mouse Genome Informatics, MGI: http://www.informatics.jax.org (*see* **Note 2**).
2. PC/MAC software: GCK, Gene Construction Kit (Texto Biosoftware); ApE-A plasmid editor (Biologylabs); SnapGeneViewer (GSL Biotech LLC) (*see* **Note 3**).

2.2 DNA Analysis

1. Kit for DNA Isolation (from ES cells from 96-well plates).
2. Kit for isolation of up to 100 μg high-molecular-weight DNA.

2.3 Nonradioactive Southern Blot

1. Special reagents: Positively charged nylon membrane; direct alkaline phosphatase labeling system; chemiluminescent detection reagent; light-sensitive film.
2. Special equipment: UV-crosslinker.
3. SSC solution: 50 mM Tris–HCl pH 7.5, 300 mM NaCl, 30 mM sodium citrate.
4. Wash buffer: 2 M Urea, 0.1% SDS, 50 mM Na phosphate pH 7, 150 mM NaCl, 1 mM $MgCl_2$, 0.2% blocking reagent.
5. Secondary wash buffer: 50 mM Tris–HCl, pH 10, 100 mM NaCl, 2 mM $MgCl_2$.

3 Methods

1. The classical design of our targeting vector from the year 2000 [6] reflects some general principles that should be considered for efficient genetic manipulations of the NMDAR channel

3.1 Targeting Vector Addressing the Ion Channel Domain Coding Exons of the Grin1 *Gene*

pore (*see* **Note 4**). Interestingly, our former colleague Dr. Schoepfer used the identical strategy for manipulating the *Grin1* gene with same success [33]. Both targeting vectors were composed of two *Grin1* gene segments encoding the essential elements of the NMDAR channel pore (Figs. 1 and 2), a short gene fragment of about 2 kb (short arm) and a larger fragment of approximately 8 kb (long arm). The short arm contained exons 11–18, including the GluN1(N598R) mutation in exon 15 (Fig. 2). In addition, a diagnostic restriction site (X) was introduced next to the GluN1(N598R) mutation in the short arm. A unique restriction site (U) has to be present at the beginning of the short arm or at the end of the long arm. This restriction site was used to linearize the targeting vector before transformation of ES cells. The exposure of at

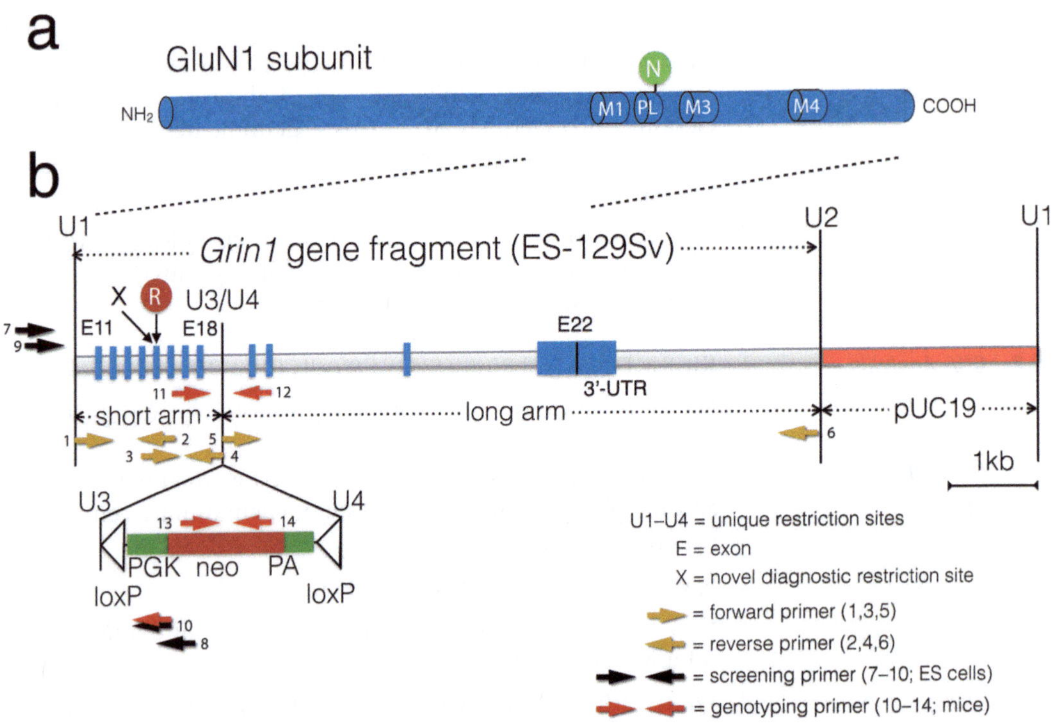

Fig. 2 General outline of the targeting vector. (**a**) Primary structure of GluN1 subunit. The position of N598 at the channel pore forming segment (Pore Loop; PL) in the GluN is indicated. **(b)** A schematic view of our targeting vector, which can be used for gene replacement targeting to introduce the point mutation GluN1 (N598R) into the mouse *Grin1* gene in ES cells. The *Grin1* gene fragment derived from ES cell of mouse line 129Sv encodes exons (E, *blue boxes*) 11–22 of the *Grin1* gene. In exon 15 the positions of the Arginine (R) mutation N598R and of the diagnostic restriction site X are indicated. Primers used for the amplification of the long and short arms of the targeting vector are shown as *yellow arrows*; during the amplifications the unique restriction sites U1–U4, X, and the mutation N598R are introduced. Positions of the plasmid backbone (pUC19) and of the loxP site flanked G418 selection marker (PGKneoPA) are indicated. Primers for the screening of ES cell clones and for the genotyping of the mice are shown as *black* and *red arrows*, respectively.

least one end of the targeting vector is expected to increase the efficiency of gene replacement by homologous recombination in ES cells [36]. However, during homologous recombination, the N598 codon of exon 15 was replaced by the R598 codon of the targeting vector only in some *Grin1* targeted ES cell clones. In other targeted ES cell clones, when the homologous recombination on the short arm took place between exon 15 and exon 18, the GluN1(N598R) mutation was not introduced into the *Grin1* gene of the ES cells. As an advantage of this targeting strategy, the successful N-to-R codon exchange can be monitored during the screening procedure of the ES cell clones by regular or nested PCR. The digest of the PCR amplified fragments with the restriction enzyme X indicates which of the targeted ES cell clones encode GluN1(N598R).

2. In Fig. 2, the important features like the length of the homology arms, the positions of the loxP flanked selection maker and of the N598R mutation, restriction sites, and primer binding sites of the targeting vector are summarized (*see* **Note 4**). For the correct incorporation of all these features in the targeting vector, the sequence information of the entire *Grin1* gene locus is required.
3. In these days, we use the nucleotide sequence information from public sequence databases to design a targeting vector and to outline the cloning strategy. Bacterial artificial chromosome (BAC) clones can be used as a source for genomic mouse DNA. For sequencing of the mouse genome, BAC libraries representing the entire mouse genome were generated and are publically available. The BAC clone spanning the entire *Grin1* gene locus is an essential DNA source for the assembly of the targeting vector described in Subheading 3.6 (*see* **Note 5**).

3.2 Sequence Information Assisted Design of Grin1 (N598R) Targeting Vector

1. You start your database search with a simple search for the GluN1 cDNA. Visit the NCBI homepage and do a nucleotide search for "mouse GluN1."
2. As first hit, you will find the nucleotide sequence of coding region for the GluN1 mRNA with the accession number, e.g., NM_008169 (Fig. 3a). Several entries for GluN1 cDNAs will show up since the *Grin1* gene encodes several isoforms. All isoforms share the same channel pore. Therefore, you can use any cDNA isoform encoded by the mouse *Grin1* gene.
3. Open the sequence and copy the GluN1 encoding cDNA sequence into the clipboard.
4. Visit the UCSC homepage and perform a "BLAT" search on the mouse genome using the GluN1 cDNA from your clipboard as the query sequence.

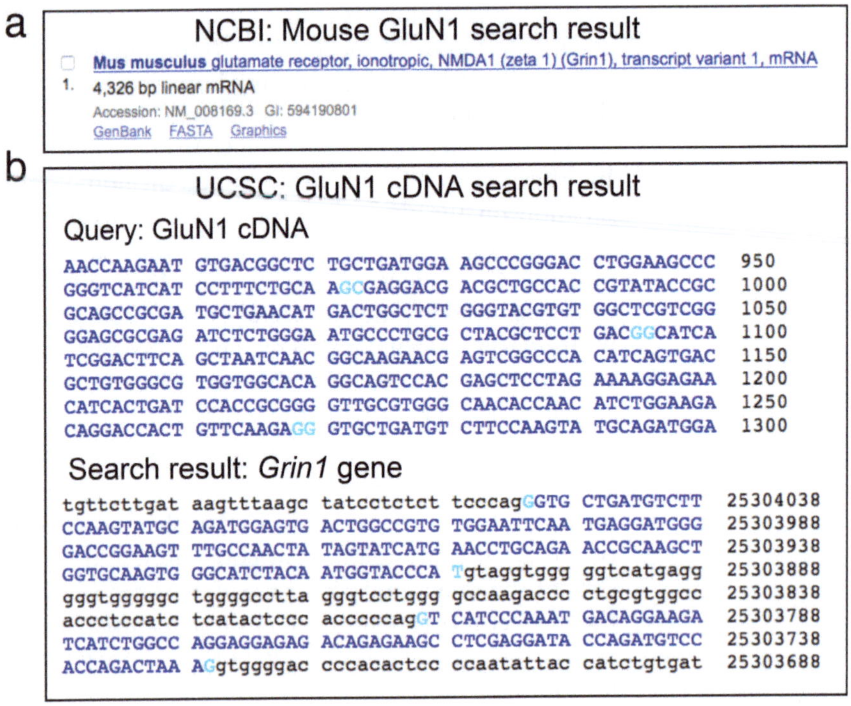

Fig. 3 Partial screen shots from the database search. (**a**) The result of the NCBI nucleotide database search for "mouse GluN1" is depicted. The accession number and description of one of the *Grin1* transcript sequences is displayed. (**b**) UCSC/BLAT search result output in the UCSC/BLAT "details" view. (**b**, upper part) Partial sequence from the GluN1 cDNA used as query sequence for the BLAT search. Paired nucleotides in *light blue* represent the last and the first nucleotide of neighboring exons (**b**, lower part). Partial *Grin1* gene sequence of the BLAT search result. Exonic and intronic sequences are displayed in *black* and *blue letters*, respectively. Note: the chromosomal positions on the right indicate a reverse orientation of the *Grin1* gene on the mouse chromosome.

5. The UCSC/BLAT search result window shows you a long list of sequences that are similar to your query sequence. Activate the first hit in the UCSC/BLAT "details" view to visualize the sequence similarity between your query sequence and the *Grin1* gene in the mouse genome (Fig. 3). In the upper part (Fig. 3a), you will see your query sequence in *blue*. Neighboring nucleotides in *light blue* indicate the position of *Grin1* introns in the mouse gene on chromosome 2. In the lower part (Fig. 3b), you will see the entire mouse *Grin1* gene with intronic nucleotides in *black* (small letters) and exonic nucleotides (capital letters) in *blue*. Copy the entire *Grin1* gene sequence and paste it into your favorite local DNA analysis program and name the file, e.g., *Grin1.seq*.
6. Now you can switch back to your UCSC search result and you can activate the "browser" view of the first hit. In the new window you will see that the *Grin1* gene is encoded on chromosome 2 in the reverse orientation (Fig. 4a, b; browser).

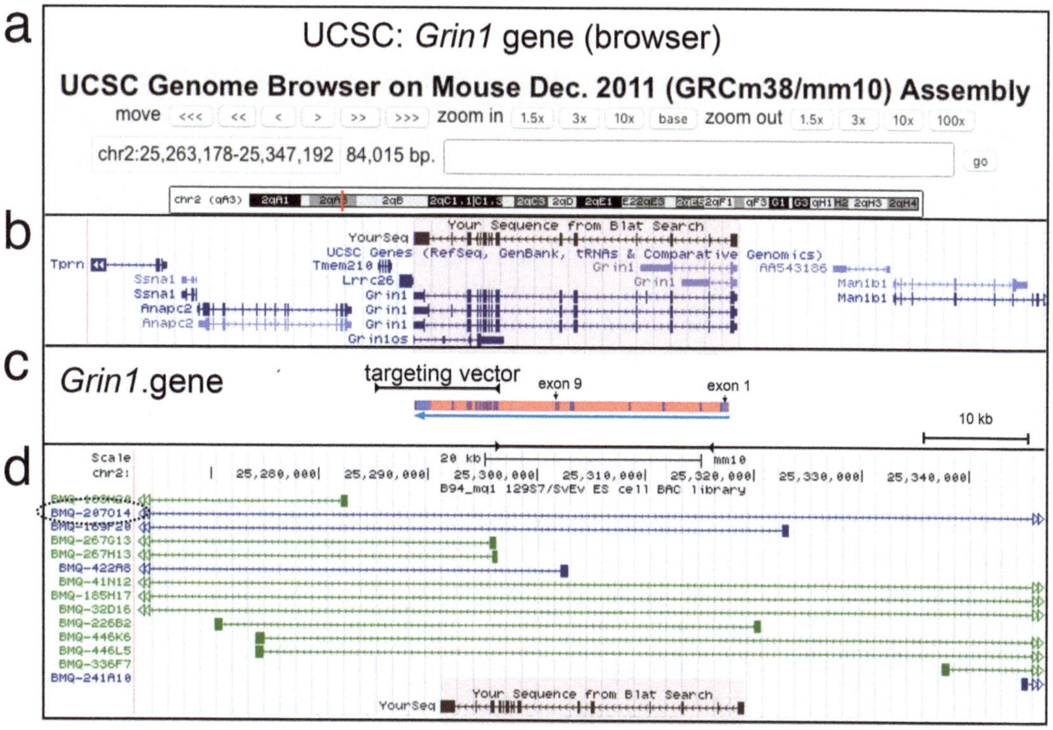

Fig. 4 Partial screen shots from the UCSC/BLAT search result output in the UCSC/BLAT "browser" view. (**a**) The position of the *Grin1* gene on chromosome 2 is indicated. (**b**) The transcripts and their orientation at the *Grin1* locus are displayed together with the query sequence. (**c**) The *Grin1* gene nucleotide sequence was copied from the USCS/BLAT search (Fig. 3) into a local sequence analysis program (GCK). The GCK display view of the *Grin1* gene mimics the exon/intron structure of the UCSC/BLAT "browser" view. The chromosomal region that should be covered by the targeting vector starts 5′-prime to exon 10 of the *Grin1* gene and contains additional 3–5 kb nucleotide sequence 3′-prime to the *Grin1* gene. Exons are given in *blue* and introns in *red*. (**d**) UCSC/BLAT "browser" view showing the BACs covering this chromosomal region. The BAC clone BMQ-207014 was selected as a DNA source for the targeting vector.

Please note that you do not need to reverse/complement your local *Grin1* gene sequence because in the UCSC search result window (Fig. 3b; details) the sequence of chromosome 2 is already displayed in the reverse/complement orientation.

7. Now you can annotate the *Grin1* gene sequence using your local DNA-analysis program. We prefer the local software GCK because it is easy to do sequence editing and annotations. Most importantly, you can switch between the graphic and sequence view with one click (*see* **Note 3**).

8. You can use the information field of the GluN1 mRNA sequence of NM_008169 for GluN1 protein accession number NP_032195 to identify the coding regions. To add annotations to your local sequence file of the *Grin1* gene, please consult the user manual of your local program.

9. You have to find the codons for LL<u>N</u>SGIG that are to be mutated to encode LL<u>R</u>SGIG (Fig. 1b). According to the information of the primer sequence given in Subheading 3.5, "LL<u>N</u>SGIG" is encoded on exon 15 of *Grin1*.
10. In your local program you should activate the "graphical" view of the sequence (Fig. 4). Copy the local map of your *Grin1* gene and compare it to *Grin1* gene transcripts in the "browser" view (Fig. 4b). Now you will see if the annotations of your local *Grin1* gene file are correct and reflect the exon distribution of the *Grin1* gene of the database (Fig. 4b, c). But you will also notice that your targeting vector (Fig. 2) of about 10 kb has to contain at least 3–5 kb genomic DNA 3′-prime to the *Grin1* gene and that the 3′-prime part of the downloaded *Grin1* gene sequence is too short (Fig. 4b).

3.3 More Information from the Grin1 Gene Locus (In Silico)

1. To generate the final targeting vector, extend the sequence environment of your *Grin1*.gcc file at the 3′-prime end using the sequence information for BAC clones at the UCSC genome browser.
2. Open again your UCSC/BLAT search result in the "browser" view.
3. Zoom out 3× to see the surrounding coding regions of neighboring genes. In the lower part of the browser page you can modify the output of the displayed information by activating or deactivating many different features.
4. In order to visualize the sequence information of BAC clones from 129/Sv ES cells (129S7/SvEv ES cell BAC library), you have to click on the button "Clone Ends" in the mapping and sequencing field. A new window will open and the available BAC libraries will be listed. The default is "C57BL/6 mouse BAC clone set". This can be changed to "B94_mql 129S7/SvEv ES cell BAC library". After you have activated the submit button, the BLAT search window will be updated and refreshed. The new window now shows the *Grin1* gene together with other regions, which are also cloned on BACs. For example, BAC clone BMQ-207O14 covers the entire *Grin1* gene locus and some neighboring genes (Fig. 4d).
5. You can click on the name BMQ-207O14 in the map view and a link to the NCBI "clone" database will open in a new window to show you the information of NCBI DB:BMQ-207O14. Now just click on NCB DB BMQ-207O14 to open the "Genome View" of this region of chromosome 2 at the NCBI server (Fig. 5a). Scroll down to visualize the regions covered by the different BACs (Fig. 5b). If you have found the region for BMQ-207O14, place the curser on the name "BMQ-207O14". A pop-up window will show you your

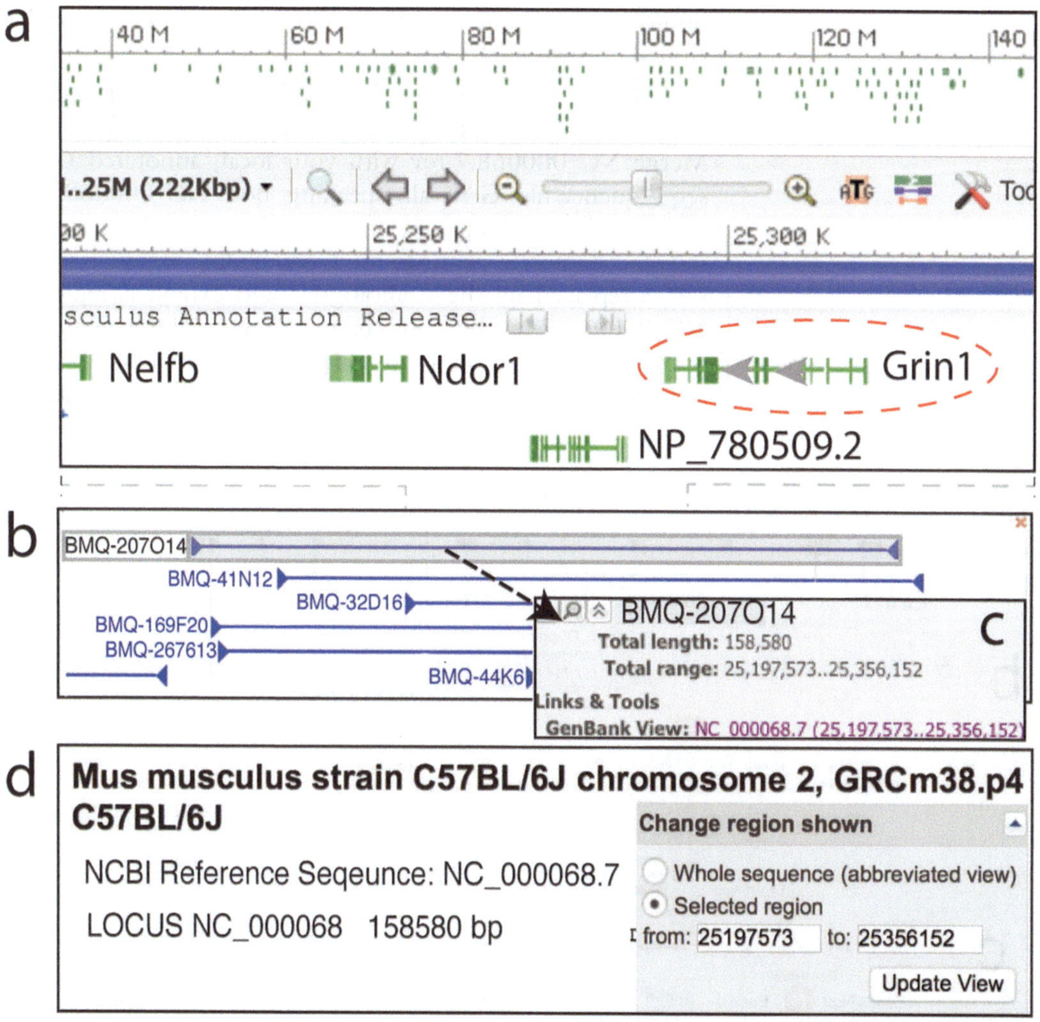

Fig. 5 Modified, partial screen shots of the *Grin1* gene locus in the NCBI Clone database. (**a**) The *Grin1* gene locus on mouse chromosome 2 is depicted. The position of the *Grin1* gene is highlighted by a *dashed red circle*. (**b**) Genetic maps of the cloned chromosome 2 fragments that are inserted in BACs. The size and the ends of the cloned fragments are indicated. The insert of BAC clone BMQ-207014 is highlighted in *grey*. It indicates that BMQ-207014 encodes the entire coding information of the *Grin1* gene (**c**). The pop-up window shows the Genbank accession number NC_000068.7 of the C57BL/6 derived sequence that is homologous to the 129/Sv derived BMQ-207014 sequence. (**d**) NCBI Genbank entry for the NC_000068.7 sequence.

options (Fig. 5c). In this pop-up window you can activate under "Links and Tools" the option Genbank view: NC_000068.7. Now the NCBI Genbank entry for NC_000068.7 will be displayed in a new region (Fig. 5d) and you can download the sequence of NC_000068.7 to your local computer (*see* **Note 6**). The sequence information NC_000068.7 represents a sequence of 158,580 bp encoded by the C57BL/6 mouse chromosome 2 fragment. The homologous fragment of the 129S7/SvEv chromosome 2 is cloned on BMQ-207014.

6. Make the reverse/complement of the downloaded NC_000068.7 sequence and name it NC_000068.7.rev. Merge NC_000068.7.rev with your local, annotated *Grin1*.seq sequence file. Generally speaking, both NC_000068.7.rev and *Grin1*.seq are opened in two different windows and the sequence region 3′-prime to the *Grin1* coding region in *Grin1*.seq is replaced by the homologous and extended region of NC_000068.7.rev by copy/paste. Now the long *Grin1*.seq sequence information (about 120 kb) is available as local sequence information (Fig. 6). This is sufficient sequence

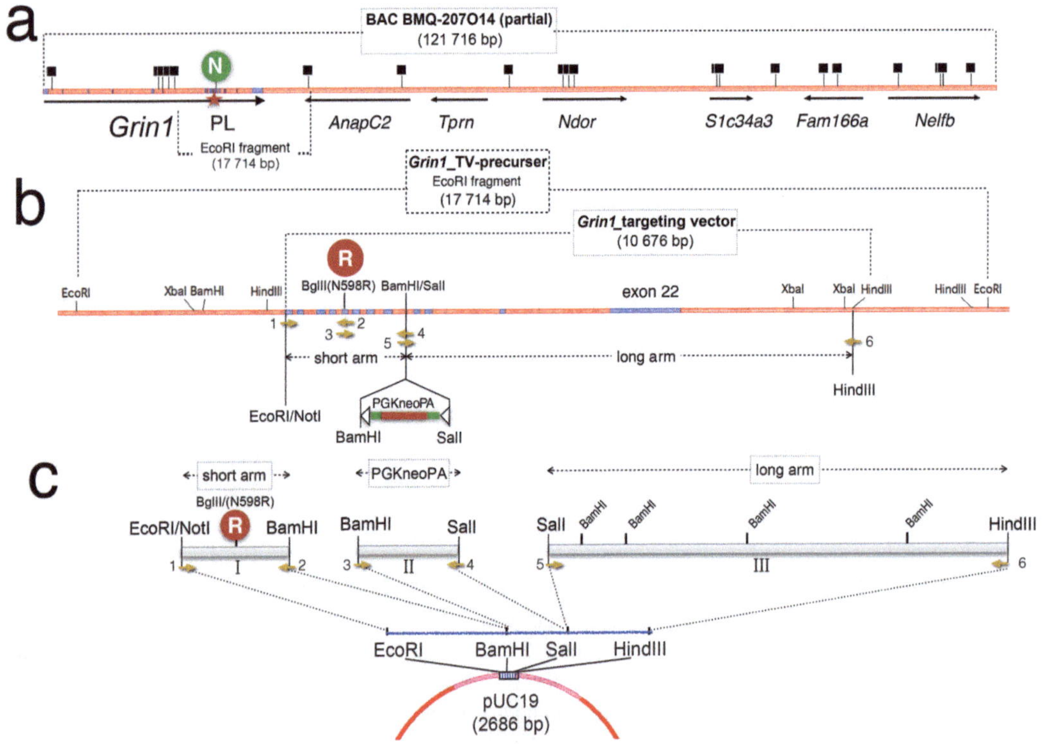

Fig. 6 Cloning strategy of the *Grin1* targeting vector (TV). (**a**) Graphical display of the *Grin1* gene sequence and coding regions of six downstream genes subcloned in the genomic BAC BMQ-207014. *Grin1* exons are shown in *blue*. The channel pore encoding *Grin1* gene region is marked by a *red star*, PL and N (*see* Fig. 1). Positions of EcoRI restriction sites are shown by *black squares*. (**b**) Graphical view of the 17,741 bp EcoRI fragment of the *Grin1* gene fragment in your local sequence *Grin1*_TV-precursor. The positions and directions of primers used for the PCR amplification of the short and the long arm of the *Grin1* targeting vector are indicated by *yellow arrows*; primers 1–6. Primers 2 and 3 introduce the N598R mutation together with the diagnostic BglII restriction site. Important restriction sites added by the PCR primers for the easy cloning of the short and long arms of the *Grin1* targeting vector from BAC BMQ-207014 PCR amplification products and from the ploxPneo plasmid are indicated (*see* Fig. 2). (**c**) DNA fragments that can be used for the cloning of the *Grin1* targeting vector in the pUC19 vector backbone are given: the short arm (I), the PGKneoPA selection marker (II), and the long arm (III).

information, which is necessary for the computer-assisted generation of the targeting vector design and the analysis of the targeted locus in ES cells.

3.4 Design of the Grin1^{N598R} Targeting Vector (In Silico)

1. First, you need to locate the regions of the short and the long arm of the targeting vector on the restriction map of the local *Grin1*.seq sequence. The *Grin1*.seq EcoRI restriction map identifies a 17.7 kb EcoRI gene fragment as the *Grin1* gene region encoding the ion channel pore of the GluN1 subunit (Fig. 6a).
2. Therefore, you should focus your analysis on this 17.7 kb EcoRI fragment (Fig. 6a, b) and copy this sequence together with about 500 bp neighboring sequences into a new local DNA file that is called, e.g., *Grin1*_TV.precursor.
3. Now the position of the short arm and the long arm of the targeting vector can be identified in detail. The short arm should encode the N598 codon (AAC) on exon15 that has to be changed into an R598 codon (CAG). Preferentially, the R598 codon should be located in the center of the short arm (SA). The short arm should not be longer than 2 kb because the PCR amplification of the short arm will be used for the screening of the homologous recombination events in ES cell after gene targeting.
4. The restricted length of the short arm and its exact position defines the insertion site of the selection marker that has to be inserted in a bigger intron away from the branch point sequence that is necessary for the formation of the lariat structure [38, 39]. The formation of the lariat structure is essential for the correct splicing of the intron and might be destroyed by the insertion of foreign DNA. In the described example intron 18 was selected as an insertion site for the selection marker (floxed PGKneoPA; loxP site flanked PGKneoPA) (Fig. 6b). Intron 20 is bigger and would be safer for the position of the selection marker, but when positioned in intron 20 the short arm fragment is oversized and hence, is not suitable for the PCR screen of targeted ES cell clones (*see* **step 1** in Subheading 3.9).
5. As 3′-prime end of the long arm, we selected the endogenous HindIII restriction site (Fig. 6b). The HindIII restriction map of the EcoRI *Grin1* gene fragment revealed that the long and the short arm for the *Grin1*^{N598R} targeting vector are localized on a single HindIII fragment (Fig. 6b). This HindIII fragment can be subcloned or amplified from BAC BMQ-207O14 that was identified in the database search described above (*see* **step 4** in Subheading 3.3).

6. In an internet search "BAC libraries BMQ-207O14", internet resources are displayed which provide the BAC BMQ-207O14 (e.g., Source Bioscience http.//www.lifesciences.sourcebioscience.com/clone-products/mammalian/genomic-clones-others/sanger-resources/mouse-bmq-bac-library).
7. The BAC clone can be ordered from this public resource and BAC DNA can be prepared using commercial chemicals.

3.5 Setting Up the Cloning Strategy for the Grin1^{N598R} Targeting Vector

1. First, you have to prepare the selection marker, PGKneoPA, for the insertion into the *Grin1* targeting vector.
2. We use a selection marker from plasmid ploxPneo1. We modified the Neo gene of ploxPneo1 according to [40] to provide a high G418 resistance to ES cells expressing ploxP-neo1 (*see* **Note** 7).
3. Other floxed PGKneoPA plasmids (e.g., PGKneotpAlox2) can also be used (available at Addgene; https://www.addgene.org as source for the PGKneoPA (Figs. 2b, 6b). But make sure that the plasmids encode the improved version of the Neo (*see* **Note** 7).
4. You have to insert the selection marker PGKneoPA in the sense orientation. The PGK termination sequence (PA) after the Neo coding sequence is used for premature transcriptional termination within the targeted gene. This premature termination is important for gene silencing by the Neo gene insertion.
5. As a golden rule, you have to introduce a unique restriction site that is absent in the short arm of the targeting vector right in front of the 5′-prime loxP site of the floxed PGKneoPA. Similarly, another unique restriction site, absent on the long arm, has to be introduced after the 3′-prime loxP site of the PGKneoPA. In our example, we introduced BamHI and SalI, respectively. This has to be done by regular cloning techniques modifying the improved ploxPneo1 plasmid (Fig. 6b).
6. Now, the primers for the amplification and the cloning can be designed in detail (Fig. 6b). The oligonucleotides can be ordered using the following sequence derived from your local *Grin1* sequence file (nucleotides that are inserted or added to the *Grin1* sequence are given in bold or small letters).
7. For amplification (gene assembly) of the short arm and to replace the Asparagine codon by an Arginine codon (LLNSGIG will be mutated to encode LL**R**SGIG) primers 1–4 are used, restriction sites are underlined: Primer 1 (N1in10Ndo, 5′-CGG AAT TCG CG GCC GCT TGG GAT TTA CTG CAG CAC-3′; NotI), primer 2 (N1LRSup, 5′-CCA ATG CCA GAT **CT**G AGC AGG ACG CC-3′; BglII), primer 3 (N1LRSdo, 5′-GGC GTC CTG CTC **AGA** TCT GGC ATT GG-3′; BglII), primer 4 (N1in18Xup, 5′-cga gga tcc TGT

GGC TGG AGC ATT GGG GAC TC-3′ BamHI) (*see* **step 2** in Subheading 3.6).

8. For the amplification of the long arm primers 5 and 6 are used: primer 5 (N1in18Xdo, 5′-ggc cgt cga cCC CTG GCC TGG CTG ACA GC-3′; SalI), primer 6 (N1i3UTRup 5′-acg ccA AGC TTT CTG TCC CAG GAT TCC TGG-3′; HindIII) (*see* **step 2** in Subheading 3.6).
9. For the screening of ES cell clones by nested PCR you can order primer pair 7/8 and primer pair 9/10 (*see* Subheading 3.9). The reverse primers 8 and 10 are encoded by the PGK promoter. The forward primers 7 and 9 are outside the targeting vector, in intron 10 of the *Grin1* gene: primer 7 (N1in10do1, 5′-GGA TCT GTC CCC AAG GGT AGC-3′), primer 8 (pgkprom1, 5′-GAA TGT GTG CGA GGC CAG AGG-3′), primer 9 (N1in10do2, 5′-CTA GCC ATG TCA GAA GGA TGT G-3′), primer10 (pgkprom, 5′-CAG ACT GCC TTG GGA AAA GCG-3′).
10. For genotyping of the gene targeted mice by PCR of tail-tip DNA primers 13 and 14 can be ordered: primer 13 (rspneo4; 5′ -GGC TAT TCG GCT ATG ACT GGG C-3′), primer 14 (rspneo5; 5′-GGG TAG CCA ACG CTA TGT CCT G-3′). The $Grin1^{N589Rneo}$ alleles will provide an amplified Neo gene fragment of 624 bp.
11. As an alternative genotyping PCR protocol we prefer a triplex PCR reaction. In the triplex PCR reaction, both the wild-type *Grin1* allele and the gene targeted $Grin1^{N589Rneo}$ allele can be amplified in the same reaction (*see* **step 3** in Subheading 3.12). This is a big advantage for the correct identification of homozygous, heterozygous, and wild-type animals. In our triplex PCR the three primers are used in the same PCR: primer 10 (pgkprom, 5′-CAG ACT GCC TTG GGA AAA GCG-3′), primer 11 (N1ex18do1, 5′-CTG GGA CTC AGC TGT GCT GG-3′), and primer 12 (N1in18up1, 5′-AGG GGA GGC AAC ACT GTG GAC-3′). Primer pair 10/11 will amplify a 372 bp gene fragment of the $Grin1^{N589R}$ allele and primer pair 11/12 will amplify a 449 bp gene fragment of wild-type $Grin1^{+}$ allele (*see* **step 3** in Subheading 3.12).
12. You can also use primer pair 11/12 to genotype mice in which Cre recombination was used to delete the PGKneoPA selection marker. Primers 11 and 12 flank the insertion site of the selection marker: primer 11 (N1ex18do1, 5′-CTG GGA CTC AGC TGT GCT GG-3′), primer 12 (N1in18up1, 5′-AGG GGA GGC AAC ACT GTG GAC-3′). After Cre-mediated PGKneoPA removal the PCR products are 449 bp and 493 bp DNA fragments for the wild-type and $Grin1^{N598R\text{-}delta\_neo}$ allele, respectively, in heterozygous mice.

3.6 Cloning of the Grin1^{N598R} Targeting Vector

1. As soon as the oligonucleotides (primers 1–6) arrive in the lab, you can perform the molecular cloning. As outlined in Fig. 6c, three different fragments have to be inserted into a plasmid backbone. As a vector, the popular cloning vector pUC19 was selected. This vector contains convenient restriction enzyme cutting sites in the multiple cloning site (MCS). We recommend doing the cloning subsequently in three cloning steps.
2. Introduce the BglII restriction site and the arginine codon by gene assembly of the short arm from BAC BMQ-207O14 (*see* **step 6** in Subheading 3.4) using primers 1–4 (*see* **step 7** in Subheading 3.5). Dilute BAC-DNA BMQ-207O14 down to 0.1 μg/μL in 200 μL H_2O. Then boil the sample for 10 min and put it on ice. Set up two PCR reactions on ice. PCR1 (left short arm): Mix 4 μL of 5× Phusion® HF/GC Buffer (New England Biolabs) with 0.4 μL of 10 mM dNTP's (chase) and 1 μL of 0.1 μg/μL BMQ-207O14. Add 1 μL of 10 μM primer 1 (N1in10Ndo), 1 μL of 10 μM primer 2 (N1LRSup), 0.6 μL of DMSO (optional), and 0.2 μL of 5 U/μL Phusion polymerase (New England Biolabs). Add H_2O to 20 μL. PCR2 (right short arm): Mix 4 μL of 5× Phusion® HF/GC Buffer with 0.4 μL of 10 mM dNTP's (chase) and 1 μL of 0.1 μg/μL BMQ-207O14. Add 1 μL of 10 μM Primer 3 (N1LRSdo), 1 μL of 10 μM primer 4 (N1in18Xup), 0.6 μL of DMSO (optional), and 0.2 μL of 5 U/μL Phusion polymerase (New England Biolabs). Add H_2O to 20 μL. Use hot start: preheat thermocycler lid, preheat thermocycler block to 98 °C, and add the tubes. Start PCR program immediately following this protocol: Pre-denaturation at 98 °C for 30 sec followed by 35 cycles amplification (Each cycle: denaturation for 10 s at 98 °C, annealing for 30 s at 55 or 60 °C and elongation for 60 s at 72 °C). The amplification is terminated after the 35 cycles by a 5 min elongation reaction at 72 °C.
3. Analyze 5 μL on 1.5% agarose gel with E-buffer. Dilute the fragments for the left and the right short arm 1:5 in 50 μL H_2O. Boil the diluted DNA fragments for 10 min and put both samples on ice. Set up the gene assembly PCR reactions on ice. Mix 4 μL of 5× Phusion® HF/GC buffer with 0.4 μL of 10 mM dNTP's (chase). Add 4 μL of the 1:5 PCR-product1 and PCR-product2 each. Add 1 μL of 10 μM primer1 (N1in10Ndo), 1 μL of 10 μM primer4 (N1in18Xup), 0.6 μL of DMSO (optional), and 0.2 μL of 5 U/μL Phusion Polymerase. Add H_2O to 20 μL. Start the same PCR program as described in **step 2** of Subheading 3.6.
4. Analyze 5 μL on 1 % agarose gel with E-buffer. Digest the agarose gel purified DNA PCR fragment with EcoRI and BamHI and subclone the fragment of the short arm (SA).

5. Insert the PCR-amplified short arm into pUC19 using the restriction enzymes EcoRI and BamHI to generate plasmid pUC19.SA-*Grin1*^{N598R}.
6. Digest several pUC19.SA-*Grin1*^{N598R} plasmids with BglII to verify the presence of the mutation. Sequence one BglII positive pUC19.SA-*Grin1*^{N598R} plasmid to confirm the sequence identity of the exon 11–18 of the cloned short arm fragment.
7. Make a Maxi-preparation of the correct plasmid pUC19.SA-*Grin1*^{N598R}. Double digest the plasmid pUC19.SA-*Grin1*^{N589R} with BamHI and SalI. Insert the BamHI/SalI DNA fragment that encodes the PGKneoPA gene of the modified plasmid ploxPneo1 (*see* **step** 2 in Subheading 3.5). Screen and identify the pUC plasmids that can be named pUC19. SAneo-*Grin1*^{N598R} because it contains both the short arm with the *Grin1* mutation (N598R) and the floxed PGKneoPA selection marker.
8. Set up the long range PCR for cloning of the long arm of the targeting vector from BAC DNA BMQ-207O14 (*see* **step 5** Subheading 3.4). Dilute the BAC DNA in H_2O down to a concentration of 0.1 μg/μL. Boil 20 μL of the BAC dilution for 10 min and put on ice. Set up PCR reaction on ice. Mix 4 μL of 5× Phusion® HF/GC Buffer with 0.4 μL of 10 mM dNTP's (chase) and 1 μL of 0.1 μg/μL of BMQ-207O14. Add 1 μL of 10 μM primer 5 (N1in18Xdo), 1 μL of 10 μM primer 6 (N1i3UTRup), 0.6 μL of DMSO (optional), and 0.2 μL of 5 U/μL Phusion polymerase. Add H_2O to 20 μL. Start PCR program immediately following this protocol: Predenaturation at 98 °C for 30 s followed by 35 cycles amplification (Each cycle: denaturation for 10 s at 98 °C, annealing for 30 s at 55 or 60 °C and elongation for 3 min at 72 °C). The amplification is terminated after the 35 cycles by a 5 min elongation reaction at 72 °C.
9. Analyze 5 μL on 1% agarose gel with E-buffer. Digest the expected agarose gel purified PCR fragment with SalI and HindIII.
10. Make a maxi-preparation of the plasmid pUC19.SAneo-*Grin1*^{N598R} and digest plasmid pUC19.SAneo-*Grin1*^{N598R} with SalI and HindIII for the insertion of the SalI/HindIII long range PCR product from BAC BMQ-207O14.
11. After the transformation of the ligation product in *E. coli* you should select the small ampicillin-resistant colonies. The expected pUC-based plasmid is about 16 kb in size. Usually, the *E. coli* cells that contain large multi copy plasmids grow more slowly. Several colonies have to be analyzed by restriction digest (e.g., using BamHI, Fig. 7) to find the correct clone.

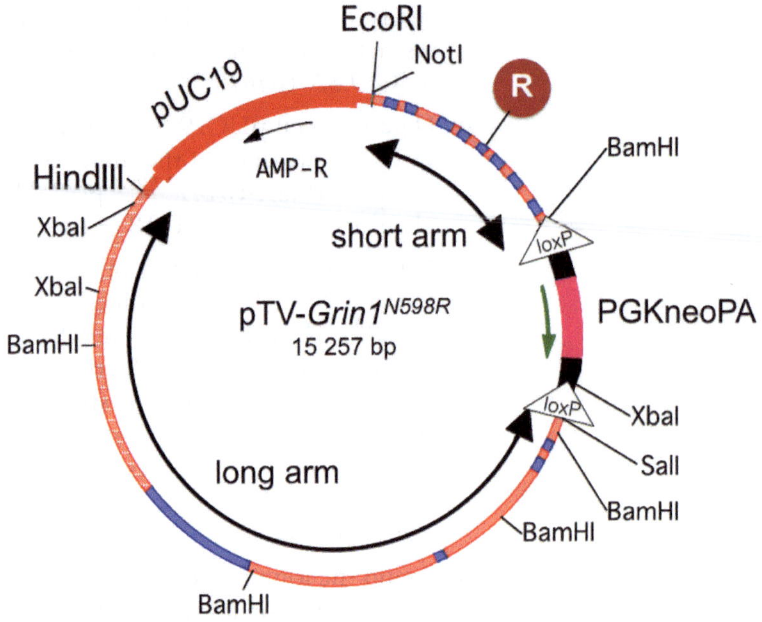

Fig. 7 Restriction map of the final targeting vector pTV-*Grin1*^{N598R}. Restriction enzymes NotI or EcoRI can be used to linearize the targeting vector for ES cell telectroporation. *Grin1* exons are given in *blue*. The coding position of the GluN1 (N598R) mutation (R) is indicated.

The final plasmid can be named targeting vector pTV-*Grin1*^{N598R} (Fig. 7).

12. For the ES cell electroporation, about 50 μg of plasmid pTV-*Grin1*^{N598R} linearized by NotI is necessary.

3.7 Transformation of ES Cells

1. The efficient electroporation of ES cells is described in detail elsewhere [36, 41]. Depending on the origin of the ES cells, the ES cell supplier will provide detailed information about how the ES cells have to be cultured [42].
2. Usually, the ES cells are grown on feeder cells (mitomycin treated primary mouse fibroblasts) and in the presence of Leukemia Inhibitory Factor (LIF), to avoid differentiation of the pluripotent cells.
3. After electroporation, ES cells are selected for their resistance to the aminoglycoside geneticin (G418), a toxic substrate, which is inactivated in ES cells expressing the PGKneoPA containing targeting vector. About 100–300 resistant ES cell colonies are picked in 96-well plates in duplicates.
4. When the cells are confluent, one 96-well plate is frozen at -70 °C in a freezing medium. The other 96-well plate is used for a crude DNA preparation of the G418-resistant ES cell colonies within the 96-well plate.

3.8 DNA Isolation from 96-Well Plates

1. Quickly invert the plate to dump the medium and to remove excess liquid by blotting on a paper towel.
2. Wash cells of each well with 150 μL PBS and dump the PBS.
3. Prepare the DNA from the 96-well plates using the DNA extraction kit. Follow the manufacturer's instruction.
4. Use 1 μL of the isolated DNA for the PCR reactions. The DNA preparation does not contain a lot of high molecular weight DNA and therefore is not useful for Southern blots.

3.9 Screening and Expanding of the G418-Resistant ES Cell Colonies

1. The DNA of the ES cell clones is screened by nested PCR using primer pairs 7/8 and 9/10 (*see* **step 9** in Subheading 3.5). After 25 PCR cycles with primer pair 7/8, 1 μL of the PCR product is diluted 1:20 and amplified in a second PCR with primer pair 9/10 for 20 more PCR cycles. PCR products of the second PCR are visualized in a 1% agarose gel. It is expected that only a few ES cell clones provide the correctly sized PCR fragment.
2. A subsequent BglII digestion of the PCR fragments provides further evidence that the correct targeting event of PCR-positive ES cell clones includes the point mutation.
3. The positive ES cell clones are revitalized from the frozen duplicate ES cell colonies (*see* **step 4** in Subheading 3.7) and expanded on feeder cell plates for frozen stock generation and on feeder-free gelatinized plates for high-quality DNA preparation.
4. The positive ES cell clones that are expanded on feeder-free 10 cm dishes covered with gelatin are grown to confluency.
5. High-quality DNA is prepared from a confluent 10 cm dish (about 10^8 ES cells). Follow the manufacturer's instruction. Use the isolated DNA first to confirm the positive targeting event in the expanded ES cell clones by the nested PCR, described above (*see* **step 1** in Subheading 3.9).
6. However, for convincing evidence of the correct targeting event by homologous recombination at the *Grin1* locus you have to analyze the isolated ES cell DNA by Southern blot.

3.10 Nonradioactive Southern Blot

1. Cut high-molecular-weight DNA from the expanded ES cell colonies by using either HindIII, EcoRI or XbaI (High fidelity restriction Endonucleases; 10–20 units/μL). Please perform a two-step digestion to ensure that DNA is fully digested. In **step 1** use 3 μL of the restriction enzyme and 3.5 μL of the respective reaction buffer. Use 10 μg of DNA with a minimal concentration of 350 ng/μL. Add H_2O to reach a final volume of 35 μL. Incubate for 16 h at 37 °C. For **step 2**, add 7.5 μL of H_2O, 1.5 μL of enzyme, and 1 μL of the buffer to the DNA

digestion mix of **step 1** and incubate at the same temperature for additional 4 h. Finally, incubate the reaction mix for 20 min at 65 °C to inactivate the restriction enzyme.

2. Next, you can perform the gel electrophoresis of the DNA fragments, depurination, denaturation, neutralization, and blotting of the DNA fragments as described for the classical Southern blot [37].
3. For effective blotting and detection we use a positively charged nylon membrane.
4. After blotting, the nylon membrane is removed and transferred to a dish containing 0.4 M NaOH such that the nylon membrane is covered with the NaOH solution.
5. Agitate the membrane in the NaOH solution in the dish on a pivoting platform for 1 min.
6. Put the membrane in 200 mL SSC solution and agitate on a pivoting platform for 1 min.
7. Transfer the membrane on a dry Whatman® paper and wait for a few minutes to remove the excess of SSC fluid.
8. UV-crosslink both sites of the membrane in the UV Stratalinker 2400 at "optimal crosslinking" settings to crosslink the DNA on the membrane.
9. Expose the membrane at RT for drying. Preheat the hybridization buffer provided with the direct alkaline phosphatase labeling system to 55 °C.
10. Place the blots into the hybridization buffer and prehybridize for at least 15 min at 55 °C in a shaking water bath (approximately 60 strokes/min) or hybridization oven. The volume of buffer should be equivalent to 0.25 mL per one cm^2 nylon membrane. The prehybridization can already be performed in sealed plastic bags or in hybridization roller bottles.

3.11 Preparation of the Labeled Probe, Hybridization, and Probe Detection

1. For the labeling of the probe with the enzyme alkaline phosphatase we use a commercial direct alkaline phosphatase labeling system from GE-Healthcare.
2. We routinely crosslink 300 ng of DNA fragment following the manufacturer's protocol. For optimal results you should add the crosslinked probe directly to the prehybridization, which you have already prepared as described in **Step 10** in Subheading 3.10.
3. Hybridization is done overnight at 55 °C in a shaking water bath or hybridization oven.
4. Prepare and preheat the wash buffer to 55 °C.

5. Transfer the nylon membrane from the hybridization to the wash buffer (about 2–5 mL per one cm^2 nylon membrane) and wash 10 min with gentle agitation.
6. Repeat the first washing step with fresh washing buffer for another 10 min.
7. Transfer the blot to a new container and wash the blot twice for 5 min in secondary wash buffer at RT.
8. Remove the blot from the secondary wash buffer and allow the excess of secondary wash buffer to drip off from the blot.
9. Place the blot on a clean tray with the sample side up. Make sure the membrane lines the bottom of the tray without any major cavities.
10. Add 4.8 mL of the chemiluminescent detection reagent CDP-Star™.
11. Pivot with your hands carefully for 5 min and make sure the CDP-Star™ runs over the complete membrane.
12. Remove with forceps and drain excess CDP-Star™ from the membrane. Do not let the membrane dry out.
13. Put membrane sample side up between two same-sized pieces of projector foil (e.g., Q-media film; Minolta) in a film cassette. Make sure that there is a small liquid film of CDP-Star™ solution on the membrane and that no air-bubbles are included.
14. Fix the sides of the foil with paper tape and wait for 1 h.
15. Move the cassette to a dark room and turn the red light on and the white light off.
16. Remove one light-sensitive film from the stock package and re-package the box carefully.
17. Put the film quickly on the membrane in the cassette and close it immediately.
18. After 1 h exposure time the film is developed in an X-ray developing machine. If signals are too week the film should be exposed for 12 h.
19. Use a flatbed scanner to digitize your result for further analysis.
20. ES cell clones that show the expected DNA fragments after enzyme digestions are now replated and used for the cellular manipulation of mouse embryos [43].

3.12 Blastocyst Injections and Generation of Chimeric and Grin1$^{+/N598Rneo}$ Mice

1. The experimental manipulation of the mouse embryo by injection of ES cells in mouse blastocysts to generate chimeric mice will not be described in detail in this chapter. Most investigators use a commercial service or the service of a local core facility. The culturing conditions, the genetic manipulation of mouse

ES cells, as well as the manipulation of the mouse embryo by blastocyst injection or morula fusion are described in vol.159 of this book series: Methods in Molecular Biology [35] in five excellent and detailed chapters [41–45].

2. As the next step (usually about 6–8 months after the ES cell manipulation that was performed by a company or service unit), you will receive tail tip biopsies from mice that might be heterozygous for the targeted *Grin1* locus. These putative heterozygous mice are offspring—F1 generation—from chimeric founders that were generated by the blastocyst injection or morula fusion [43, 45]. Chimeric founders are identified by a mixed coat color, mostly with agouti fur derived from ES cells and, e.g., black fur derived from the recipient C57BL/6 embryo. However, even a highly chimeric animal (usually male animals, since most ES cells are derived from male animals) is not a relevant indication that the ES cells transmitted to the embryo participated in the production of germ cells. If this first offspring derives from a breeding to the same background as the recipient embryo, germline transmission to the F2 animals is indicated by the agouti coat color derived from the ES cells. But still, the F1 offspring has to be analyzed for the presence of the targeted allele. This is done first by the analysis of crude tail-tip DNA preparations. The DNA of this crude preparation is usually heavily fragmented and the average fragment size might be between 1000 and 3000 bp. Therefore, the expected PCR product should not be too long. Primers 10–12 can be used for genotyping. Alternatively, you can simply amplify a 300–400 bp fragment of the Neo gene using own primers (Fig. 2b).

3. Genotype mice tails by putting the tail tip (about 0.2 mm) in 300 μL of 50 mM NaOH in a 1.5 mL tube. Incubate the sample at 99 °C at 80 g for 1 h in a heatable shaker. Put the tube on ice and add 30 μL of 1 M Tris–HCl pH 8. Use 1 μL for PCR and freeze the rest of the sample. PCR-Mix for 10 tails: Add 25 μL of 10× PCR-buffer to 10 μL of 50 mM $MgCl_2$ and 10 μL of 20 mM chase (dNTP). Add 5 μL of 10 μM primer 10, 5 μL of 10 μM primer 11, 5 μL of 10 μM primer 12 (*see* **step 11** in Subheading 3.5) and 189 μL of H_2O. Add 1 μL of 5 U/μL Taq polymerase. Put 10 μL of PCR-mix +1 μL of tail DNA in a PCR tube and start the PCR amplification: Pre-denaturation at 96 °C for 1 min followed by 35 cycles amplification (Each cycle: denaturation for 20 s at 94 °C, annealing for 30 s at 60 °C and elongation for 1 min at 72 °C). The amplification is terminated after the 35 cycles by a 5 min elongation reaction at 72 °C. Primer pair 10/11 will amplify a 372 bp gene fragment of the *Grin1*$^{(N589R)}$ allele and primer pair 11/12 will amplify a 449 bp gene fragment of wild-type *Grin1*$^{+}$ allele.

4. Analyze 10 μL for each PCR reaction on a 1.5% agarose gel in E-buffer and freeze the rest of the sample.
5. Tail-PCR positive mice are used for further breeding to provide animals of the F2 generation. Male mice are used for brother sister matings and backcrosses to two C57BL/6 females in one cage (e.g., one male $Grin1^{+/N598Rneo}$ X one female $Grin1^{+/N598Rneo}$ and two C57BL/6 females).
6. Pregnant females are separated and tail-PCR positive F2 offspring of the backcrossed $Grin1^{+/N598Rneo}$ F2 are sacrificed for further genetic analysis. The Southern blot has to confirm the presence of the targeted locus in tail-PCR positive F1 animals.
7. Fresh liver of F2 $Grin1^{+/N598Rneo}$ and of wild-type littermates can be used for the preparation of high molecular weight DNA. High-quality DNA can be prepared using the appropriate kit. Follow the manufacturer's instructions. You can use the isolated DNA for Southern blot analysis, as described in Subheadings 3.10 and 3.11.
8. Pregnant females of $Grin1^{+/N598Rneo}$ X $Grin1^{+/N598Rneo}$ matings have to be observed carefully. Some pups of the $Grin1^{+/N598Rneo}$ females are expected to have no milk in the stomach at postnatal day 0.5. Those pups are unable to suckle and to receive milk from the weaning mother. The genotyping of the dead pups will confirm that the pups are homozygous [6].
9. If the results are as expected for heterozygous $Grin1^{+/N598Rneo}$ and homozygous $Grin1^{N598Rneo/N598Rneo}$ mice, the targeting experiment was successful and you can go on to use the 2loxP $Grin1^{+/N598Rneo}$ mice to generate conditional and cell type specific $Grin1^{N598R}$ mice.
10. However, before you set up matings with Cre driver lines (*see* Subheading 3.13), some backcrosses to C57BL/6 animals should be performed if 129/Sv-derived ES cells were used for the targeting. For backcrossing, it is important that you use heterozygous F1 $Grin1^{+/N598Rneo}$ male mice to set up the first backcross with C57BL/6 females. All male offspring will carry the X chromosome derived from the C57BL/6 female.
11. Before the conditional NMDAR mice can be generated, at least 10 backcrosses to C57BL/6 mice should be performed (20 backcrosses are necessary to generate a pure inbred mouse line with the name C57BL6/$Grin1^{+/N598Rneo}$). C57BL/6 inbred mice were found to be most suitable for the analysis of some cognitive and social functions [46, 47]. In case you prefer to work with F1 or F2 hybrids of different genetic backgrounds, you should follow the recommendation of the Banbury conference in 1997 [48].

3.13 Generation of Conditional Grin1$^{+/N598R}$ Mice

1. To generate mice with expression of GluN1(N598R) in specific brain areas, the *Grin1*^{N598Rneo} mice have to be crossed with Cre expressing mice. A huge collection of Cre expressing mice is available to the public at the "Mouse Genome Informatics CrePortal". Each of the more than 2500 mouse lines is described at the CrePortal. However, many Cre driver strains display characteristics, like ectopic expression, that were not obvious in the initial design. Consistency and precision of the Cre recombination activity of a particular Cre line is crucial for the interpretation of experimental results. It will have a major impact on the number and the age of mice that need to be examined before a conclusion is drawn [17]. The detailed information about Cre recombinase activity for individual Cre driver lines is displayed in the JAX Cre Repository (currently 365 strains). The Jackson lab analyzed Cre-mediated recombination at four time points (E10.5, E15.5, P7, and P56) in a wide range of organs, tissues, and cell types. The data are permanently archived at the Mouse Genome Informatics CrePortal [17]. Nevertheless, it is obligatory to reevaluate the Cre expression pattern of a publicly available Cre line to make sure that the Cre expression pattern is as described and that there was no accidental mixup. Most laboratories use Cre reporter mice for the evaluation of Cre activity. These lines express beta-galactosidase [49] or fluorescent proteins [50], which can easily be analyzed in the lab. After the confirmation of the Cre expression you can use the Cre mice for the generation of conditional NMDAR mutants. Depending on the NMDAR mutation, the mating schema for generating experimental cohorts has to be designed. As mentioned above, mice heterozygous for dominant NMDAR mutation might die early or might not be able to reproduce. This has to be considered in the selection of the genotypes of breeders used to generate experimental cohorts of test and control animals. The genetic description of Cre expressing mice and mice with a targeted mutations (TM) of the *Grin1* gene that is silenced be the Neo insertion (*Grin1*^{TMneo}) is as follows: the presence of the Cre transgene (*Tg*$^{+/Cre}$) is indicated as "*Grin1*$^{+/TMneo/+/Cre}$" or "*Grin1*$^{+/Cre}$" in *Grin1* targeted and *Grin1* wild-type mice, respectively.
2. Breeding strategies for generating conditional *Grin1* mutant mice from the Neo gene silenced, dominant NMDAR targeted mutations (TMneo) using the Cre mice: *Grin1*$^{+/TMneo}$ X *Tg*$^{+/Cre}$ mating provide offspring with four different genotypes: the test animals (*Grin1*$^{+/TMneo/+/Cre}$) and the controls (*Grin1*$^{+/TMneo}$), (*Grin1*$^{+/Cre}$), and (*Grin1*$^{+/+}$).
3. Breeding strategies for generating conditional *Grin1* mutant mice from the Neo gene silenced, dominant NMDAR

mutations (TMneo) using the doxycycline controlled Cre system. The Cre expression of the transgene Tg^{LC1} is controlled by the doxycycline controlled transcription factor tTA encoded by transgene Tg^{CN12} [22]. First you have to establish two lines with complex genotypes: mouse line $Grin1^{+/TMneo/+/tTA}$ and mouse line $Grin1^{+/TMneo/+/LC1}$. Then you have to set up intercrosses between the two lines. You have to add 20 mg of doxycycline and 5% sugar into the drinking water of pregnant females to suppress the Cre expression of Tg^{LC1} because the doxycycline bound to tTA protein from Tg^{CN12} is inactive [21, 22]. At the day when the animals are born you have to remove the doxycycline from the drinking water to activate the tTA- controlled Cre expression in the offspring. The genotyping will show that the offspring of ($Grin1^{+/TMneo/+/tTA}$) X ($Grin1^{+/TMneo/+/LC1}$) are composed of the following genotypes: Test animals ($Grin1^{TMneo/TMneo/tTA/LC1}$), ($Grin1^{+/TMneo/tTA/LC1}$), and control animals ($Grin1^{+/TMneo/+/tTA}$), ($Grin1^{+/TMneo/+/LC1}$), ($Grin1^{+/TMneo/+/LC}$), ($Grin1^{+/+}$), ($Grin1^{+/LC1}$), and ($Grin1^{+/TMneo}$).

3.14 Release of the Grin1$^{+/N598Rneo}$ Mice to the Public

1. Once you have analyzed and published your NMDAR mutant mouse line, please make the mouse line available for public research by submitting the data of the new NMDAR mutant mouse line to the European Mouse Mutant Archive: (EMMA, INFRAFONTIER: https://www.infrafrontier.eu/emma/publicSubmission/submissionForm.emma) or to the international database resource for the laboratory mouse (Mouse Genome Informatics, MGI: http://www.informatics.jax.org/mgihome/submissions/recombinase_submission.cgi).

4 Notes

1. The presented method is an example how genes for NMDAR subunits can be silenced and reactivated by Cre recombinase. In our hands, this method was very efficient for several genes. Other researchers used a very similar, but more complex strategy for the conditional reactivation of the NMDAR point mutation GluN2A(N595Q) of the *Grin2a* gene [10].
2. The presented method is based on sequence information obtained from the sequence database at the National Center for Biotechnology Information (NCBI) and at the Genome Browser of the University of California Santa Cruz (UCSC). The sequence and nucleotide information is also available from the joint project between EMBL-European Bioinformatics Institute and the Wellcome Trust Sanger Institute (Ensembl). The web-based resources are under constant development.

Please understand that the presented approach is based on the current web profile, which might be different when you access the presented links. However, since the sequence and mouse databases are essential for genome research, they will never be erased.

3. ApE: A plasmid editor (Biologylabs, Utah) and SnapGen-eViewer (GSL Biotech LLC) are freeware for MACs and PCs. Commercial programs are also available. GCK: Gene Construction Kit is our favorite for keeping our plasmid library in shape.
4. A targeting vector has to provide the following features: (1) high targeting efficiency, (2) successful selection of transfected ES cells, (3) fast and simple screening procedure of ES cell colonies, and (4) a unique restriction enzyme cutting site. We use a traditional gene-targeting vector of the first or second generation, which permits a conditional expression of the mutated NMDAR gene and a PCR based screening of the gene targeted ES cell colonies. The targeting vector contains three essential elements: (1) a gene fragment that is sequence identical to the NMDAR gene segment that should be replaced, (2) a selector gene that can express a selection marker in ES cells, and (3) a plasmid backbone. Usually, the gene for the resistance to aminoglycoside antibiotics, the gene for neomycin resistance (Neo) from Transposon TN5 is used. Some labs prefer the puromycin resistance gene since ES cells that do not express the puromycin resistance gene die faster. The selector genes are usually flanked by two loxP sites. The loxP sites can be used later to remove the selector gene. The loxP flanked selector gene has to be inserted into the NMDAR gene segment such that the NMDAR gene fragment splits up in two parts: a short fragment of 1.5–2 kb and a long fragment of 8–15 kb. To achieve a reasonably good targeting efficiency, the length of the NMDAR gene fragment is of major importance. For the facilitation of homologous recombination, the targeting vector has to be linearized before the electroporation of mouse ES cells. We usually use a restriction site at the 5′-prime end of the short arm for linearization. This site should be a unique restriction site and should be part of the NMDAR subunit gene or only a few nucleotides away.
5. In our experiments, long-range PCRs over several kb genomic DNA always failed. It only worked with genomic DNAs inserted in BACs.
6. When downloading the BMQ-207O14 BAC encoded *Grin1* gene fragment, please make sure that you just download the sequence information covered by BMQ-207O14 and not the entire sequence of mouse chromosome 2. Please note that the

BAC BMQ-207O14 derived from a BAC library form ES cell clone (B94_mq1 129S7/SvEv) isolated from 129/Sv mice. However, the sequence entry from BAC MQ-207O14 sequence, which is given in the Genbank entry NC_000068.7, is derived from C57BL/6 genomic DNA. Therefore, you might notice some minor sequence polymorphisms between the *Grin1* of C57BL/6 and 129/Sv mice in your cloning work.

7. A mutation on many Neo genes of several commonly used selection vectors encodes a mutation that reduces the resistance of transformants expressing the mutated Neo gene. For efficient selection of ES cells, this mutation has to be repaired [40]. According to the sequence information available, the Addgene distributed plasmid PGKneotpAlox2 still encodes the mutation.

Acknowledgments

We thank Annette Herold and Markus Hüser for providing and testing the most recent protocols for genotyping genomic DNA from mice. The gene targeting of NMDAR and AMPAR subunit genes was supported by the Max Planck Society and the German Research Foundation (SFB636/A4 to R.S.).

References

1. Burnashev N, Szepetowski P (2015) NMDA receptor subunit mutations in neurodevelopmental disorders. Curr Opin Pharmacol 20:73–82. doi:10.1016/j.coph.2014.11.008
2. Endele S, Rosenberger G, Geider K, Popp B, Tamer C, Stefanova I, Milh M, Kortüm F, Fritsch A, Pientka FK, Hellenbroich Y, Kalscheuer VM, Kohlhase J, Moog U, Rappold G, Rauch A, Ropers H-H, von Spiczak S, Tönnies H, Villeneuve N, Villard L, Zabel B, Zenker M, Laube B, Reis A, Wieczorek D, Van Maldergem L, Kutsche K (2010) Mutations in GRIN2A and GRIN2B encoding regulatory subunits of NMDA receptors cause variable neurodevelopmental phenotypes. Nat Genet 42(11):1021–1026. doi:10.1038/ng.677
3. Hsu PD, Lander ES, Zhang F (2014) Development and applications of CRISPR-Cas9 for genome engineering. Cell 157(6):1262–1278. doi:10.1016/j.cell.2014.05.010
4. Sander JD, Joung JK (2014) CRISPR-Cas systems for editing, regulating and targeting genomes. Nat Biotechnol 32(4):347–355. doi:10.1038/nbt.2842
5. Inui M, Miyado M, Igarashi M, Tamano M, Kubo A, Yamashita S, Asahara H, Fukami M, Takada S (2014) Rapid generation of mouse models with defined point mutations by the CRISPR/Cas9 system. Sci Rep 4:5396. doi:10.1038/srep05396
6. Single FN, Rozov A, Burnashev N, Zimmermann F, Hanley DF, Forrest D, Curran T, Jensen V, Hvalby Ø, Sprengel R, Seeburg PH (2000) Dysfunctions in mice by NMDA receptor point mutations NR1(N598Q) and NR1 (N598R). J Neurosci 20(7):2558–2566
7. Niewoehner B, Single FN, Hvalby Ø, Jensen V, Meyer zum Alten Borgloh S, Seeburg PH, JNP R, Sprengel R, Bannerman DM (2007) Impaired spatial working memory but spared spatial reference memory following functional loss of NMDA receptors in the dentate gyrus. Eur J Neurosci 25(3):837–846. doi:10.1111/j.1460-9568.2007.05312.x
8. Nakazawa K, Quirk MC, Chitwood RA, Watanabe M, Yeckel MF, Sun LD, Kato A, Carr CA, Johnston D, Wilson MA, Tonegawa S (2002) Requirement for hippocampal CA3 NMDA

receptors in associative memory recall. Science 297(5579):211–218. doi:10.1126/science.1071795

9. McHugh TJ, Jones MW, Quinn JJ, Balthasar N, Coppari R, Elmquist JK, Lowell BB, Fanselow MS, Wilson MA, Tonegawa S (2007) Dentate gyrus NMDA receptors mediate rapid pattern separation in the hippocampal network. Science 317(5834):94–99. doi:10.1126/science.1140263
10. Hayashi Y, Nabeshima Y, Kobayashi K, Miyakawa T, Tanda K, Takao K, Suzuki H, Esumi E, Noguchi S, Matsuda Y, Sasaoka T, Noda T, J-i M, Mishina M, Funabiki K, Nabeshima Y (2014) Enhanced stability of hippocampal place representation caused by reduced magnesium block of NMDA receptors in the dentate gyrus. Mol Brain 7(1):44–17. doi:10.1186/1756-6606-7-44
11. Lewandoski M (2001) Conditional control of gene expression in the mouse. Nat Rev Genet 2 (10):743–755. doi:10.1038/35093537
12. Birling MC, Gofflot F, Warot X (2009) Site-specific recombinases for manipulation of the mouse genome. In: Cartwright EJ (ed) Transgenesis techniques, Methods in molecular biology, vol 561. Humana Press, New York, pp 245–263. doi:10.1007/978-1-60327-019-9_16
13. Deng C-X (2011) Chapter 2: The use of Cre--loxP technology and inducible systems to generate mouse models of cancer. In: Green JE, Ried T (eds) Genetically engineered mice for cancer research: design, analysis, pathways, validation and pre-clinical testing, vol 2. Springer New York, New York, NY, pp 17–36. doi:10.1007/978-0-387-69805-2_2
14. Nagy A (2000) Cre recombinase: the universal reagent for genome tailoring. Genesis 26 (2):99–109
15. Sauer B (1998) Inducible gene targeting in mice using the Cre/lox system. Methods 14 (4):381–392. doi:10.1006/meth.1998.0593
16. Wang X (2009) Cre transgenic mouse lines. In: Cartwright EJ (ed) Transgenic techniques, Methods in molecular biology, vol 561. Humana Press, New York, pp 265–273. doi:10.1007/978-1-60327-019-9_17
17. Heffner CS, Pratt CH, Babiuk RP, Sharma Y, Rockwood SF, Donahue LR, Eppig JT, Murray SA (2012) Supporting conditional mouse mutagenesis with a comprehensive cre characterization resource. Nat Commun 3:1218. doi:10.1038/ncomms2186
18. Harris JA, Hirokawa KE, Sorensen SA, Gu H, Mills M, Ng LL, Bohn P, Mortrud M, Ouellette B, Kidney J, Smith KA, Dang C, Sunkin S, Bernard A, Oh SW, Madisen L, Zeng H (2014) Anatomical characterization of Cre driver mice for neural circuit mapping and manipulation. Front Neural Circuits 8:76. doi:10.3389/fncir.2014.00076
19. Indra AK, Warot X, Brocard J, Bornert JM, Xiao JH, Chambon P, Metzger D (1999) Temporally-controlled site-specific mutagenesis in the basal layer of the epidermis: comparison of the recombinase activity of the tamoxifen-inducible Cre-ER(T) and Cre-ER (T2) recombinases. Nucleic Acids Res 27 (22):4324–4327
20. Shimshek DR, Kim J, Hübner MR, Spergel DJ, Buchholz F, Casanova E, Stewart AF, Seeburg PH, Sprengel R (2002) Codon-improved Cre recombinase (iCre) expression in the mouse. Genesis 32(1):19–26
21. Sprengel R, Hasan MT (2007) Tetracycline-controlled genetic switches. In: Feil R, Metzger D (eds) Conditional mutagenesis: an approach to disease models, vol 178. Springer, Berlin, pp 49–72
22. Bannerman DM, Bus T, Taylor A, Sanderson DJ, Schwarz I, Jensen V, Hvalby Ø, Rawlins JNP, Seeburg PH, Sprengel R (2012) Dissecting spatial knowledge from spatial choice by hippocampal NMDA receptor deletion. Nat Neurosci 15(8):1153–1159. doi:10.1038/nn.3166
23. Bannerman DM, Sprengel R, Sanderson DJ, McHugh SB, Rawlins JNP, Monyer H, Seeburg PH (2014) Hippocampal synaptic plasticity, spatial memory and anxiety. Nat Rev Neurosci 15(3):181–192. doi:10.1038/nrn3677
24. Brusa R, Zimmermann F, Koh DS, Feldmeyer D, Gass P, Seeburg PH, Sprengel R (1995) Early-onset epilepsy and postnatal lethality associated with an editing-deficient GluR-B allele in mice. Science 270(5242):1677–1680
25. Feldmeyer D, Kask K, Brusa R, Kornau HC, Kolhekar R, Rozov A, Burnashev N, Jensen V, Hvalby Ø, Sprengel R, Seeburg PH (1999) Neurological dysfunctions in mice expressing different levels of the Q/R site-unedited AMPAR subunit GluR-B. Nat Neurosci 2 (1):57–64. doi:10.1038/4561
26. Krestel HE, Shimshek DR, Jensen V, Nevian T, Kim J, Geng Y, Bast T, Depaulis A, Schonig K, Schwenk F, Bujard H, Hvalby Ø, Sprengel R, Seeburg PH (2004) A genetic switch for epilepsy in adult mice. J Neurosci 24 (46):10568–10578. doi:10.1523/JNEUROSCI.4579-03.2004
27. Burnashev N, Schoepfer R, Monyer H, Ruppersberg JP, Gunther W, Seeburg PH,

Sakmann B (1992) Control by asparagine residues of calcium permeability and magnesium blockade in the NMDA receptor. Science 257 (5075):1415–1419

28. Kuner T, Wollmuth LP, Karlin A, Seeburg PH, Sakmann B (1996) Structure of the NMDA Receptor Channel M2 segment inferred from the accessibility of substituted Cysteines. Neuron 17(2):343–352. doi:10.1016/S0896-6273(00)80165-8
29. Lee C-H, Lu W, Michel JC, Goehring A, Du J, Song X, Gouaux E (2014) NMDA receptor structures reveal subunit arrangement and pore architecture. Nature 511 (7508):191–197. doi:10.1038/nature13548
30. Kuner T, Seeburg PH, Guy HR (2003) A common architecture for K+ channels and ionotropic glutamate receptors? Trends Neurosci 26 (1):27–32
31. Sprengel R, Aronoff R, Völkner M, Schmitt B, Mosbach R, Kuner T (2001) Glutamate receptor channel signatures. Trends Pharmacol Sci 22(1):7–10
32. Malenka RC, Nicoll RA (1993) NMDA-receptor-dependent synaptic plasticity: multiple forms and mechanisms. Trends Neurosci 16 (12):521–527
33. Chen PE, Errington ML, Kneussel M, Chen G, Annala AJ, Rudhard YH, Rast GF, Specht CG, Tigaret CM, Nassar MA, Morris RG, Bliss TV, Schoepfer R (2009) Behavioral deficits and subregion-specific suppression of LTP in mice expressing a population of mutant NMDA receptors throughout the hippocampus. Learn Mem 16(10):635–644. doi:10.1101/lm.1316909
34. Rudhard Y, Kneussel M, Nassar MA, Rast GF, Annala AJ, Chen PE, Tigaret CM, Dean I, Roes J, Gibb AJ, Hunt SP, Schoepfer R (2003) Absence of whisker-related pattern formation in mice with NMDA receptors lacking coincidence detection properties and calcium signaling. J Neurosci 23(6):2323–2332
35. Cartwright EJ (2009) In: Cartwright EJ (ed) Transgenesis techniques. Principles and protocols, Methods in molecular biology, vol 561, 3rd edn. Springer/Humana Press, Totowa, NJ. doi:10.1007/978-1-60327-019-9
36. Joyner AL (2003) Gene targeting: a practical approach, 2nd edn. Oxford University Press, Oxford
37. Green MR, Sambrook J (2012) Molecular cloning: a laboratory manual, 4th edn. Cold Spring Harbor Laboratory Press, New York
38. Gao K, Masuda A, Matsuura T, Ohno K (2008) Human branch point consensus sequence is yUnAy. Nucleic Acids Res 36 (7):2257–2267. doi:10.1093/nar/gkn073
39. Harris NL, Senapathy P (1990) Distribution and consensus of branch point signals in eukaryotic genes: a computerized statistical analysis. Nucleic Acids Res 18(10):3015–3019
40. Yenofsky RL, Fine M, Pellow JW (1990) A mutant neomycin phosphotransferase II gene reduces the resistance of transformants to antibiotic selection pressure. Proc Natl Acad Sci U S A 87(9):3435–3439
41. Friedel RH (2009) Targeting embryonic stem cells. In: Cartwright EJ (ed) Transgenesis techniques, Methods in molecular biology, vol 561. Humana Press, New York, pp 185–197. doi:10.1007/978-1-60327-019-9_12
42. Barbaric I, Dear TN (2009) Culture of murine embryonic stem cells. In: Cartwright EJ (ed) Transgenesis techniques, Methods in molecular biology, vol 561. Humana Press, New York, pp 161–184. doi:10.1007/978-1-60327-019-9_11
43. Pluck A, Klasen C (2009) Generation of chimeras by microinjection. In: Cartwright EJ (ed) Transgenesis techniques, Methods in molecular biology, vol 561. Humana Press, New York, pp 199–217. doi:10.1007/978-1-60327-019-9_13
44. Pluck A, Klasen C (2009) Surgical techniques for the generation of mutant mice. In: Cartwright EJ (ed) Transgenesis techniques, Methods in molecular biology, vol 561. Humana Press, New York, pp 231–243. doi:10.1007/978-1-60327-019-9_15
45. Pluck A, Klasen C (2009) Generation of chimeras by morula aggregation. In: Cartwright EJ (ed) Transgenesis techniques, Methods in molecular biology, vol 561. Humana Press, New York, pp 219–229. doi:10.1007/978-1-60327-019-9_14
46. Crawley JN (1996) Unusual behavioral phenotypes of inbred mouse strains. Trends Neurosci 19(5):181–182
47. Wehner JM, Silva A (1996) Importance of strain differences in evaluations of learning and memory processes in null mutants. Dev Disabil Res Rev 2(4):243–248. doi:10.1002/(SICI)1098-2779(1996)2:4<243::AID-MRDD8>3.0.CO;2-R
48. Silva AJ, Simpson EM, Takahashi JS, Lipp H-P, Nakanishi S, Wehner JM, Giese KP, Tully T, Abel T, Chapman PF, Fox K, Grant S, Itohara

S, Lathe R, Mayford M, Mcnamara JO, Morris RJ, Picciotto M, Roder J, Shin H-S, Slesinger PA, Storm DR, Stryker MP, Tonegawa S, Wang Y, Wolfer DP (1997) Mutant mice and neuroscience: recommendations concerning genetic background. Neuron 19(4):755–759

49. Soriano P (1999) Generalized lacZ expression with the ROSA26 Cre reporter strain. Nat Genet 21(1):70–71. doi:10.1038/5007

50. Madisen L, Zwingman TA, Sunkin SM, Oh SW, Zariwala HA, Gu H, Ng LL, Palmiter RD, Hawrylycz MJ, Jones AR, Lein ES, Zeng H (2010) A robust and high-throughput Cre reporting and characterization system for the whole mouse brain. Nat Neurosci 13 (1):133–140. doi:10.1038/nn.2467

Chapter 12

Electrophysiological Investigation of NMDA Current Properties in Brain Slices

Alexandre Pons-Bennaceur and Natalia Lozovaya

Abstract

Analysis of electrophysiological properties of NMDARs and NMDAR-mediated synaptic transmission in identified neurons and synapses in brain slices is a major step in understanding their function in normal and pathological neuronal brain networks. In many central synapses excitatory postsynaptic currents (EPSCs) are mediated by excitatory neurotransmitter glutamate that activates colocalized AMPAR and NMDAR generating a complex EPSC. Here, we describe the methods commonly used in brain slices to study the electrophysiological properties of NMDAR-mediated component of spontaneous or evoked EPSCs by extracellular stimulation or by stimulating synaptically connected neurons. This approach is based on whole-cell patch-clamp recordings, pharmacological tools, and the analysis of the difference in temporal parameters between the AMPA and NMDA receptors. It allows pinpointing of the basic functional properties of NMDARs that are specific to identified brain regions, neurons, and synapses of wild-type or genetically manipulated mice.

Key words *N*-Methyl-D-aspartate receptor, Slices, Patch-clamp, Neurons, Synapse, EPSC

1 Introduction

N-Methyl-D-aspartate receptors (NMDARs) belong to the family of ionotropic glutamate receptors and are cationic ligand-gated ion channels that play a critical role in excitatory neurotransmission. NMDAR-mediated signaling is involved in normal brain development, synaptic plasticity, learning, memory, and high cognitive functions [1, 2]. NMDARs play an important role in temporal integration of neuronal network activity and long-term alterations in synaptic structure and function. NMDAR dysfunctions, on the other hand, are often associated with various pathological conditions such as ischemia [3], hyperalgesia [4], excitotoxicity [5], epilepsy [6, 7], and neurodegenerative diseases in the central nervous system (CNS) [8]. Important basic electrophysiological properties of NMDARs include the use of co-agonist (glycine, D-serine) for activation, relatively slow current kinetics, voltage-dependent

Nail Burnashev and Pierre Szepetowski (eds.), *NMDA Receptors: Methods and Protocols*, Methods in Molecular Biology, vol. 1677, DOI 10.1007/978-1-4939-7321-7_12, © Springer Science+Business Media LLC 2017

block by extracellular Mg^{2+} and high Ca^{2+} permeability. In general, much of the variation in function between NMDAR subtypes has been attributed to the identity of the modulatory GluN2 (GluN3) subunits in the channel assembly [9]. The GluN2 (GluN3) subunits have different temporal and spatial expression patterns in the brain and NMDAR subtypes also vary according to the cell types and subcellular localization. Different neuronal types usually express distinct combinations of NMDAR subunits, thus providing a diversity of cell-specific and even input-specific synaptic responses. In addition, mutations in NMDARs genes may have pathological consequences in several brain diseases [10, 11]. Therefore, relation of a particular mutation to the corresponding alteration of NMDARs function may provide an avenue to the targeted therapy for the pharmacological treatment of disorders. Several transgenic animals with knockin (knockout) of different NMDAR subunits as well as mice with genetically manipulated NMDARs appeared allowing us to study the functional significance of NMDAR molecular alterations in identified synapses and neurons in normal and pathological brains. Thus, it is of importance to investigate synaptic NMDARs function in living slices using electrophysiological approaches. We present here the methodological approaches suitable for the detection of the NMDAR component of excitatory postsynaptic current (EPSC) either from spontaneous or evoked EPSC in identified synapses in acute brain slices.

2 Materials

2.1 Solutions

Prepare all solutions from powder and distilled water as precisely as possible. Solutions have to be carbo-oxygenated for at least 5 min before use and osmolarity has to be set at 300 mOsmol. Three different extracellular solutions are used to keep slices in the optimal conditions: slicing solution to cut slices, recovering Artificial Cerebrospinal Fluid (ACSF) to keep slices before recordings and promote recovering after slicing and typical recording ACSF to keep slices during recordings. One pipette intracellular solution that mimics intracellular ionic composition is used.

1. Low chloride pipette potassium gluconate solution: 130 mM K-Gluconate, 10 mM Na-Gluconate, 4 mM NaCl, 4 mM MgATP, 4 mM Phosphocreatine, 10 mM HEPES and 0.3 mM GTP (pH 7.3, 320–330 mOsm). Biocytin (final concentration 0.3–0.5%) can be added to the pipette solution to label the neurons from which recordings were obtained for further morphological analysis.
2. Slicing solution: 118 mM Choline chloride, 2.5 mM potassium chloride, 1.2 mM NaH_2PO_4, 34 mM $NaHCO_3$, 8 mM D-

glucose, 7 mM $MgCl_2$ and 0.5 mM $CaCl_2$ oxygenated with 95% O_2 and 5% of CO_2.

3. Recovering ACSF: 125 mM NaCl, 3.5 mM KCl, 0.7 mM $CaCl_2$, 3 mM $MgCl_2$, 1.25 mM NaH_2PO_4, 26 mM $NaHCO_3$, and 10 mM glucose.
4. Recording ACSF: 125 mM NaCl, 3.5 mM KCl, 2 mM $CaCl_2$, 1 mM $MgCl_2$, 1.25 mM NaH_2PO_4, 26 mM $NaHCO_3$, and 10 mM glucose (*see* **Notes 1** and **2**).

2.2 Micropipettes and Microelectrodes

1. Patch clamp recording electrodes. We use fire-polished borosilicate glass pipettes with a diameter of 1.5 mm pulled with micropipette puller system to obtain pipette resistance about 7 MΩ in ACSF for whole-cell recordings (*see* **Note 3**).
2. Extracellular stimulating electrodes can be either commercially available concentric bipolar microelectrodes or homemade electrodes prepared by twisting two coated nichrome wires glued together to obtain a straight rigid wire. These electrodes are connected to an isolation unit and a stimulator to generate electromagnetic field between the two tips of electrodes.

2.3 Vibratome

For slicing different models of vibratomes are available, we usually use Leica VT1200S model that contains associated cooler system.

2.4 Electrophysiological Setup

For this kind of investigation, a basic conventional slice electrophysiological setup with the following components is adapted: motorized stage, upright microscope equipped with water-immersion, infrared optics and camera, two holders controlled by micromanipulators, a recording chamber, two-channels patch clamp amplifier.

2.5 Slices Chamber

We keep slices in homemade chambers built from 70 μM pore cell strainers and plastic dishes designed for cell culture (*see* **Note 4**).

3 Methods

3.1 Brain Slice Preparation (See Note 5)

1. Prepare vibratome, adjusting adapted razor blade position with vibrocheck function to obtain a perfect horizontally positioned blade.
2. Fill vibratome tank with oxygenated slicing solution and cool it with a connected cooler system to keep solution fresh about 4 °C. Alternatively, an ice-cold slicing solution can be used.
3. Decapitate rodent head, put it quickly in iced slicing solution for removing of brain from the skull, and glue the brain to the vibratome platform.

4. Connect this platform to the vibratome tank filled with 4 °C oxygenated (95% O_2 and 5% CO_2) slicing solution.
5. Set vibratome parameters: the thickness and velocity of slicing (for rodents we cut 300 μm slices at 0.04 mm/s).
6. Transfer the slices carefully one by one from the vibratome tank to a first chamber containing the recovering ACSF solution. Wait 5 min to wash from the slicing solution and put slices in a second dish with the recovering ACSF solution. This prevents possible contamination by remaining choline that may alter the normal activity of slices.
7. Depending on the animal age and brain structure, adapt the time of slices recovery. For cortical slices from young animals (P10-20) wait at least 1 hour before transferring slices to the setup recording chamber filled with classical ACSF solution for recordings. For adult animals (>P40) wait at least 2 h before starting experiment.

3.2 Whole-Cell Patch Clamp Recordings

1. Place slices into the recording chamber mounted on the stage of an upright microscope and fix the slice with a U-shaped platinum-wire with a grid of nylon threads. During recordings, the chamber has to be continuously perfused with the carbogen-saturated recording ACSF solution.
2. Visualize slices with the objective of low magnification (e.g., 5×), identify structure of interest, and switch to a higher magnification (40× or 60×) to visualize the cells.
3. Select cells with a strong membrane contrast and homogenous smooth shape, indicators of a good cellular health (*see* **Note 6**).
4. Apply positive pressure inside the patch pipette (50–150 hPa depending on age and location of brain tissues) before immerging it in ACSF. Before touching the tissue, compensate liquid junction potential.
5. Put pipette tip on the membrane of the selected cell to visualize a membrane deflection. Reach cell-attached configuration by removing the positive pressure, the membrane should be sucked inside the pipette tip. At this point, the access resistance should reach more than 1 GΩ (*see* **Note 7**).
6. Open the cell by applying a short pulse of negative pressure to the pipette to switch in whole-cell configuration. Keep the cell in voltage clamp close to the resting membrane potential around −70 mV (adjust according to the cell type) (*see* **Notes 8** and **9**).

3.3 Recordings of NMDA Component of Evoked EPSCs

1. Position an extracellular stimulating electrode in the area of interest in the slice close to the recording cells. Set the stimulus duration to 100 μs and adjust the stimulus magnitude to 50%

of the maximal response to stimulate tissue. The interval between subsequent stimulus should be 10–20 s to allow complete recovery of the synapses (*see* **Notes 10** and **11**).

2. Record evoked composite EPSC (mixed AMPAR/NMDAR-mediated response) from selected cell in voltage clamp mode at −60 mV in Mg^{2+}-free recording ACSF in the presence of 50 μM of bicuculline to block GABAR-mediated currents. At least ten responses should be recorded.
3. Record composite EPSCs from several stimulations in the same solution in the presence of 10 μM of NBQX to block AMPAR-mediated current to obtain isolated NMDAR-mediated current.

3.4 Analysis of Isolated Synaptic NMDAR-Mediated Currents

Several parameters of synaptic NMDAR-mediated currents can be estimated from these recordings, such as: (1) AMPA/NMDA ratio reflecting relative contribution of NMDAR to EPSC; (2) decay kinetics; (3) strength and voltage dependence of the Mg^{2+} block.

1. By mathematical subtraction of NMDA current from total current pure AMPA current can be estimated (*see* Fig. 1) and ratio of AMPA and NMDA currents amplitude (AMPA/NMDA ratio) can be calculated (*see* **Note 12**).
2. For kinetics analysis, fit the decaying phase of the NMDAR-mediated current with the sum of two exponentials: $I(t) = A_1$ exp. $(-t/\tau_1) + A_2$ exp. $(-t/\tau_2)$, where A is the current amplitude; τ is the time constant.
3. For the estimation of the voltage dependence and strength of the Mg^{2+} block of NMDAR channels, record NMDAR currents at different potentials in Mg^{2+}-free recording ACSF as

Fig. 1 Isolation of NMDAR-mediated component of composite eEPSC in Mg^{2+}-free ACSF. (**a**) Traces of averaged eEPSCs recorded in patch clamp at −60 mV in Mg^{2+}-free ACSF. Trace recorded in control condition (*black*) represents composite eEPSC; *red* trace recorded in the presence of 10 μM NBQX represents NMDAR-mediated component. (**b**) Traces of averaged eEPSCs recorded in patch clamp at −60 mV in Mg^{2+}-free ACSF. Traces recorded in the presence of NBQX (NMDAR-mediated current, *red*) and AMPAR-mediated current (*blue*) obtained by the subtraction of NMDA trace from the composite eEPSC

described above (*see* **step 3** in Subheading 3.3). Average traces recorded at the same potential (at least ten responses for each holding potential) and measure current amplitudes. Measure NMDAR-mediated currents at the same holding potentials, but in the presence of Mg^{2+} in recording ACSF. Plot the measured current amplitudes versus corresponding voltages to obtain the current-voltage (I-V) relations for the measured voltage range in Mg^{2+}-free and Mg^{2+} containing ACSF. The I-V relation in Mg^{2+} containing ACSF shows the voltage dependence of Mg^{2+} block. The strength of Mg^{2+} block can be estimated as a ratio of the NMDAR current amplitudes measured in the presence of Mg^{2+} to the amplitude values obtained in the Mg^{2+}-free solution.

3.5 Paired Recordings of Synaptically Connected Neurons

1. Identify and select pair of necessary neighboring cells under visual control. Importantly, use biocytin added to the intracellular solution to confirm identity of both cell types by morphology after the end of the experiment.
2. Establish cell-attached configuration with both cells simultaneously.
3. Reach access resistance higher than 1 GΩ, then switch to the whole-cell configuration in both cells.
4. Check action potential firing patterns for both cells by applying depolarizing current steps.
5. Keep presynaptic cell in current clamp configuration and the postsynaptic cell in voltage clamp configuration.
6. In synaptically connected neurons, suprathreshold intracellular stimulation of the presynaptic cell in current clamp evokes composite EPSCs in the postsynaptic cell. Typically, presynaptic cells can be stimulated with short 5 ms suprathreshold depolarizing current pulses (≅800 pA).
7. After recording 30–50 baseline composite EPSCs (control) evoked at 0.1 Hz, the EPSCs should be averaged. Isolate and analyze NMDAR-mediated component as described above (*see* Subheadings 3.3 and 3.4).
8. If cells are not synaptically connected, you can re-patch another neighboring postsynaptic cell, keeping the presynaptic cell in whole-cell mode. The use of triple or multipatch setups significantly increases probability of success to find connected pairs.

3.6 Recordings of NMDA Component of Spontaneous Miniature EPSCs (mEPSCs)

1. Add 1 μM of tetrodotoxin in ACSF to abolish action potentials mediated postsynaptic currents and 10 μM of bicuculline to block GABA-activated current.
2. Record spontaneous mEPSCs in whole-cell configuration for 15 min in voltage clamp mode at −50 mV with normal ACSF

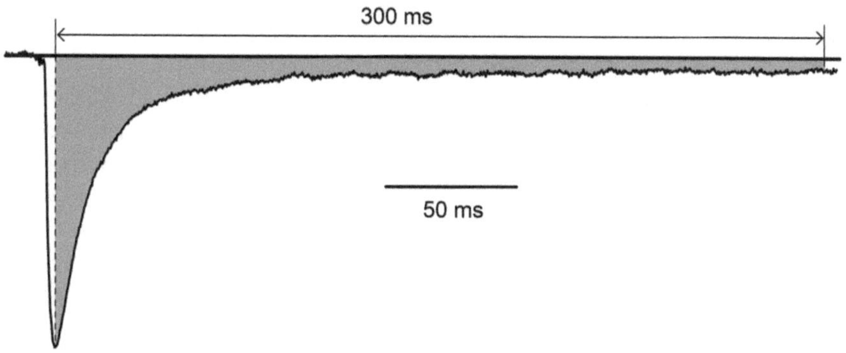

Fig. 2 Estimation of contribution of NMDAR-mediated current to composite mEPSC in physiological Mg^{2+} containing ACSF. Example trace of averaged composite mEPSCs recorded at −50 mV in Mg^{2+} containing recording ACSF. Charge transfer was calculated by integrating the area shown in *gray* under the current waveform in the interval of time between peak of sEPSC and 300 ms after the peak

solution. At this potential the block of NMDAR by Mg^{2+} is largely relieved so that inward current is a composite EPSC.

3. For analysis only events that do not show any signs of multiple peaks (that is contamination of rise or decay phases by subsequent events) should be selected for subsequent analyses of the kinetics and for exponential fitting. Events that do not show a typical synaptic waveform should be rejected.
4. For each neuron, selected traces of individual events (100–200) should be aligned based on the beginning of their rising phases and averaged. These averaged traces from individual experiments should be normalized by peak amplitudes.
5. Relative contribution of NMDAR-mediated component in composite mEPSC can be quantified by measuring charge transfer of mEPSC normalized by the peak. Measurement of charge transfer can be achieved by integrating the area under the current waveform after the peak (*see* Fig. 2) or even more precisely for pure NMDAR-mediated component, after (50–60) ms from the peak. A larger charge transfer corresponds to a slower decay kinetics and increased contribution of the NMDAR-mediated component and vice versa.
6. In addition, to estimate the contribution of NMDAR-mediated component, the weighted time constant of mEPSC decay (τ_w) can be calculated using Eq. 1:

$$\tau_W = (I_{fast}/(I_{fast} + I_{slow}))^* \tau_{fast} + (I_{slow}/(I_{slow} + I_{fast}))^* \tau_{slow}, \quad (1)$$

where I is the amplitude of the fast or slow component, and τ is the respective decay time constant.

4 Notes

1. With this solution, at −50 mV small depolarization will partially relieve magnesium block and you will record complex AMPA/NMDA/GABA currents. GABA-activated currents can be easily separated from glutamate activated currents based on the direction of the peak (upward for GABA, downward for glutamate). However, you can pharmacologically prevent this GABA contamination using 10 μM of bicuculline. The NMDAR-mediated component can be isolated by using selective AMPAR antagonists.
2. To potentiate spontaneous and evoked NMDA currents, 10–15 μM of glycine can be added to the ACSF solution and/or $MgCl_2$ can be removed from this solution.
3. By increasing the resistance of pipettes, you increase the probability of getting gigaseal but decrease the ability to get optimal voltage clamp conditions.
4. *Getting high quality tissue*: For young slices (P0-P10) simple plastic dishes are sufficient. Older slices are more sensitive to ischemia and tissue is less permeant. To avoid any anoxic-related damage of the slices, prefer to use the pore strainers to provide slices oxygenation from both the sides.
5. Minimize animal stress before slicing by avoiding keeping pups far from their mother for too long, avoiding letting them in a bright and noisy room, and reducing as much as possible the time out of the nursery. Velocity of your surgery is crucial, the more time you will use to open the skull and cut the brain, the longer will be the anoxic context and stronger will be the damages in tissue. Cutting in protective isotonic solutions (choline, sucrose, NMDG, K-Gluconate based, etc.) will prevent cellular damage.
6. *Choosing healthy cells*: Watching the cell with interferential contrast gives you some clues to identify good ones. You should identify healthy cells with the intensity of contrast between membrane and extracellular matrix. Cells contour has to be smooth and regular whatever its shape. The inside of the cell has to be bright with as less as possible vacuoles that could reflect a bad cellular state. An easy way to identify dead cells is the appearance of their nuclei that reflects nuclear fragmentation, a sign of apoptosis.
7. *Reaching good seal*: Even if in theory, gigaseal is obtained when your access resistance reaches 1 GΩ, a higher resistance can improve the quality of your recordings by reducing leak currents. In order to increase the seal quality, wait a few minutes before opening the cell.

8. *Gentle sucking*: Certain softwares propose a special function (ZAP), which is a soft alternative solution to go to the whole-cell configuration. By applying a small voltage step (400–800 mV) for a short time (100 μs), you will destabilize membrane lipids and help the opening of the membrane. In our hands, it is a good way to minimize failure of opening.
9. *Waiting for stabilization*: It can take a few minutes for the cell to reach stability: time for the diffusion of the intracellular solution, time for the stabilization of the conductances when you hold potential far from the resting membrane potential. Analyzing stable part of recordings after this stabilization period is a good way to avoid any compensatory mechanisms interfering with the data.
10. External stimulation in the Shaffer collateral and recording of CA1neurons in hippocampus for example or external stimulation in the thalamus and recording of cortical neurons.
11. Adjust the stimulation intensity (usually 50% of the maximal amplitude, unless you need stimulus-dependent protocol) to obtain reliable post-synaptic currents but avoid the appearance of action potentials in your evoked responses.
12. To exclude GABAergic currents use bicuculline (10 μM).

References

1. Collingridge GL, Volianskis A, Bannister N, France G, Hanna L, Mercier M et al (2013) The NMDA receptor as a target for cognitive enhancement. Neuropharmacology 64:13–26. doi:10.1016/j.neuropharm.2012.06.051
2. Vyklicky V, Korinek M, Smejkalova T, Balik A, Krausova B, Kaniakova M et al (2014) Structure, function, and pharmacology of NMDA receptor channels. Physiol Res 63:S191–S203
3. Simon RP, Swan JH, Griffiths T, Meldrum BS (1984) Blockade of *N*-methyl-D-aspartate receptors may protect against ischemic damage in the brain. Science 226(4676):850–852. doi:10.1126/science.6093256
4. Petrenko AB, Yamakura T, Baba H, Shimoji K (2003) The role of *N*-methyl-D-aspartate (NMDA) receptors in pain: a review. Anesth Analg 97(4):1108–1116
5. Gupta K, Hardingham GE, Chandran S (2013) NMDA receptor-dependent glutamate excitotoxicity in human embryonic stem cell-derived neurons. Neurosci Lett 543:95–100. doi:10.1016/j.neulet.2013.03.010
6. Croucher MJ, Collins JF, Meldrum BS (1982) Anticonvulsant action of excitatory amino acid antagonists. Science 216(4548):899–901
7. Lozovaya N, Gataullina S, Tsintsadze T, Tsintsadze V, Pallesi-Pocachard E, Minlebaev M et al (2014) Selective suppression of excessive GluN2C expression rescues early epilepsy in a tuberous sclerosis murine model. Nat Commun 5:4563. doi:10.1038/ncomms5563
8. Zhou Q, Sheng M (2013) NMDA receptors in nervous system diseases. Neuropharmacology 74:69–75. doi:10.1016/j.neuropharm.2013.03.030
9. Wyllie DJ, Livesey MR, Hardingham GE (2013) Influence of GluN2 subunit identity on NMDA receptor function. Neuropharmacology 74:4–17. doi:10.1016/j.neuropharm.2013.01.016
10. Traynelis SF, Wollmuth LP, McBain CJ, Menniti FS, Vance KM, Ogden KK et al (2010) Glutamate receptor ion channels: structure, regulation, and function. Pharmacol Rev 62(3):405–496. doi:10.1124/pr.109.002451
11. Endele S, Rosenberger G, Geider K, Popp B, Tamer C, Stefanova I et al (2010) Mutations in GRIN2A and GRIN2B encoding regulatory subunits of NMDA receptors cause variable neurodevelopmental phenotypes. Nat Genet 42(11):1021–1026. doi:10.1038/ng.677

Chapter 13

Analysis of Functional NMDA Receptors in Astrocytes

Frank Kirchhoff

Abstract

Neuronal *N*-methyl-D-aspartate (NMDA) receptors are well known for their pivotal role in memory formation. Originally, they were thought to be exclusive to neurons. However, numerous studies revealed their functional expression also on various types of glial cells in the nervous system. Here, the methodology how to study the physiology of NMDA receptors selectively on astrocytes will be described in detail. Astrocytes are the main class of neuroglia that control transmitter and ion homeostasis, which link cerebral blood flow and neuronal energy demands, but also affect synaptic transmission directly.

Key words NMDA receptor, Astrocyte, GFAP-EGFP transgenic mice, Patch-clamp recording, Ca^{2+} imaging, Neuron–glial interaction, Mouse, Cortex

1 Introduction

Astrocytes are in intimate contact with neurons in gray matter, particularly at synapses. Several recent studies indicate that there is an intense cross-talk between these two cell types. For that purpose astrocytes express the prerequisite to participate in neuronal communication: numerous glutamate-binding molecules, such as glutamate transporters, α-amino-3-hydroxy-5-methylisoxazole-4-propionic acid (AMPA) receptors, kainate receptors, *N*-methyl-D-aspartate (NMDA) receptors, and metabotropic glutamate receptors.

Astroglial expression of the NMDA receptor appears to be the most enigmatic one. While neuronal NMDA receptors have been studied abundantly and their role in memory formation is undisputable [1], their analysis on glial cells is still sparse. Early evidence came from the immunohistochemical analysis of NMDA receptor expression in the rat cortex, where respective immunolabel could be localized to perisynaptic processes of astrocytes [2]. More recent work, which addressed the physiological role, was facilitated by the use of various transgenic mouse models. We now learn more and

Nail Burnashev and Pierre Szepetowski (eds.), *NMDA Receptors: Methods and Protocols*, Methods in Molecular Biology, vol. 1677, DOI 10.1007/978-1-4939-7321-7_13, © Springer Science+Business Media LLC 2017

more about the functional properties of NMDA receptors on glial cells, on astrocytes but also on oligodendrocytes [3–5].

The combination of mouse genetics, single-cell electrophysiology, and imaging did not only confirm the localized expression of NMDA receptors on the distal processes of the astrocytes, it also uncovered apparent biophysical and pharmacological differences between glial and neuronal receptors [3, 4, 6]. To encourage further research and to provide a methodological guideline to uncover the physiological properties of astroglial NMDA receptors, technical approaches that have been proven successful in the past will be described in detail:

- Electrophysiological characterization of NMDA receptor-mediated currents in slices using either direct drug application or neuronal stimulation.
- Analysis of the kinetic properties of NMDA receptor-mediated currents in acutely isolated, single astrocytes by fast drug application and visualization of NMDA receptor localization by Ca^{2+} imaging using confocal laser-scanning microscopy.

2 Materials

2.1 Solutions

1. Preparation solution for vibratome slices of the mouse brain. Use ultrapure deionized water and dissolve to final concentrations: 87 mM NaCl, 3 mM KCl, 25 mM $NaHCO_3$, 1.25 mM NaH_2PO_4, 3 mM $MgCl_2$, 0.5 mM $CaCl_2$, 75 mM sucrose and 25 mM glucose. $MgCl_2$ and $CaCl_2$ are added from 1 M stock solutions. Before use, the preparation solution is cooled on ice and oxygenated with carbogen (5% CO_2/95% O_2) giving pH 7.4.
2. Bath solution for maintenance and perfusion during patch-clamp recording of vibratome slices. Use ultrapure deionized water and dissolve the various salts at final concentrations: 126 mM NaCl, 3 mM KCl, 25 mM $NaHCO_3$, 15 mM glucose, 1.2 mM NaH_2PO_4, 2 mM $CaCl_2$, and 2 mM $MgCl_2$. By gassing the solutions with carbogen (5% CO_2/95% oxygen), the pH is adjusted to 7.4–7.5. Use 1 M stock solutions to add $CaCl_2$ and $MgCl_2$.
3. External solution for the analysis of acutely isolated astrocytes. Use ultrapure water to prepare external solution for isolated cell experiments containing the following: 135 mM NaCl, 2.7 mM KCl, 2.5 mM $CaCl_2$, 1 mM $MgCl_2$, 10 mM HEPES, 1 mM NaH_2PO_4, and 15 mM glucose. Adjust pH with NaOH to 7.3.
4. Pipette solution for whole-cell patch-clamp recordings. To prepare the pipette solution dissolve at final conventrations:

130 mM KCl, 0.5 mM $CaCl_2$, 3 mM $MgCl_2$, 5 mM Na_2-ATP, 5 mM EGTA, and 10 mM HEPES. The Ca^{2+} activity of the pipette solution will be approximately 11 nM, pH 7.4. Please note that also the Mg^{2+} is chelated by EGTA. The free intracellular Mg^{2+} concentration will be about 140 μM.

5. Drug solutions for pharmacological characterization. Agonists and antagonists of receptors, transporters, and ion channels should be dissolved first in pure water to generate 100 × stock solutions. The solutions should be stored frozen and light protected. Dilute the stock solutions immediately prior to the experiment at final concentrations in bath solution. Drugs final concentrations: 100 μM for glutamate, 100 μM for NMDA, 100 μM for kainate, 50 μM for L-*trans*-pyrrolidine-2,4-dicarboxylic acid (PDC), 20 μM for 6-cyano-7-nitroquinoxaline-2,3-dione (CNQX, disodium salt), 20 μM for 2,3-Dioxo-6-nitro-1,2,3,4-tetrahydrobenzo[f]quinoxaline-7-sulfonamide (NBQX, disodium salt), 100 μM for $CdCl_2$, 20 μM for (+)MK-801, 1 μM for tetrodotoxin citrate (TTX), 200 μM for dihydrokainic acid, and 20 μM for D,L-*threo*-hydroxyaspartate.
6. Solutions for cell labeling and Ca^{2+} imaging. To visualize intracellular Ca^{2+} transients evoked by activation of NMDA receptors on astrocytes, cells have to be perfused by respective indicator dyes. Dissolve Fluo4 pentapotassium salt in Ca^{2+}- and EGTA-free pipette solution for the analysis of astrocytes at a final concentration of 100 μM. Use a long-range Ca^{2+} indicator such as Calcium Orange tetrapotassium salt (200 μM) for the analysis of green fluorescent protein-expressing astrocytes, e.g., of GFAP-EGFP transgenic mice (*see* **Note 1**).

2.2 Transgenic Mice to Study Astrocytes

Use transgenic mice with astrocyte-specific expression of fluorescent proteins. TgN(GFAP-EGFP) mice in which the enhanced green fluorescent protein (EGFP) is under the control of the human glial fibrillary acidic protein (GFAP) promoter have been widely used [3, 4, 7–10]. Already, heterozygous offspring display a robust bright EGFP expression in astrocytes of all brain regions visible by conventional fluorescence microscopy (*see* **Note 2**).

3 Methods

3.1 Preparation of Brain Slices

1. Anesthetize mice with genetically labeled astrocytes (e.g., TgN (GFAP-EGFP) mice, *see* **Note 3**) of defined age (preferentially of 1–4 weeks old) by isoflurane. Decapitate and remove the brain from the skull.

2. Dissect and immerse the brain immediately in an ice-cold preparation solution (**step 1** in Subheading 2.1) oxygenated with carbogen (95% O_2, 5% CO_2). Cut the hemispheres into 150–300 μm-thick slices in frontal orientation using a vibratome.
3. Transfer slices for recovery into a nylon basket slice holder immersed in a warm bath solution (30 °C) for 30 min at room temperature. Subsequently, maintain the slices in an oxygenated bath solution at room temperature.

3.2 Electrophysiology of Astroglial Whole-Cell Membrane Currents In Situ

1. Place the slices for electrophysiological characterization into an observation chamber mounted on the stage of an upright microscope with 40× objective and equipped with water-immersion, infrared optics, and epifluorescence illumination using EGFP and TexasRed filter sets (Fig. 1a–f).
2. Fix the slice with a U-shaped platinum-wire with a grid of nylon threads. Perfuse the chamber continuously with the carbogen-saturated bath solution at a flow rate of 2.5 mL/min.
3. Prepare patch pipettes with a resistance of 7–9 MΩ from borosilicate capillaries (outer diameter: 1.5 mm, inside diameter: 0.86 mm; Sutter, USA) using a Micropipette Puller (Model P-97, Sutter Instrument Co., CA). Apply positive pressure to the recording pipette filled with the pipette solution (*see* **step 3** in Subheading 2.1) while lowering under microscopic control.
4. Identify cell somata of cortical astrocytes using water immersion optics (40×). Blow cellular debris aside and place pipette tip onto the soma surface.
5. Use the EGFP channel to excite EGFP at 488/490 nm and visualize astrocytes at 510 ± 10 nm. Cells will be found located about 10–20 μm beyond the surface or deeper.
6. Record whole-cell membrane currents by amplifier, apply low-pass filter at 3 kHz, and control acquisition by electrophysiology software. Measure resting membrane potential within 30 s after establishing whole-cell recording. Set holding potential in voltage clamp mode to −80 mV.
7. Application of a series of depolarizing and hyperpolarizing voltage steps from a holding potential of −80 mV will evoke passive currents, with nearly linear I–V relationship, as an electrophysiological signature characteristic of astrocytes (Fig. 1b).
8. Evoke NMDA receptor currents by the bath application of NMDA (100 μM) for 30 s. Change the bath perfusate first, to bath solution containing cocktail of inhibitors blocking glial glutamate transporters, AMPA receptor currents, or neuronal activity, and second, to a bath solution containing NMDA in addition (Fig. 1c) (*see* **Note 4**).

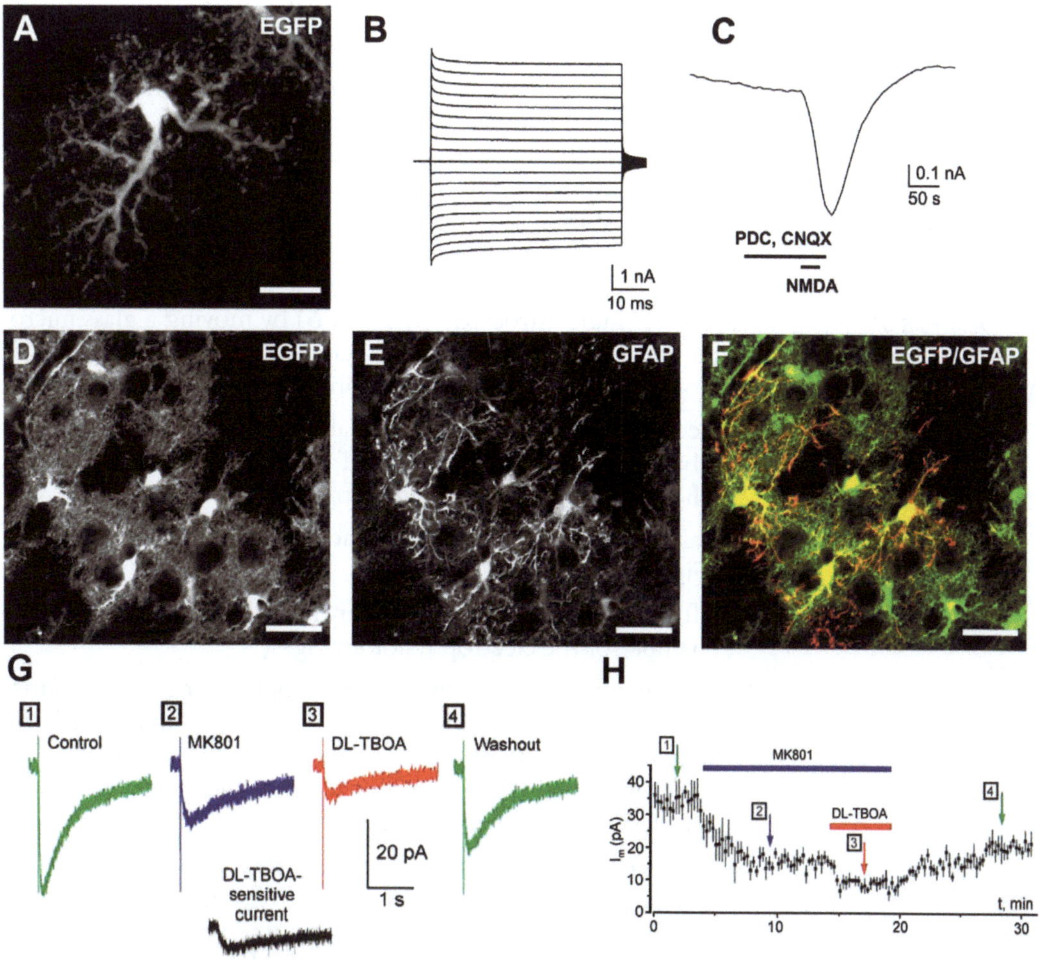

Fig. 1 Morphological and electrophysiological properties of EGFP-positive astrocytes in the slice. (**a**) Confocal laser-scanning microscope image of an astrocyte from a cortical slice expressing the EGFP under the control of the GFAP promoter. Astrocytes under study are characterized by their numerous, highly branched processes. (**b**) Whole-cell currents of the cell shown in (**a**). The cell is characterized by a symmetrical pattern of noninactivating outward and inward currents elicited by depolarizing and hyperpolarizing voltage steps (from −170 to +10 mV). The membrane was clamped at −80 mV. (**c**) NMDA was applied (100 μM) to the same cell in the presence of 50 μM PDC and 50 μM CNQX. (**d**) Micrograph of green fluorescent cells expressing EGFP in a brain slice; (**e**) the same area, in which GFAP expression is visualized by anti-GFAP antibodies; (**f**) merge. The horizontal bars represent 15 μm in A and 30 μm in **d–f**. (**g**, **h**) Synaptic currents mediated by NMDA receptors in astrocytes. Astrocytes in layer II of the slice were identified by EGFP fluorescence; electrical stimulation was in layer IV. (**g**) Synaptically evoked currents are inhibited by MK-801 (10 μM), and the residual current is partially blocked by DL-TBOA (100 μM). The DL-TBOA-sensitive component obtained by subtraction is shown in the inset. The boxed numbers 1–4 are recordings obtained at the respective time points shown in H. (**h**) Time course of recording with pharmacological intervention as indicated. Figure modified from [3, 4]

3.3 Analysis of NMDA Receptor Currents in Astrocytes After Neuronal Stimulation

1. Add 100 μM picrotoxin to the bath solution to reduce the inhibitory drive and allow for maximal stimulation.
2. Locate an astrocyte in cortical layer 2 or 3 and position a concentric bipolar microelectrode in layer V close to the border of layer IV. Set the stimulus duration to 300 μs at 0.25–0.5 Hz and stimulus magnitude three to four times higher than the minimal stimulus required to activate an adjacent pyramidal neuron by a single-axon response [4] (Fig. 1g, h) (*see* **Note 5**).

3.4 Analysis of Acutely Isolated Astrocytes

1. Isolate single astrocytes (*see* **Note 6**) by moving a glass micropipette with a small glass ball at its tip (diameter 200 μm) slowly along the surface of an acutely prepared slice.
2. Keep a distance of 10–50 μm above the slice surface, while vibrating at 100 Hz with lateral displacements of 20–30 μm. The micropipette mills single astrocytes out of the slice.
3. Take care to keep the morphology of cells largely intact (Fig. 2a) (*see* **Note 7**).
4. After single cell isolation, transfer the cells into the observation chamber of the electrophysiology rig.
5. After breakthrough of the gigaseal, apply a series of depolarizing and hyperpolarizing voltage steps (±90 mV in 10-mV steps) from the holding potential of −80 or −70 mV to elicit the characteristic passive currents with a linear current to voltage (I/V) relationship (Fig. 2b).
6. Let recordings start 10 min after getting whole-cell access to ensure equilibration between the pipette solution and the cytosol. The series resistances will vary between 4 and 12 MΩ, and input resistances between 50 and 100 MΩ for astrocytes.
7. Use a fast concentration jump method (*see* **Note 8**) to activate NMDA receptors and reveal their biophysical properties such as rise and decay times of receptor currents (Fig. 2c–f). A solution exchange should occur in less than 20 ms.

3.5 Ca^{2+} Recordings of Astrocytes In Situ

Based on their Ca^{2+} permeability NMDA receptors can be selectively localized on the complex astrocyte morphology in situ by a high-resolution confocal laser-scanning microscope connected to an electrophysiology rig (Fig. 3) [3].

1. Use a cortical brain slice obtained with transgenically labeled astrocytes (e.g., TgN(GFAP-EGFP) mice) and patch-clamp a chosen astrocyte. Use Ca^{2+} and EGTA-free pipette solution containing Calcium Orange tetrapotassium salt (200 μM) or one of the more recently developed red-shifted Ca^{2+} indicators such as CalciumRuby-Nano [11, 12] at the same concentrations.

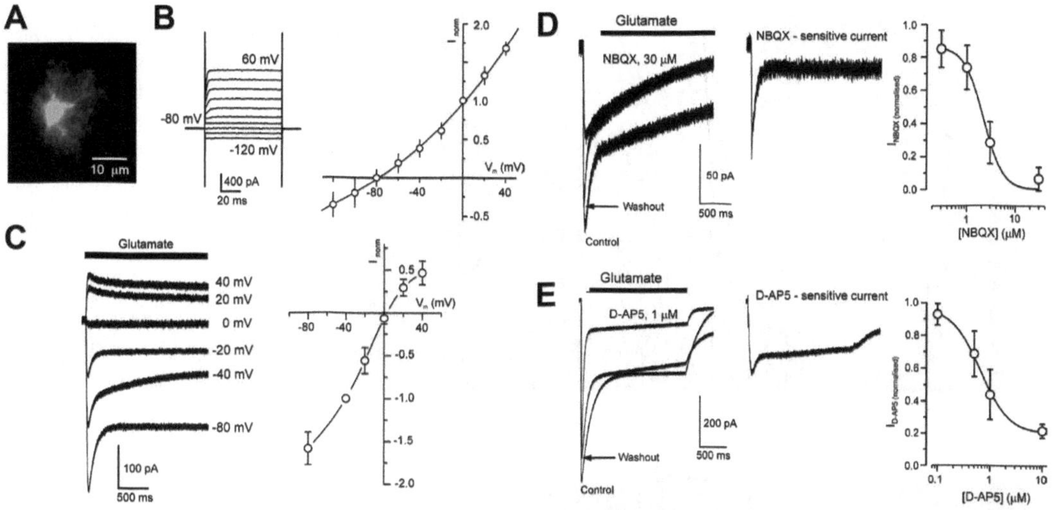

Fig. 2 Electrophysiological and pharmacological characterization of acutely isolated astrocyte. (**a**) Epifluorescence image of an isolated astrocyte. (**b**) Representative whole-cell currents recorded from the astrocyte shown in (**a**) in response to hyperpolarizing and depolarizing steps from −120 to +40 mV (step interval, 20 mV), from the holding potential of −80 mV. Current–voltage relationship of membrane currents was determined from isolated cortical astrocytes. Amplitudes of currents were normalized to the value measured at 0 mV (I_{norm}), data are mean ± SD for 20 cells. (**c**) Voltage dependence of receptor currents evoked by glutamate in an isolated astrocyte. The left panel shows currents induced by application of glutamate (10 μM, 2 s) recorded at different holding potentials. The right panel shows current–voltage relationship constructed from 11 independent experiments. The amplitudes of currents were normalized to the value measured at −40 mV (I_{norm}). (**d**, **e**) Selective AMPA and NMDA antagonists differentially suppress glutamate-induced currents in single astrocytes. (**d**) NBQX inhibits the fast component of glutamate-induced current. Representative traces illustrate the current before, during, and after the application of 30 μM NBQX (*left*) and the NBQX-sensitive current obtained by the subtraction (*middle*). The right panel shows the concentration dependence of the block of the fast component for four cells (IC_{50} 2.2 ± 0.4 μM; Hill coefficient, 1.9). (**e**) D-AP-5 inhibits the slow component of glutamate-induced current. Representative traces demonstrating the effect of 1 μM D-AP-5 (left) and the D-AP-5-sensitive component obtained by subtraction (*middle*) are shown. The right panel shows the concentration dependence of the block for five cells (IC_{50} 0.64 ± 0.1 μM; Hill coefficient, 1.6). Error bars represent SD. Figure modified from [4]

2. Wait for 10 min of dialysis and the cells will be sufficiently filled with the indicator dye. Use Fluo-4 (100 μM) for red-fluorescently labeled (e.g., expressing tdtomato or mRFP1) astrocytes.
3. Induce Ca^{2+} transients either by the bath application of NMDA (100 μM) or by electrical stimulation of adjacent neuronal fibers. Record images of 256 × 256 or 512 × 512 pixels at 1–5 Hz before, during, and after agonist application.
4. Excite Fluo-4 or Calcium Orange at 488 or at 514 nm, and detect emission with 510 nm or 600 nm long pass filters, respectively.
5. Display Ca^{2+} concentration changes as fluorescent intensity ratios F/F0 with the resting fluorescence value F0 determined at the beginning of each experiment.

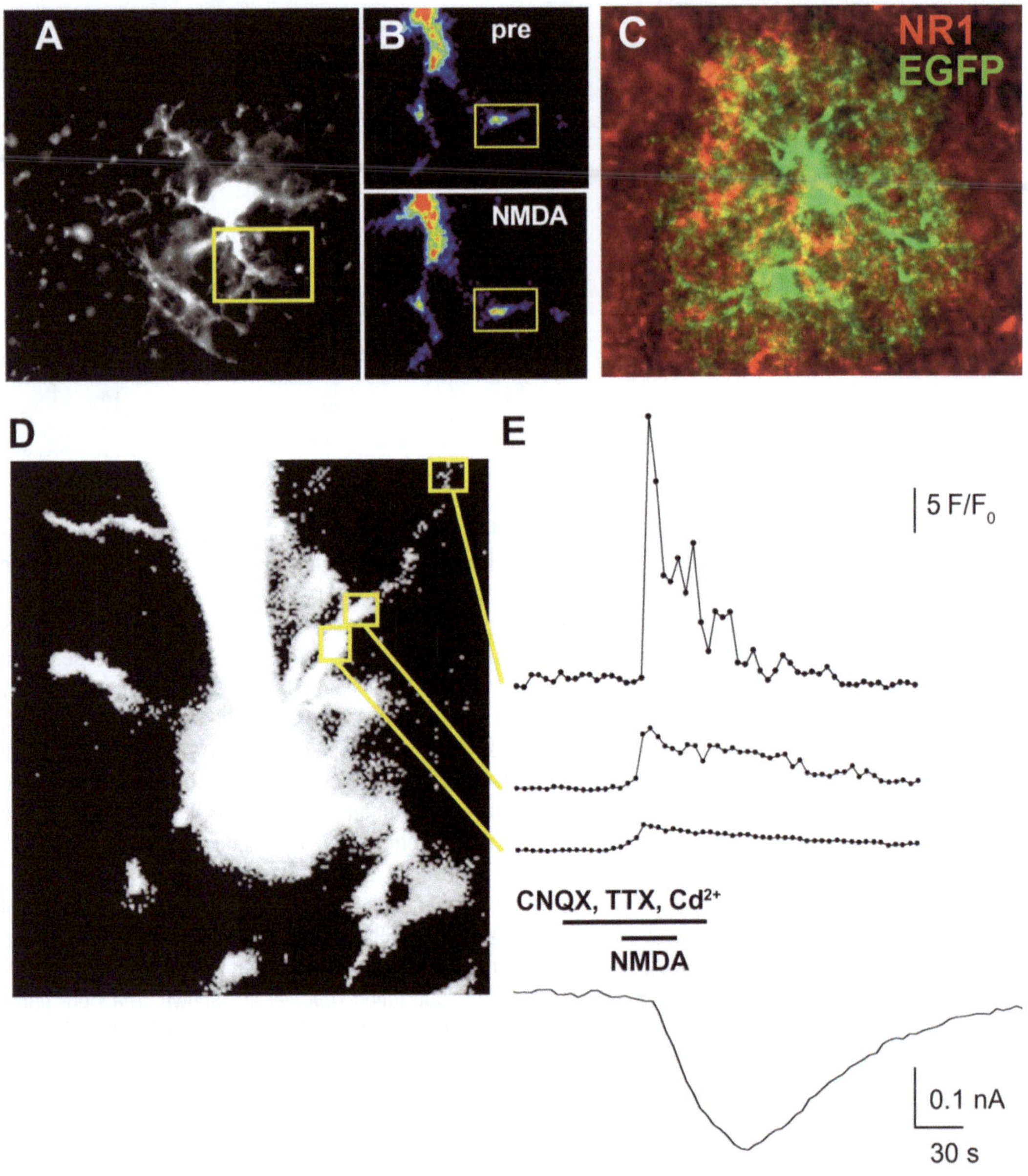

Fig. 3 NMDA triggers an increase in $[Ca^{2+}]_i$. (**a**) The confocal laser-scanning micrograph displays the EGFP fluorescence of an astrocyte in a cortical slice. (**b**) Magnified view from A and displaying the fluorescence of the Ca^{2+} indicator Calcium *Orange* prior to (upper micrograph) and during NMDA application (lower micrograph). Note the increase in fluorescence at the processes (*yellow box*). (**c**) Immunohistochemical localization of the NMDA receptor subunit NR1 (*red*) at the surface of an EGFP-labeled astrocyte (*green*). Note the localization at processes (*yellow*). (**d**) Micrograph of Fluo-4 fluorescence an unlabeled astrocyte of a non-transgenic mouse. Note the recording pipette approaching the cell from the top. Fluorescence changes (F/F_0) were analyzed in the areas denoted by the squares and are displayed in the top three traces in (**e**). The lower trace is the simultaneously recorded current response. NMDA (100 μM) and CNQX, TTX, Cd^{2+} were applied as indicated by bars. Note the larger F/F_0 ratio at the most distal process indicating highly localized expression of NMDA receptor complexes. Figure modified from [3]

4 Notes

1. Alternatively, you may test the novel Ca^{2+} indicator CaRuby-Nano (200 μM) [3, 11, 12].
2. Alternative mouse lines are available. Aldh1l1-EGFP have been generated using bacterial artificial chromosomes [13, 14]. In GFAP-CreERT2 or GLAST-CreERT2 × ROSA26-floxed-STOP-tdTomato mice the astroglial expression of the red fluorescent protein tdTomato has to be induced by intraperitoneal tamoxifen injection [15–17].
3. Cortical astrocytes of wild-type mice or other species without fluorescent protein expression can be labeled by the red fluorescent dye sulforhodamine 101 (SR101, TexasRed) that is selectively transported by the multispecific thyroid hormone transporter OATP1C1 into the cytosol [18]. To label astrocytes incubate brain slices in a bath solution containing 1 μM sulforhodamine 101 for 20 min at 30 °C. Subsequently remove excess dye by changing to a fresh bath solution.
4. The analysis of NMDA receptor-mediated inward currents in astrocytes can be performed by the direct bath application of NMDA [3]. This approach will reveal almost the total amount of membrane currents. However, for this purpose, other currents that could be indirectly elicited in astrocytes must be blocked. Such currents could be glutamate uptake or astroglial AMPA receptor currents elicited by glutamate release from adjacent neurons or P2X receptor-mediated currents activated by neuronal ATP release. An effective blocking solution is composed of 50 μM PDC to block glutamate transport, 20 μM CNQX or NBQX (block of AMPA receptors), 100 μM $CdCl_2$ and 1 μM tetrodotoxin to block neuronal voltage-gated ion channels, thereby preventing neuronal transmitter release. Alternative effective glutamate transport blockers are dihydrokainate and TBOA. However, a disadvantage of receptor stimulation via bath perfusion is the slow activation kinetic that is largely affected by flow-rate and drug diffusion into the brain slice. An alternative approach is the direct intrinsic stimulation of astroglial NMDA receptors via evoked neuronal glutamate release.
5. Note that the currents evoked in the astrocytes will vary substantially. Depending on the path of the stimulated axon different numbers of astroglial NMDA receptors will be activated. The average amplitude will normally reach only 10–20% of the maximal current evoked by a saturating stimulus, in general tenfold larger than a minimal stimulus. The specificity of astroglial NMDA receptors can further be substantiated by blocking the astroglial NMDA receptor current with MK801 at 1 μM.

6. Acutely isolated single astrocytes from a vibratome brain slice are the best model system to study astroglial membrane currents without modulatory input from adjacent neurons. For that purpose the "vibrating ball" technique has been developed [19] (Vorobjev, 1991; Pankratov et al., 2002, 2003; Akaike and Moorhouse, 2003).
7. In contrast to an enzymatic isolation procedure (e.g., as used for FACsorting), the ectodomains of cell surface proteins, and therefore transmitter receptors such as the NMDA receptor, stay fully functional.
8. A "square-pulse" concentration jump device has originally been developed by Oleg Krishtal [20] and was later modified by others to study glial cells [4, 21, 22].

Acknowledgments

The author is particularly grateful to the long-term collaboration with Alexei Verkhratsky (University of Manchester, UK), Ulyana Lalo and Yuriy Pankratov (University of Warwick, Coventry, UK).

The author's lab received grant support from Deutsche Forschungsgemeinschaft DFG SPP 1757 and SFB 894; Fondation pour l'Aide a la Recherche sur la Sclerose En Plaques and Association Française contre les Myopathies (ARSEP-AFM); European Commission FP7-202167 NeuroGLIA and FP7-People ITN-237956 EdU-Glia.

References

1. Tsien JZ, Huerta PT, Tonegawa S (1996) The essential role of hippocampal CA1 NMDA receptor-dependent synaptic plasticity in spatial memory. Cell 87(7):1327–1338
2. Conti F, DeBiasi S, Minelli A, Melone M (1996) Expression of NR1 and NR2A/B subunits of the NMDA receptor in cortical astrocytes. Glia 17(3):254–258. doi:10.1002/(SICI)1098-1136(199607)17:3<254::AID-GLIA7>3.0.CO;2-0
3. Schipke CG, Ohlemeyer C, Matyash M, Nolte C, Kettenmann H, Kirchhoff F (2001) Astrocytes of the mouse neocortex express functional *N*-methyl-D-aspartate receptors. FASEB J 15(7):1270–1272
4. Lalo U, Pankratov Y, Kirchhoff F, North RA, Verkhratsky A (2006) NMDA receptors mediate neuron-to-glia signaling in mouse cortical astrocytes. J Neurosci 26(10):2673–2683. doi:10.1523/JNEUROSCI.4689-05.2006
5. Saab AS, Tzvetavona ID, Trevisiol A, Baltan S, Dibaj P, Kusch K, Mobius W, Goetze B, Jahn HM, Huang W, Steffens H, Schomburg ED, Perez-Samartin A, Perez-Cerda F, Bakhtiari D, Matute C, Lowel S, Griesinger C, Hirrlinger J, Kirchhoff F, Nave KA (2016) Oligodendroglial NMDA receptors regulate glucose import and axonal energy metabolism. Neuron 91 (1):119–132. doi:10.1016/j.neuron.2016.05.016
6. Palygin O, Lalo U, Pankratov Y (2011) Distinct pharmacological and functional properties of NMDA receptors in mouse cortical astrocytes. Br J Pharmacol 163 (8):1755–1766. doi:10.1111/j.1476-5381.2011.01374.x
7. Nolte C, Matyash M, Pivneva T, Schipke CG, Ohlemeyer C, Hanisch UK, Kirchhoff F, Kettenmann H (2001) GFAP promoter-controlled EGFP-expressing transgenic mice: a tool to visualize astrocytes and astrogliosis in living brain tissue. Glia 33(1):72–86. doi:10.1002/1098-1136(20010101)33:1<72::AID-GLIA1007>3.0.CO;2-A. [pii]

8. Malatesta P, Hack MA, Hartfuss E, Kettenmann H, Klinkert W, Kirchhoff F, Gotz M (2003) Neuronal or glial progeny: Regional differences in radial glia fate. Neuron 37 (5):751–764. doi:10.1016/s0896-6273(03)00116-8
9. Matthias K, Kirchhoff F, Seifert G, Hüttmann K, Matyash M, Kettenmann H, Steinhäuser C (2003) Segregated expression of AMPA-type glutamate receptors and glutamate transporters defines distinct astrocyte populations in the mouse hippocampus. J Neurosci 23 (5):1750–1758. doi:23/5/1750 [pii]
10. Hirrlinger J, Hulsmann S, Kirchhoff F (2004) Astroglial processes show spontaneous motility at active synaptic terminals in situ. Eur J Neurosci 20(8):2235–2239. doi:10.1111/j.1460-9568.2004.03689.x
11. Collot M, Wilms CD, Bentkhayet A, Marcaggi P, Couchman K, Charpak S, Dieudonne S, Hausser M, Feltz A, Mallet JM (2015) CaRuby-Nano: a novel high affinity calcium probe for dual color imaging. Elife 4. doi:10.7554/eLife.05808
12. Oheim M, van 't Hoff M, Feltz A, Zamaleeva A, Mallet JM, Collot M (2014) New red-fluorescent calcium indicators for optogenetics, photoactivation and multi-color imaging. Biochim Biophys Acta 1843(10):2284–2306. doi:10.1016/j.bbamcr.2014.03.010
13. Doyle JP, Dougherty JD, Heiman M, Schmidt EF, Stevens TR, Ma G, Bupp S, Shrestha P, Shah RD, Doughty ML, Gong S, Greengard P, Heintz N (2008) Application of a translational profiling approach for the comparative analysis of CNS cell types. Cell 135 (4):749–762. doi:10.1016/j.cell.2008.10.029
14. Winchenbach J, Duking T, Berghoff SA, Stumpf SK, Hulsmann S, Nave KA, Saher G (2016) Inducible targeting of CNS astrocytes in Aldh1l1-CreERT2 BAC transgenic mice. F1000Res 5:2934. doi:10.12688/f1000research.10509.1
15. Mori T, Tanaka K, Buffo A, Wurst W, Kuhn R, Gotz M (2006) Inducible gene deletion in astroglia and radial glia—a valuable tool for functional and lineage analysis. Glia 54 (1):21–34. doi:10.1002/glia.20350
16. Madisen L, Zwingman TA, Sunkin SM, Oh SW, Zariwala HA, Gu H, Ng LL, Palmiter RD, Hawrylycz MJ, Jones AR, Lein ES, Zeng H (2010) A robust and high-throughput Cre reporting and characterization system for the whole mouse brain. Nat Neurosci 13 (1):133–140. doi:10.1038/nn.2467
17. Hirrlinger PG, Scheller A, Braun C, Hirrlinger J, Kirchhoff F (2006) Temporal control of gene recombination in astrocytes by transgenic expression of the tamoxifen-inducible DNA recombinase variant CreERT2. Glia 54 (1):11–20. doi:10.1002/glia.20342
18. Schnell C, Shahmoradi A, Wichert SP, Mayerl S, Hagos Y, Heuer H, Rossner MJ, Hulsmann S (2015) The multispecific thyroid hormone transporter OATP1C1 mediates cell-specific sulforhodamine 101-labeling of hippocampal astrocytes. Brain Struct Funct 220 (1):193–203. doi:10.1007/s00429-013-0645-0
19. Vorobjev VS (1991) Vibrodissociation of sliced mammalian nervous tissue. J Neurosci Methods 38(2-3):145–150
20. Krishtal OA, Marchenko SM, Pidoplichko VI (1983) Receptor for ATP in the membrane of mammalian sensory neurones. Neurosci Lett 35(1):41–45
21. Lalo UV, Pankratov YV, Arndts D, Krishtal OA (2001) Omega-conotoxin GVIA potently inhibits the currents mediated by P2X receptors in rat DRG neurons. Brain Res Bull 54 (5):507–512
22. Seifert G, Steinhauser C (1995) Glial cells in the mouse hippocampus express AMPA receptors with an intermediate Ca2+ permeability. Eur J Neurosci 7(9):1872–1881

Chapter 14

GluNs Detection and Functions in Microglial Cells

Stéphane Peineau, Vincent Degos, Catherine Verney, and Pierre Gressens

Abstract

Proving endogenous GluN presence and functions in microglia require complementary steps to demonstrate (1) that GluN genes are transcripted and translated, (2) their cellular localization, (3) that the GluN are functional, and (4) the role of the functional GluN. The complete demonstration is performed by using mRNA detection technics, western blots, immunofluorescence, electrophysiology, calcium imaging, morphology studies, multiplex immunoassay together with conditional microglial Knock-Out mice and brain lesion models.

Key words Microglia, Brain, NMDA receptor, GluN, Western blots, Immunofluorescence, Electrophysiology, Calcium imaging, Morphology studies, Multiplex immunoassay

1 Introduction

Our lab is focused on post-natal brain injury and two hot topics were studied in parallel: excitotoxicity with glutamate excitation and neuroinflammation with microglial activation. In 2005, Catherine Verney incidentally described that microglial cells were expressing GluN1 subunit receptors in human fetuses. At this time, the glutamate system was supposed to be dedicated for polarized cells and we aim to understand if this in vivo expression was just an artifact or if it was associated with a function. Proving GluN presence and function in a cell type require complementary methods. First is the detection of GluN mRNAs which allow demonstrating that GluN genes are transcripted. Second is the detection of the GluN proteins to demonstrate that mRNAs are translated. Third is the cellular localization of the GluN. Usually, cellular localization is performed with fluorescence techniques. However, it is tricky to fully demonstrate with such techniques whether the receptors are localized at the plasma membrane or in a sub-membrane compartment. Fourth is the demonstration of the presence of functional receptors with electrophysiology and calcium imaging approaches. Fifth is the role of the functional GluN for the studied cell type.

Nail Burnashev and Pierre Szepetowski (eds.), *NMDA Receptors: Methods and Protocols*, Methods in Molecular Biology, vol. 1677, DOI 10.1007/978-1-4939-7321-7_14,

The fifth step will depend on the cell type. For the microglia, the most relevant demonstration is to show that GluNs are able to activate them. Activation of microglia can be evaluated by quantifying cytokines/chemokines release, detecting ROS and NO production, and studying cell morphology.

Using relatively simple biological tools with a dedicated strategy, our group was allowed (1) to demonstrate GluN presence (GluN1/GluN2B/GluN2C mRNAs and proteins) in human and murine microglial cells and (2) that microglial GluNs are functional, (3) to show that microglial GluN activation is associated with changes in microglia phenotype and with microglial activation. Finally, we generate a conditional microglia GluN1 knockout mouse line to demonstrate an in vivo deleterious physiological role of microglia GluNs activation on neuronal survival.

2 Materials

2.1 Prepare Dedicated Biological Materials

2.1.1 Primary Microglial Cultures

1. Dissect cortices of newborn *Swiss* mice (P0) in PBS 1×, 6% glucose, and 2% penicillin-streptomycin (PS).
2. Remove the meninges, and chop the cortices into small pieces to subsequently mechanically dissociate the tissue with Pasteur pipette.
3. Suspend the cells in precooled low glucose Dulbecco's modified Eagle's minimum essential medium (DMEM) supplemented with 10% fetal calf serum (FCS) and 0.01% PS.
4. Isolate microglial cells from primary mixed glial cultures on day in-vitro 14 (DIV14) by vigorous shaking parafilm-sealed culture dishes for 20 min at room temperature (RT) and repeat rinsing with their medium, followed by smooth centrifugation.
5. Resuspend the pellet and culture the cells in DMEM supplemented with 10% FCS at a concentration of 5×10^5 cells/mL.
6. Culture microglia for further 3 days in 500 μL of DMEM supplemented with 10% FCS and 0.01% penicillin-streptomycin.
7. Verify the purity of primary microglia cultures by immunostaining using cell-type specific antibodies against tomatolectine (microglia), GFAP (astrocytes), and NeuN (neurons), respectively.

2.1.2 Primary Neuronal Cultures

1. Gather embryos from pregnant Swiss mice at E14.4 sacrificed through elongation.
2. Dissect the cortices and remove of the meninges, then mince the cortices into small pieces and chemically dissociate with 0.25% trypsin and 1% DNAse at 37 °C for 20 min.

3. Stop the reaction by the addition of 0.001% horse serum, and subsequently dissociate mechanically cortex pieces.
4. Suspend the cells in 12-well culture plates (300×10^3 cells per well) precoated with 15 µg/mL poly-DL-ornithine (PO) or 96-well culture plates (7×10^4 cells per well) precoated with 30 µg/mL PO and culture in Neurobasal Medium supplemented with 2% B27, 1% glutamine 100×, and 0.05% PS.
5. Treat the cultured cells with 1 mM AraC on DIV4.
6. Exchange one third of the culture medium for fresh solution twice a week.

2.1.3 Human Tissues and Tissue Processing

1. Obtain approvals from the appropriate local and National Ethics committees.
2. Obtain fetal and adult brain specimens within 48 h postmortem.
3. Cut blocks of cortical areas of 1 cm thickness and fix in 4% PFA in 0.12 M PB for 6–30 h.
4. After cryoprotection, freeze the blocks, then collect coronal cryostat sections of 15 µm thickness on Superfrost plus slides, and store at −80 °C.
5. Brain sections can also be provided by BNE BrainNet Europe (http://www.brainnet-europe.org/).

2.1.4 Conditional GluN1 Microglia

1. Bread $Nr1^{flox/flox}$ mice with $LysM^{cre/cre}$ mice resulting in heterozygous $Nr1^{flox/-}$ mice with $Nr1^{flox/flox}$ mice.
2. Rebread the next generations among each other to obtain $Nr1^{flox/flox}$ $LysM^{cre/cre}$ mice (KO).
3. Use $Nr1^{flox/flox}$ $LysM^{-/-}$ as controls.
4. Keep mice in a 12-h light–12-h dark cycle under enriched environment. Conditions and experimental protocols must be performed in accordance with the local guidelines following the approval of the institutional review committee.

2.1.5 Preparation of Brains from In Vivo Excitotoxic Brain Lesion Models

1. Perform an intracerebral injection of 10 µg of glutamate analogue ibotenate (5 µg/µL) on P5, respectively, as described previously [1] into the neopallial parenchyma of neonatal mice with a 50 µL Hamilton syringe mounted on a calibrated microdispenser attached to a mechanically rigid holder.
2. Insert the needle 2 mm under the external surface of scalp skin in the frontoparietal area of the right hemisphere, 2 mm from the midline in the lateral–medial plane, and 3 mm in the rostro–caudal plane from the junction between the sagittal and lambdoid sutures.

3. Inject two boluses of 1 μL at an interval of 30 s and keep the needle in place for an additional 30 s.
4. After the injections, allow pups to recover from anesthesia, and return them to their dams.
5. For adult mice, inject 10 μg of ibotenate intracranially following a trepanation (0.8 mm hole) under general anesthesia with 150 μL 5% w/v chloral hydrate and antibiotic treatment with 1 mg of penicillin.
6. Suture the cutaneous lesion and allow animals to recover from anesthesia at 37 °C.
7. Sacrify pups by decapitation (1 and 5 days after ibotenate injection).
8. Sacrify adult mice after isoflurane inhalation (5 days after ibotenate injection) followed by intracardial perfusion with 4% paraformaldehyde (PFA) in phosphate buffer (PB; 0.12 M TPO4, pH 7.4).
9. Remove brains and postfix in the same fixative solution for five further days at room temperature.

2.1.6 Preparation of cDNAs

1. Extract total RNA from primary microglial or neuronal cell cultures (*see above*), and purify with DNAseI treatment, as described previously [2, 3].
2. Assess for RNA quality and concentration by spectrophotometry and capillary electrophoresis on RNAstdsens biochips.
3. Design sense and antisense NR subunits, iNOS and SOD1 primers.
4. Generate reverse transcription.

2.1.7 Preparation of Proteins

1. Isolate protein extracts from primary microglia or neuron cultures (*see above*) by homogenization in Hepes buffer containing protease inhibitors (Cell Lysis Buffer).
2. Quantify protein concentrations by spectroscopy (Bradford protein assay).
3. Denaturate protein extracts in Laemmli sample loading buffer.
4. Separate proteins by sodium dodecyl sulfate polyacrylamide gel electrophoresis (SDS-PAGE) in a NuPageTM 10% gel.
5. Electrotransfer in transfer buffer to a PVDF membrane.
6. Block in buffer (TBS 1× with TWEEN20 0.1% v/v).
7. Visualize immunoreactive protein using an enhanced chemiluminescence system and picture with radiographic films.

2.1.8 Preparation of Biological Material for Immunocytochemistry

1. Perform sagittal cryostat sections of infant (P5) and adult Swiss mice (P45) animals as well as fixed primary microglia and neuronal cultures (*see above*) with PFA 4% in 0.12 M phosphate buffer (PB).

2.1.9 Preparation of Biological Material for Electrophysiology and for Calcium Imaging

1. Prepare primary microglia (*see above*) in 6-well-plates with 60×10^4cells/well with coverslip.

2.1.10 Preparation of Biological Material for Multiplex Immunoassay

1. Prepare primary microglia (*see* above) in 24-well plates with 15×10^4 cells/well.

2.2 Solutions

2.2.1 Preparation of Solutions for Electrophysiology

1. Prepare all the solutions either with ultrapure water or natural spring water (*see* **Note 1**).
2. Prepare Phosphate Buffer Saline (PBS) solution at 1× and keep it at 0 °C.
3. Prepare NMDA solution. Solve 50 mg of NMDA in 3.398 mL water. Solution is at 100 mM. Do 30 μL aliquot of NMDA solution 100 mM. Store at −20 °C.
4. Prepare D-serine solution. Solve 2.1018 mg D-serine in 1 mL water. Solution is at 20 mM. Do 10 μL aliquot of D-Serine solution 20 mM. Store at −20 °C.
5. Prepare MK-801 solution. Solve 10 mg MK-801 in 2.964 mL water. Solution is at 10 mM. Do 10 μL aliquot of MK-801 solution 10 mM. Store at −20 °C.
6. Prepare 1 M $CaCl_2$ solution. Add 29.404 g of NaCl in 200 mL water. It can be stored at +4 °C for weeks.
7. Prepare Artificial Cerebrospinal Fluid (aCSF) solution. In a beaker, add 7.246 g of NaCl (124 mM), 0.298 g of KCl (4 mM), 0.172 g of NaH_2PO_4, $1H_2O$ (1.25 mM), 2.184 g of $NaHCO_3$ (26 mM) and 0.901 g of D-Glucose (5 mM). Add 1 L of water. Stir. Add 2 mL of 1 M $CaCl_2$ solution. Adjust to pH = 7.30 by adding HCl. Control osmolarity is close to 300 mOsm. Store at room temperature. It is to be used on the day of preparation.
8. Prepare an intracellular Patch-clamp solution. In a beaker, add 889 mg of $CsMeSO_4$ (130 mM), 71.49 mg of HEPES (10 mM), 14.03 mg of NaCl (8 mM), 5.71 mg of EGTA (0.5 mM), 60.86 mg of Mg-ATP (4 mM), 4.71 mg of Na-GTP (0.3 mM), and 50 mg of lidocaine N-ethyl bromide (QX-314, 5 mM). Add 30 mL of water. Stir. Adjust to PH = 7.2 by adding CsOH (in solution). Control osmolarity is close to 285 mOsm. Do 400 μL aliquot of intracellular Patch-Clamp solution. Store at −20 °C.

9. Unfreeze 1 aliquot of NMDA solution and 1 aliquot of D-serine solution. Add 30 μL of NMDA 100 mM and 10 μL of D-Serine 20 mM–10 mL of aCSF to obtain a final concentration of 300 μM NMDA/ 20 μM D-Serine.
10. Add 1 MK-801 aliquot at 10 mM–10 mL of aCSF to prepare MK-801 solution.
11. Add 1 MK-801 aliquot at 10 mM–10 mL of NMDA/D-serine solution.

2.2.2 Preparation of Solutions for Calcium Imaging

1. Prepare recording solution. Add 20 mM of Hepes/Tris (pH 7.4), 1.3 mM of $CaCl_2$, 5 mM of D-Glucose in 1 L PBS 1× buffer without Mg^{2+}.
2. Prepare imaging solution. Add 2 μM of FURA 2-AM and 0.02% of pluronic F-127 to recording solution.
3. Add NMDA and D-Serine from stock solution in the culture dish to obtain 300 μM NMDA/ 20 μM D-Serine final concentration.

3 Methods

3.1 GluN mRNAs Detection

1. Remove smoothly media from cell culture.
2. Clean out cell cultures were three times with iced PBS 1×.
3. Remove cells from the plastic by using 100 μL of a mixture with the reagent from extraction mRNA silica-membrane spin columns associated with 1% betamercapto-ethanol.
4. Directly freeze at −80 °C.
5. Load 900 ng of total RNA to reverse transcription.
6. Run RT-PCRs with dedicated primers for *GRIN1*, *GRIN2A*, *GRIN2B*, *GRIN2C*, *GRIN2D*, *GRIN3A*, *iNOS*, and *SOD1*.

3.2 GluN Protein Detection

1. Denaturate protein extracts (30 μg per sample) and a biotinylated molecular weight marker at 95 °C for 10 min.
2. After the transfer, incubate the membrane for 1 h at RT in the blocking buffer.
3. Incubate overnight at 4 °C with rabbit anti-GluN1 and mouse anti-beta-actin, respectively.
4. Incubate for 45 min with anti-rabbit horseradish peroxidase-labeled secondary antibody.
5. Perform serial exposures with radiographic films.
6. Run Western blot experiments in triplicate.

GluN1 antibody specificity was tested with the GluN1-antibody corresponding peptide (20 μg for 0.1 μg of antibody).

3.3 GluN Protein Cellular Localization in Brain

1. Incubate brain sections (sagittal cryostat sections of infant (P5), adult Swiss mice animals for example) in 10% normal donkey serum (NDS) for 30 min at room temperature.
2. GluN1 cellular localization with rabbit polyclonal anti-GluN1 antibody. Visualize with biotinylated anti-rabbit Ig followed by streptavidin- CY3 (1:1000). Then fix brain sections with PFA 4% for 30 min and repeated rinsing. Incubate in rabbit anti-Iba1 and visualize with Alexa Fluor® 488 anti-rabbit antibody (*see* **Note 2**).
3. GluN2B cellular localization with mouse polyclonal anti-GluN2B antibody. Add biotinylated tomatolectin (1:1000) for 2 days at 4 °C in a humidity chamber (*see* **Note 3**). Alternatively, human brain sections can be incubated with mouse anti-CD68 instead of applying tomatolectin with an additional step of incubation with biotinylated anti-mouse Ig. Then incubate subsequently in secondary AlexaFluor® 488 donkey anti-rabbit IgG antibody or donkey anti-mouse IgG antibody and Streptavidin-Cy3 for 2 h at RT.
4. GluN2D cellular localization with rabbit polyclonal anti-GluN2D. Add biotinylated tomatolectin for 2 days at 4 °C in a humidity chamber (*see* **Note 3**). Alternatively, the human brain sections can be incubated with mouse anti-CD68 instead of applying tomatolectin with an additional step of incubation with biotinylated anti-mouse Ig. Incubate subsequently in secondary AlexaFluor® 488 donkey anti-rabbit IgG antibody or donkey anti-mouse IgG antibody and Streptavidin-Cy3 for 2 h at RT.
5. Label nuclei with 4′,6- diamidino-2-phenylindole (DAPI).

3.4 GluN Protein Cellular Localization in Culture

1. Fix primary microglia cultures at 4 °C for 20 min.
2. Rinse with PBS.
3. Pre-block in 10% NDS for 30 min.
4. Incubate fixed cells on coverslips with rabbit polyclonal anti-GluN1 antibody, mouse polyclonal anti-GluN2B antibody, or rabbit polyclonal anti-GluN2D and biotinylated tomatolectin overnight at 4 °C and subsequently in secondary antibodies at RT as specified in Subheading 3.3.
5. Label nuclei with DAPI (*see* **Notes 4 and 5**).

3.5 Define Cell Phenotype After GluN1 Excitation with Luminex

1. Isolate microglial cells from astrocyte (*see* above in Subheading **2.1.1**).
2. Culture microglia plated for one further day in 500 μL of DMEM supplemented with 10% FCS and 0.01% penicillin-streptomycin.

3. Treat the cultures with PBS 1× (control), NMDA 30 μM, NMDA 300 μM, NMDA 300 μM and MK801 10 μM, MK801 10 μM, respectively, by using stock solutions of NMDA 100 mM and MK801 10 mM. Add the volume of NMDA or MK801 according to the volume of medium present in the culture dishes.
4. Harvest the medium 12 h after treatment initiation.
5. Freeze at −80 °C.
6. Process for Luminex analysis.
7. Run all samples ($n = 3$ minimum) in duplicates.

3.6 Define GluN Functionality

Studying GluN functionality is performed by two complementary technical approaches.

3.6.1 Electrophysiology

One of the most accurate electrophysiology techniques to identify NMDA current is Patch-Clamp in whole cell configuration (Voltage Clamp mode).

1. Continuously bubble all solutions with carbogen (5% CO_2, 95% O_2).
2. All whole-cell recordings are performed at Room Temperature (22–26 °C).
3. Identify microglia with a microscope (×40) and patch the soma using patch electrodes (5–7 MΩ) that contain intracellular Patch-Clamp solution.
4. Once patch is performed, start the record.
5. Record cells continuously at a holding potential of −70 mV.
6. Once baseline is stable (*see* **Note 6**), bath apply NMDA/D-serine solution during 30 s through the rapid solution perfusion system (*see* **Note** 7) of the electrophysiology setup.
7. To control that the recorded current is pure NMDA current, first add MK801-solution to the perfusion system and once baseline is stable, apply the NMDA/D-serine/MK-801 solution during 30 s through the rapid solution perfusion system of the electrophysiology setup. No current must be recorded.

3.6.2 Calcium Imaging

1. Wash microglial cells cultured on glass coverslips twice in a recording solution and incubate for 1 h at 37 °C with an imaging solution.
2. Wash with the recording solution at 37 °C.
3. Add the recording solution at 37 °C and install the culture dish in a micro-incubation system to maintain it at 37 °C all time.
4. Start calcium imaging record to perform baseline.

5. To perform MK-801 experiment, add 10 μM MK801, 5 min before adding NMDA/D-serine.
6. Apply Thapsigargin 0.5 μM at the end of the experiment.
7. Acquire images of microglial cells at intervals of 2 s at 340 and 380 nm and integrate in real time by an image processor.
8. DIC images are taken alternating with the fluorescent images.
9. Calculate Intracellular Ca^{2+} changes from the ratio of fluorescence intensities at 340 and 380 nm taken after background subtraction and normalize on the average of ratios of the recording first minute.
10. Analyze cells individually using dedicated software.
11. Cells presenting less than 5% fluorescence increase compared to baseline were discarded.

3.7 Definition of GluN Role with CreLox Mice: Brain Lesion Size Quantification (See Note 8)

1. After fixation, dehydrate the brains in alcohol.
2. Embed brains in paraffin.
3. Cut 15-mm-thick coronal serial sections.
4. Stain with cresyl violet.

Each brain is completely and serially sectioned from the frontal pole to the occipital lobes. Neocortical and white matter lesions are defined by the maximal length of three orthogonal axes: the lateral–medial axis (in a coronal plane), the radial axis (also in a coronal plane, from the pial surface to the lateral ventricle), and the fronto–occipital axis (in a sagittal plane).

The entire brain is sectioned serially in the coronal plane. This sectioning permits an accurate and reproducible determination of the maximal sagittal fronto–occipital diameter of the lesion, which is equal to the number of sections where the lesion is present multiplied by 15 mm, and represents an index of its volume. Two investigators, blinded to the treatment groups, determine the size of the lesion in each brain.

4 Notes

1. Volvic water is highly recommended as other natural spring water have been tested by others and are less efficient.
2. The specificity of the immunoreactions was ensured in control sections treated with the corresponding blocking peptide of the NR1 antibody (10 μg peptide for 1 μg of antibody) as well as in sections incubated only in the secondary antibodies.
3. Immunostainings of microglia using the following NMDA receptor subunit-specific antibodies at various concentrations were found to give no specific signal in our hands: rabbit anti-GluN2A (Sigma, M264), rabbit anti-GluN2C (Sigma, M266), rabbit anti-GluN3A (Millipore, 07-356), rabbit anti-GluN3B

(Millipore, 07-351), mouse monoclonal anti-GluN2A/B (Abcam, ab2639).

4. The specificity of the immunoreactions was ensured in control specimen incubated only in the secondary antibodies. To verify the purity of primary microglia cultures, immunocytological staining was performed using antibodies anti-tomatolectin for microglia (1:1000), mouse anti- GFAP for astrocytes (1:1000; Clontech), and mouse anti-NeuN for neurons (1:5000; Chemicon).
5. To further confirm the specificity of the applied anti-GluN1 antibodies, we analyzed the colocalization of the signal obtained following immunostaining with that of the NR1-EGFP signal following transfection through confocal microscopy.
6. Long (10 min) stable records of microglia and strong enough to remain stable during rapid drug perfusion are really difficult to obtain and require many trials (our success rate was of 2–3%).
7. Rapid solution changers are either commercially available (usually expensive but precise when high-speed perfusions are required) or can be home made with solenoid valves.
8. Preparation of the CRELOX mice: $Nr1^{flox/flox}$ mice have one *loxP* sequence in the intron that lies between exons 10 and 11 and a second loxP sequence as well as neomycin-resistance gene introduced 3 kb downstream of the last *NMDAR1* exon; thus, loxP sites flank *NMDAR1* gene region that encodes the four transmembrane domains and entire C-terminal sequence. $LysM^{cre/cre}$ mice express the Cre recombinase from the endogenous M lysozyme locus common to the myeloid lineage (donated by Irmgard Förster, PhD, University Düsseldorf). Perform PCR for genotyping with DNA extracted from tissue lysate from tail biopsy specimen. A total of 35 cycles of 30 s at subsequently 96 °C, 60 °C, and 72 °C were performed with *Taq* DNA polymerase, and the PCR reactions were subjected to electrophoresis on 1.5% agarose gels.

   The *LoxP* primers used were:

5′- GTG AGC TGC ACT TCC AGA AG -3′ (oIMR3742),

5′- GAC TTT CGG CAT GTG AAA TG -3′ (oIMR3743),

5′- CTT GGG TGG AGA GGC TAT TC -3′ (oIMR6916),

5′- AGG TGA GAT GAC AGG AGA TC -3′ (oIMR6917),

which generate a 280 bp (mutant) and 180 bp (wild type) product.

The *Cre* primers used were:

5′- CCC AGA AAT GCC AGA TTA CG -3′ (Cre8),

5′- CTT GGG CTG CCA GAA TTT CTC -3′ (Mlys1),

5′- TTA CAG TCG GCC AGG CTG AC -3′ (Mlys2),

which generate a 350-bp (wild type) and a 700 bp and 1.7 kb (*LysM* Cre recombinase) product.

To verify the floxed region on the DNA level, we performed a PCR with the primers:

5′- GGA GAA GGG AGA GGA GGT TT -3′.

5′- GGG AGC AGC CAC AAG TTG AG -3′.

References

1. Gressens P, Marret S, Hill JM, Brenneman DE, Gozes I, Fridkin M, Evrard P (1997) Vasoactive intestinal peptide prevents excitotoxic cell death in the murine developing brain. J Clin Invest 100:390–397
2. Lelievre V, Hu Z, Byun JY, Ioffe Y, Waschek JA (2002) Fibroblast growth factor-2 converts PACAP growth action on embryonic hindbrain precursors from stimulation to inhibition. J Neurosci Res 67:566–573
3. Chomczynski P, Sacchi N (2006) The single-step method of RNA isolation by acid guanidinium thiocyanate-phenol-chloroform extraction: twenty-something years on. Nat Protoc 1:581–585

Chapter 15

NMDA Receptor Activity in Circulating Red Blood Cells: Methods of Detection

Asya Makhro, Lars Kaestner, and Anna Bogdanova

Abstract

Abundance and activity of N-methyl-D-aspartate (NMDA) in circulating red blood cells contributes to the maintenance of intracellular Ca^{2+} in these cells and, by doing that, controls red cell volume, membrane stability, and O_2 carrying capacity. Detection of the NMDA receptor activity in red blood cells is challenging as the number of its copies is low and shows substantial cell-to-cell heterogeneity. Receptor abundance is reliably assessed using the radiolabeled antagonist ([^{3}H]MK-801) binding technique. Uptake of Ca^{2+} following the NMDA receptor activation is detected in cells loaded with Ca^{2+}-sensitive fluorescent dye Fluo-4 AM. Both microfluorescence live-cell imaging and flow cytometry may be used for fluorescence intensity detection. Automated patch clamp is currently used for recording of electric currents triggered by the stimulation of the NMDA receptor. These currents are mediated by the Ca^{2+}-sensitive K^+ (Gardos) channels that open upon Ca^{2+} uptake via the active NMDA receptor. Furthermore, K^+ flux through the Gardos channels induced by the NMDA receptor stimulation in red blood cells may be detected using unidirectional K^+($^{86}Rb^+$) influx.

Key words N-methyl-D-aspartate receptor, Red blood cells, Calcium, Gardos channel

1 Introduction

Functional N-methyl-D-aspartate receptors (NMDARs) were recently discovered in human and rat red blood cells (RBCs) [1, 2]. Activation of these receptors results in Ca^{2+} uptake by RBCs. Ca^{2+} accumulation is immediately followed by the activation of Ca^{2+}-sensitive K^+ (Gardos) channels, changes in cell volume and density. Further destructive processes triggered by the persisting transient Ca^{2+} accumulation-extrusion bouts include oxidation and proteolysis [1–3]. The number of receptors per cell varies between the young, mature, and senescent RBCs [2]. An abnormally high abundance of erythroid NMDA receptors (eNMDARs) is characteristic for the RBCs of patients with sickle cell disease [4]. Due to the low number of eNMDARs per cell and substantial cell-to-cell heterogeneity [1, 2], only methods allowing screening of a large

Nail Burnashev and Pierre Szepetowski (eds.), *NMDA Receptors: Methods and Protocols*, Methods in Molecular Biology, vol. 1677, DOI 10.1007/978-1-4939-7321-7_15, © Springer Science+Business Media LLC 2017

number (hundreds to thousands) of RBCs may be used to reliably assess their activity in RBCs. Throughput of the classical patch-clamp approach is clearly too low and introduces certain bias in the selection of cells [5]. Automated electrophysiological techniques validated for RBCs so far [6, 7] were used for the detection of eNMDARs. However, the current measured was mediated not only by the receptor channels, but largely by the Ca^{2+}-sensitive K^+ Gardos channels. Direct monitoring of the Ca^{2+} uptake by the eNMDARs is performed by means of microfluorescence live-cell imaging or flow cytometry using the Ca^{2+}-sensitive fluorescent dye Fluo-4 AM. Single-cell readouts are required for the characterization of "responsive" and "non-responsive" RBC fractions, an important feature due to the high degree of heterogeneity of basal Ca^{2+} levels as well as RBC responsiveness to the stimulation with NMDAR agonists [1, 2, 8]. We present here the methodological approaches suitable for the detection of the eNMDAR activity in RBC fractions of low, medium, and high density, or in single cells. The assessment of the eNMDAR activity in fractions is performed using radiolabeled antagonist ([^{3}H]MK-801) binding assay. Ca^{2+} uptake mediated by the eNMDARs is monitored in individual cells using the fluorescent dye Fluo-4 AM. Screening for the cellular response to the eNMDAR stimulation is performed by automated patch-clamp. Bulk measurement of the secondary Ca^{2+}-induced activation of Gardos channels upon the eNMDAR stimulation in RBC suspensions may also be performed ba means of the radiolabeled tracer detection technique, in which $^{86}Rb^+$ is used as a tracer for K^+. The detection of K^+ accumulation in the extracellular medium following the activation of the Gardos channel using K^+ electrodes, although technically not demanding, is prone to artifacts as the readouts may be severely affected by hemolysis induced in response to the eNMDAR activation.

2 Materials (*See* Note 1)

2.1 [^{3}H]MK-801 Binding in Light, Medium, and Dense RBC Fractions

1. RBC plasma-like medium: 140 mM NaCl, 4 mM KCl, 0.75 mM $MgSO_4$, 10 mM glucose, 0.015 mM $ZnCl_2$, 0.2 mM glycine, 0.2 mM glutamate, 0.1 mM arginine, 0.6 mM glutamine, 0.2 mM alanine, 20 mM HEPES-imidazole (pH 7.4 at room temperature), and 0.1% bovine serum albumin. Medium is prepared in 10× stock, for 1 L: 81.76 g of NaCl, 2.982 g of KCl, 1.8486 g of $MgSO_4 \cdot 7H_2O$, 18 g of glucose, 0.02 g of $ZnCl_2$, 0.15 g of glycine, 0.37 g of glutamate-Na, 0.21 g of arginine, 0.876 g of glutamine, 0.178 g of alanine, 47.66 g of HEPES. In 1× diluted plasma-like medium pH is adjusted to 7.4 and 0.1% BSA added just prior to use.

2. [^{3}H]MK-801 (1 mCi/mL, 27.5 Ci/mM, Perkin Elmer).
3. 90% Percoll density gradient is prepared from the stock: nine parts of Percoll (GE Healthcare; density 1.130 g/mL) are mixed with 1 part of 10× plasma-like medium. Then nine parts of this stock are mixed with 1 part of 1× plasma-like medium containing 0.1% bovine serum albumin (BSA). 90% gradient allows separating RBC fractions; however, to reach better resolution for a particular fraction, other percentages of Percoll mixture can be used [2, 4].
4. Lysis buffer: 5 mM NaH_2PO_4/Na_2HPO_4, 1 mM EDTA, pH 7.5 RT. For 1 L: 0.6 g of NaH_2PO_4, 0.372 g of EDTA.
5. Narrow transparent tubes (15 mL) for high-speed centrifugation (e.g., polycarbonate flanged tubes from Sorvall for SM-24 rotor).
6. 15 mL plastic tubes (Falcon or similar) for RBC washing.
7. Glass capillaries for hematocrit (Hct) determination and hematocrit centrifuge.
8. Angle rotor centrifuge capable of developing 50,000 × *g*.
9. Scintillation liquid (Quicksafe A, Zinsser Analytic or similar) and a Beta-counter.

2.2 Microscopic Assessment of Intracellular Ca^{2+} Levels and Cell Morphology

1. RBC plasma-like medium (*see* above).
2. 1 M solution of $CaCl_2$.
3. 1 mM Fluo-4 AM stock solution is prepared on Dimethyl sulfoxide (DMSO).
4. Inverted fluorescent microscope equipped with 100× oil objective and a high light sensitivity camera controlled by the software for single-image and time-laps modules.
5. 1 mL volume cell chamber for imaging with a coverslip serving as its bottom.

2.3 Flow Cytometry

1. RBC plasma-like medium (*see* above) containing 2 mM of $CaCl_2$ is prepared from the 10× stock solution.
2. 1 mM Fluo-4 AM stock solution is prepared on DMSO.
3. Flow cytometer equipped with FITC filters (525/30 BP).
4. Software for measurements and analysis.

2.4 Automated Patch-Clamp Recordings

1. RBC plasma-like medium (*see* above).
2. "Pipette" solution: 50 mM KCl, 10 mM NaCl, 60 mM KF, 20 mM EGTA, 10 mM HEPES (pH adjusted to 7.2 with KOH).

3. Seal enhancing solution: 80 mM NaCl, 3 mM KCl, 10 mM $MgCl_2$, 35 mM $CaCl_2$, 10 mM HEPES (pH adjusted to 7.3 with NaOH).
4. Bath solution in: 140 mM NaCl, 4 mM KCl, 1 mM $MgCl_2$, 2 mM $CaCl_2$, 5 mM glucose and 10 mM HEPES (pH adjusted to 7.4 with NaOH).
5. Stock solutions of NMDA, glycine (aqueous, buffered) and MK-801 (on DMSO).
6. Automated patch-clamp setup (*see* **Note 2**).
7. Patch-clamp chips for the automated setup.

2.5 Unidirectional K^+($^{86}Rb^+$) Influx Mediated by Gardos Channel

1. Incubation medium contains: 145 mM $NaCH_3SO_4$, 5 mM KCH_3SO_4, 0.15 mM $MgSO_4$, 1 mM Ca-gluconate (or 0.1 mM EGTA in Ca^{2+}-free incubation medium), 10 mM glucose, 10 mM sucrose, and 10 mM HEPES-Tris (pH 7.4 at 37 °C), 0.1 mM L-arginine, 0.1 mM ouabain.
2. Washing medium contains 100 mM $Mg(NO_3)_2$ and 10 mM imidazole-HNO_3 (pH = 7.4 when on ice).
3. 5% solution of trichloracetic acid (TCA) in distilled water.
4. 30 mM NMDA, 30 mM glycine in distilled H_2O (pH adjusted by saturated Tris-OH solution to 7.4).
5. 100 mM ouabain stock in DMSO.
6. 10 mM TRAM-34 stock in DMSO.
7. Stock solution of $^{86}RbCl$, 0.1 mCi (3.7 MBq)/mL.
8. Distilled water.
9. Shaking water bath (37 °C).
10. Centrifuge for Eppendorf tubes (1.5–2 mL) with cooling.
11. Scintillation beta-counter.
12. Hematocrit centrifuge, hematocrit capillaries, sealing clay.

3 Methods

3.1 [^{3}H]MK-801 Binding and Fractionation

1. Separate blood from plasma by centrifugation (*see* **Note 3**) at 1700 × *g*, 5 min. Wash packed RBCs three times in the plasma-like medium under the same conditions, then resuspend the cells in the plasma-like medium to 40–50% hematocrit.
2. Add 5 μL of [^{3}H]MK-801 to 1 mL of RBC suspension and incubate for 1 h at RT.
3. Pipet 1 mL of RBC suspension carefully on the top of the 13 mL Percoll mixture in a centrifuge tube. Centrifuge the tubes for 15 min at 48,000 × *g* without brakes (*see* **Note 4**).

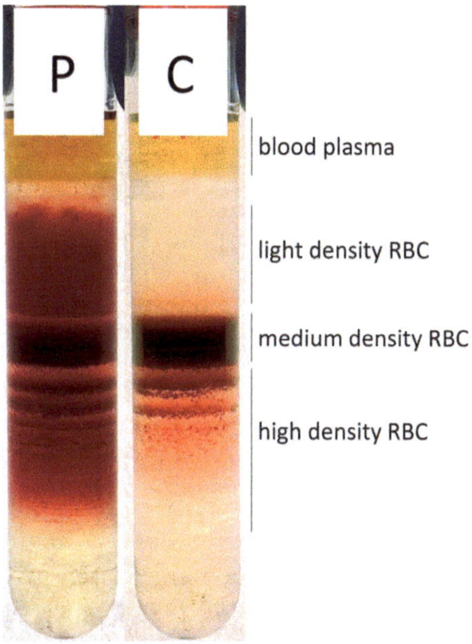

Fig. 1 Distribution of RBCs within the Percoll density gradient showing low, medium, and high density cell fractions in blood sample of a healthy subject C and a patient with sickle cell disease P

4. Examples of separation for healthy subject and patient with sicklec ell disease are shown in Fig. 1. To obtain RBCs of selected fractions, aspirate upper part with buffer and white blood cells. Then harvest RBC forming low, medium and high density fractions and transfer into a 15 mL tube for washing.
5. Wash RBC three times under the same conditions as in **step 1** of Subheading 3.1.
6. After the last washing aspirate the supernatant and adjust volume of RBC suspension to 1 mL with medium.
7. Estimate Packed RBC volume in suspension by hematocrit measurements. To do this, gently mix cells with media by slow pipetting and measure hematocrit. Correction for the trapped fluid (1–5%) may be introduced. The average RBC volume for healthy subject is 90 fL; thus, the number of RBC is calculated by dividing the total packed RBC volume to the average volume of a single cell (*see* Eq. 1). Alternatively, Sysmex or other RBC counter may be used to estimate the number of RBCs per volume suspension.

$$N_{e\mathrm{NMDAR}s} = \frac{A_{\mathrm{cells}}}{N_A \times N_{\mathrm{cells}} \times A_{\mathrm{sp}\ ([3H]\mathrm{MK-801})}} \qquad (1)$$

In the equation $N_{eNMDARs}$ stands for the number of NMDA receptors per cell, A_{cells} is [^{3}H]MK-801 activity associated with 1 L of packed RBCs in counts per minutes (CPM); N_A is the Avogardo number (6.02 × 10^{23} L/mole), N_{cells} is the number of packed RBCs per L of cell suspension, A_{sp} is specific activity (CPM/mole).

3.2 Microscopic Assessment of Intracellular Ca^{2+} Levels and Cell Morphology

1. Resuspend 1 μL of fresh whole blood in 1 mL of plasma-like medium containing 1 μM of Fluo-4 AM with or without 2 mM $CaCl_2$ and incubate for 2 h at RT (*see* **Note 5**).
2. As soon as the loading of RBCs with the dye is completed, place RBC suspension into the microscopy chamber and leave cells to precipitate to the bottom of the chamber for 5–10 min before image acquisition (*see* **Note 6**).
3. Measure basal intracellular Ca^{2+} levels in samples resuspended in the medium containing 2 mM of $CaCl_2$ as a green fluorescence with excitation wavelength of 488 nm with exposure time of 2–5 s. Since plasma-like medium contains saturating concentration of NMDA receptor agonists (NMDA and glycine, 300 μM each), basal Fluo-4 fluorescence represents steady-state intracellular Ca^{2+} levels in cells with fully active NMDA receptors. Ca^{2+}-rich intracellular compartments occurring in anemic patients (Fig. 3) also can be detected in that setting.
4. Detect Ca^{2+} uptake by the RBCs in the absence or presence of eNMDAR antagonist (memantine or MK-801) in two sets of triplicated samples. Load cells with Fluo-4 AM in plasma-like Ca^{2+}-free medium supplemented with NMDA and glycine (300 μM each) in the absence (control) or presence of 50 μM MK-801 or memantine as described in **step 1**.
5. Allow RBCs to precipitate to the bottom of the chamber and detect basal of Fluo-4 fluorescence intensity (F_0): take bright field images to assess RBC morphology (Fig. 2, baseline).
6. Set the software controlling the camera to the "time-laps" recording mode and program to take 13 frames of image couples (one in fluorescence mode and one in bright filed) at a rate of one frame per minute. Take the first frame for the "basal" Ca^{2+}-free conditions and then initiate Ca^{2+} uptake by adding a 2 μL aliquot from an aqueous 1 M stock of $CaCl_2$ carefully to 1 mL of the extracellular medium (cells at the bottom of the chamber should not be disturbed). The resulting free Ca^{2+} concentration is 1.8 mM due to the presence of 0.1 mM of EGTA in Ca^{2+}-free medium.
7. Record Ca^{2+} uptake over time and morphological responses of the individual RBCs to "Ca^{2+} pulse" for 12 min (*see* **Note 7**) (Fig. 2).

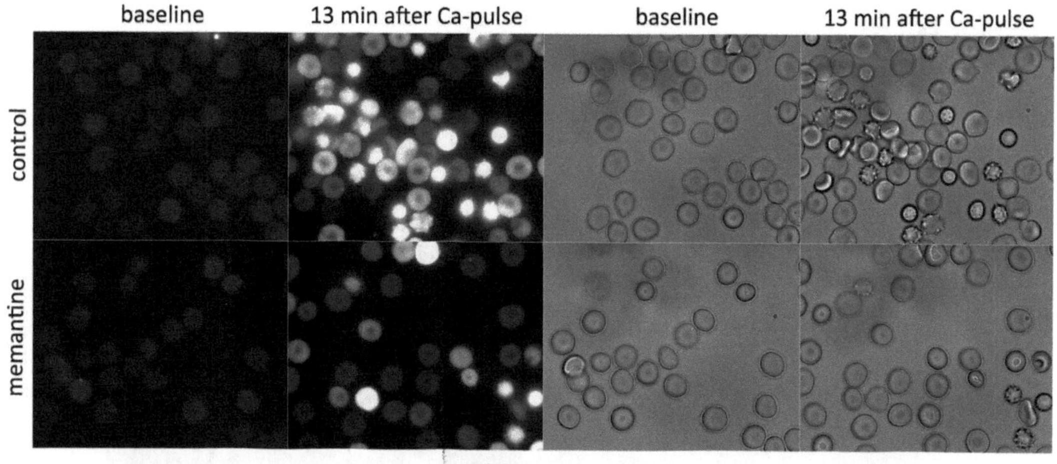

Fig. 2 Representative couples of images obtained for the "Ca^{2+} pulse" approach. *Left panels* are from the fluorescence channel showing Ca^{2+} levels at baseline and by the end of Ca^{2+} uptake measurements (frame 13) in control and antagonist (memantine)-supplemented blood samples. *Bright field images* shown in the *right* set of panels are from the same fields as fluorescence images. They are useful for the analysis of morphological changes occurring in cells upon Ca^{2+} uptake

8. Analyze the obtained images for the following parameters: (1) mean fluorescence intensity and its changes over time of incubation with Ca^{2+} (Fig. 3b, c); (2) RBC morphology including the appearance of echinocytes, stomatocytes, and other cell shape types as well as parameters related to cell projection (e.g., sphericity and anisotropy (ratio of the longest to shortest diameter, *see* Fig. 3a); (3) number and size of intracellular Ca^{2+}-filled vesicles (Fig. 3a).

Fluo-4 is not a ratiometric dye and exact absolute Ca^{2+} levels cannot, therefore, be estimated [2], Ca^{2+} uptake may be assessed from the linear part of the plots of relative fluorescence intensity (F/F_0) against the time of incubation with Ca^{2+} (Fig. 3b, c). Shrinkage introduces certain degree of underestimation into the measurements at the later time points. However, Ca^{2+} uptake always precedes the shape changes and reflects Ca^{2+} uptake fairly accurately. Furthermore, shrinkage is also associated with increase in the intracellular hemoglobin contentration, and, therefore, increase in quenching of the fluorescent signal outbalancing increase in fluorescence of the concetrated dye. Dehydration and increase in anisotropy following Ca^{2+} uptake may also serve as a secondary measure of Ca^{2+} influx. Sphericity disturbance and angular projection shapes produced during RBC dehydration are a hallmark of progressive loss of integrity of cytoskeletal proteins (Fig. 3a).

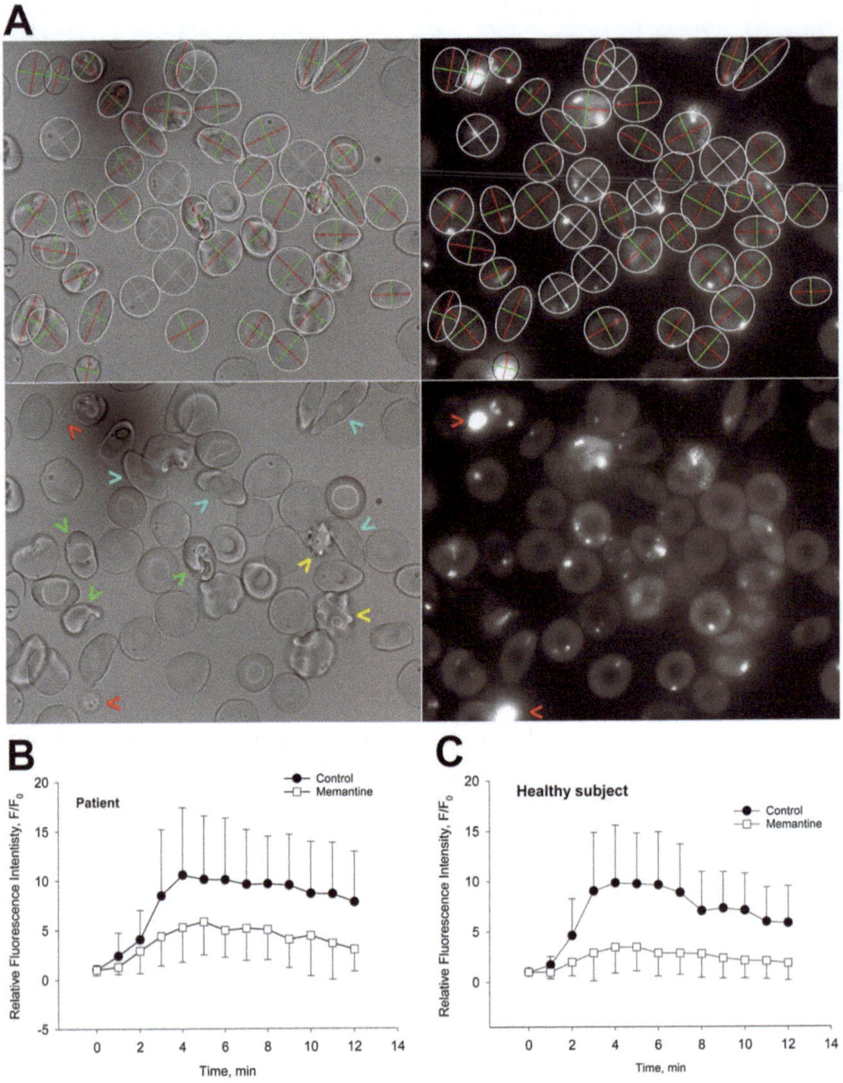

Fig. 3 Microfluorescence cell imaging technique: examples of analysis. (**a**) Morphological evaluation and assessment basal pseudo-steady state intracellular Ca^{2+} levels in RBCs. Image analysis was performed using CellFinder software. *Left* set of images are bright field images of cells. *Arrows* indicate cells with damaged cytoskeleton and angular shape (*green*), echinocytes (*yellow*), cells with sickled appearance (*blue*), and platelets (*red*). *Right set of images* shows the fluorescence intensity of Fluo-4. Ca^{2+}-filled internal vesicles are clearly seen. Ca^{2+} levels in platelets are substantially higher than those in RBCs. Shown in *white outlines* are the surface projection areas. Highlighted in *green* and *red* are the smallest and largest diameters respectively. *Arrows* indicate the cells with angular shape signifying cytoskeletal damage. (**b**) Quantification of Ca^{2+} uptake (Ca^{2+} pulse) by the untreated RBCs and those pretreated with eNMDAR antagonist memantine (50 μM) in patient with hereditary hemolytic anemia (**c**) Quantification of Ca^{2+} uptake by RBCs of a healthy subject in the presence or absence of 50 μM memantine

3.3 Flow Cytometry (FACS)

1. Incubate 2 μL of fresh whole blood for 1–2 h in the plasma-like medium containing 2 mM of $CaCl_2$ and 1 μM of Fluo-4 AM at 22–24 °C in the darkness. An unstained blood sample is used as a negative control.
2. Prepare each sample in triplicate. Samples include: (1) NMDA/Glycine-free medium, 2 mM $CaCl_2$ (2) 300 μM NMDA, 300 μM Glycine -containing medium, 2 mM $CaCl_2$ (3) 300 μM NMDA, 300 μM Glycine-containing medium supplemented with 50 μM MK-801.
3. Perform Fluo-4 fluorescence detection using 488 nm excitation/525 nm emission filters.
4. Assess fluorescence intensity of 100,000 cells at "low" flow rate to avoid lysis and artifacts caused by shear stress (*see* **Note 8**). Transient bout of Ca^{2+} uptake is associated with the release of Ca^{2+}-filled vesicles (Fig. 4B and C).
5. Intracellular Ca^{2+} varies between RBCs. High Ca^{2+} fraction (gate H) is formed by three types of cells/cell fragments: high-Ca^{2+} RBCs that do not lose phosphatidylserine asymmetry of the membrane (gate F), RBCs exposing PS to the external membrane leaflet (gate E), and Ca^{2+}-filled vesicles (gate D) (Fig. 4B).
6. Within the high Ca^{2+} RBC fraction (F gate) and upper half of the main RBC fraction some cells were CD71-positive reticulocytes (gate G) and young RBCs (Fig. 4B and C).
7. Acute administration of memantine (Fig. 4D) results in time and dose-dependent decrease in the intracellular Ca^{2+} in RBC in suspension or in whole blood (ex vivo and in vivo) due to the inhibition of eNMDARs.

3.4 Automated Patch-Clamp Recordings

1. Separate blood from the plasma by centrifugation (*see* **Note 3**) at 1700 × *g*, 5 min.
2. Wash packed RBCs three times in the plasma-like medium, and resuspend to a hematocrit of 0.5% after the final washing round (*see* **Note 9**).
3. Then transfer 1 mL of RBC suspension into the "cell-hotel" of the Patchliner.
4. All further steps occur within the automated patch-clamp setup. Initially, the "pipette solution" is filled into the chip as the internal solution and "seal enhancing solution" as the external solution. Then RBCs are transferred from the "cell hotel" to the chips and the routine for seal formation is initiated. Seals should have a resistance of at least 1 GΩ, better between 5 and 10 GΩ.
5. The whole-cell configuration is reached by applying negative pressure suction pulses between −45 and −150 mbar and its

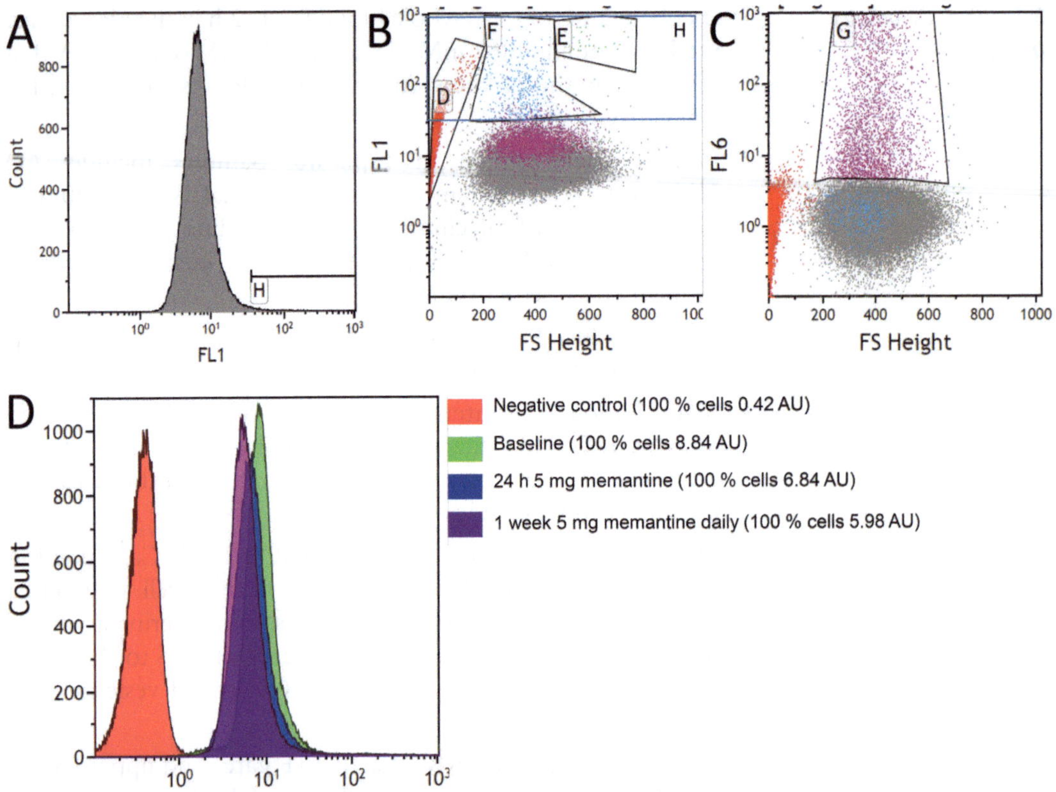

Fig. 4 Flow cytometry for the detection of the impact of eNMDAs into pseudo-steady state intracellular Ca^{2+} levels in RBCs. Measurements were performed for RBCs of an asymptomatic patient with sickle cell disease using FACS Gallios flow cytometer, BD. (**a**) Histogram for fluorescence intensity of Ca^{2+}-sensitive dye Fluo-4 in 100,000 cells. Cells with high Ca^{2+} levels are gated by gate H. (**b**) *Dot plot* showing the heterogeneity of Ca^{2+} distribution in RBCs and RBC-derived vesicles. RBCs with high Ca^{2+} are gated by the F (*blue*) and E (*green*) gates. Gate D (*red*) indicates the vesicles. Shown in *pink* are the cells within the ungated RBC fraction positive for transferrin receptor (CD71). (**c**) *Dot plot* showing cells positive for the CD71 (monoclonal anti-human CD71 antibody, APC-conjugated, clone OKT9, Ref. 17–0719 eBiosciences, San Diego, CA, USA). CD71+ – cells are gated (Gate G, *pink*). Measurements are done using 635 nm excitation 669/20 BP emission filter. (**d**) Histogram of Ca^{2+} levels (all cells in population) at baseline, and after 24 h and 1 week of 5 mg Memantin Mepha administration daily. The treatment was performed within the MemSID clinical trial (ClinicalTrials.gov Identifier: NCT02615847 approved by SWISSMEDIC (# 2015DR2096) and Cantonal ethic committee of canton Zurich (#2015-0297))

formation can be judged by the appearance of sharp capacitive transients of approximately 1 pF. For a typical success rate, please refer to [6]. After reaching whole cell configuration the "seal enhancing solution" is replaced by the "bath solution." RBC should be kept at a holding potential between −10 and −40 mV.

6. Then the experimental recordings as outlined in Fig. 5 can start. A typical scenario is the repetitive application of voltage ramps as depicted in Fig. 5b resulting in currents as exemplified

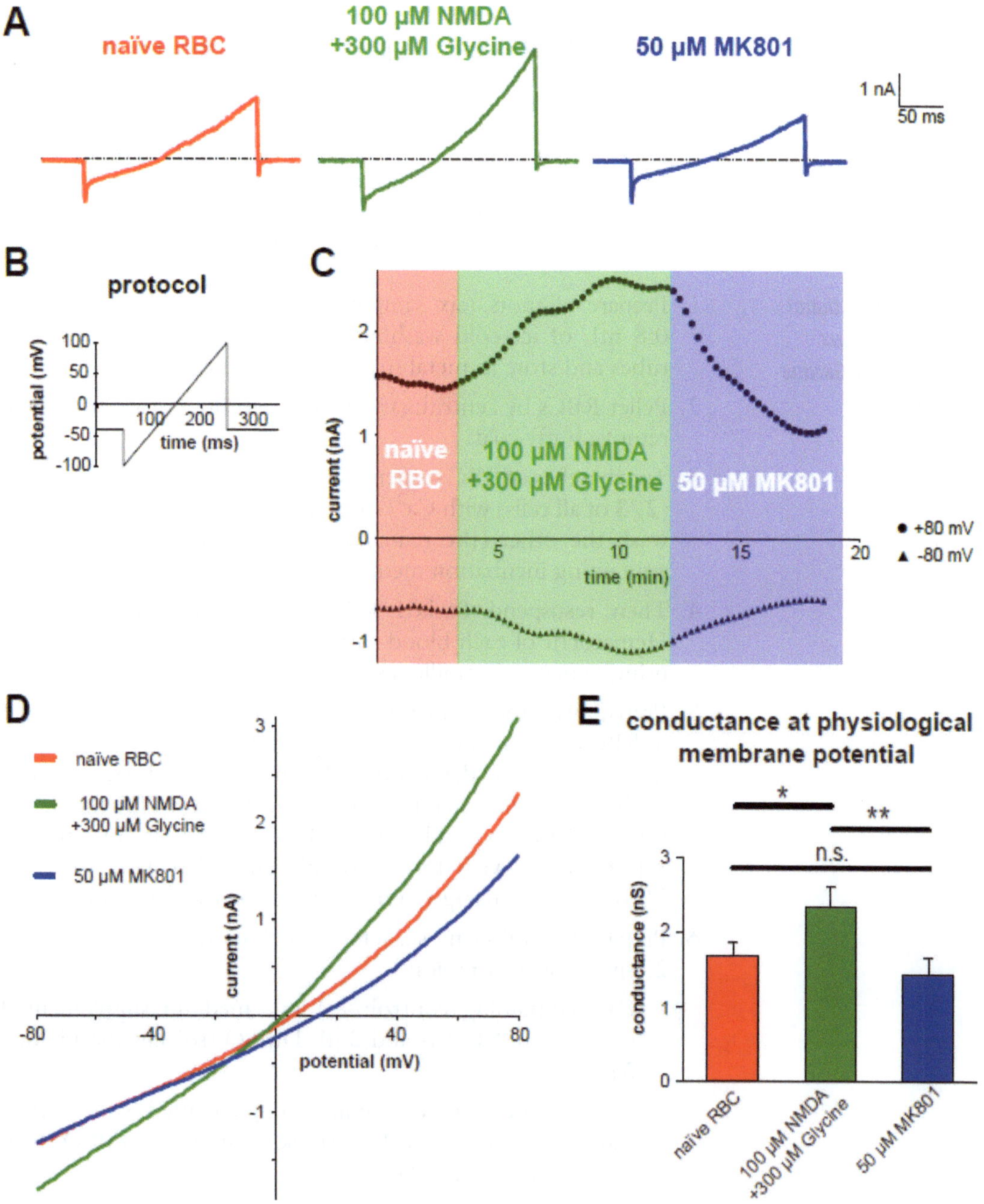

Fig. 5 Whole-cell currents of RBC under activation and blocking of the eNMDARs. Measurements as shown in panel (**a**) were performed on 56 gigasealed RBCs. However, for an observation period of 15–20 min only 34 cells were stable enough to allow analysis. 29% of the analyzed RBCs were classified as "responders." Responding cells were the RBCs showing an increase in conductance with the application of NMDA (100 μM) and glycine (300 μM) and a decrease when MK-801 was added. Panel (**a**) shows the original current traces of a RBC under the conditions presented above the traces as a response to the voltage protocol given in panel (**b**). The I-t diagram in panel (**c**) depicts the temporal distribution of the currents at +80 mV and −80 mV over the experimental course starting with naive RBC, application of NMDA and glycine for NMDAR activation, and

in Fig. 5a. An eNMDAR activating "bath solution" containing 100 μM NMDA and 300 μM glycine is pipetted as the external solution and currents recorded for several minutes (*see* **Note 10**).

7. To verify the identity of the currents as eNMDAR mediated add a "bath solution" containing 50 μM of the NMDAR inhibitor MK-801 and apply further voltage ramps. Figure 5c–e represents a possible analysis of the recorded data.

3.5 Unidirectional K^+ ($^{86}Rb^+$) Influx Mediated by Gardos Channels

1. Prepare aliquots (six samples × five time points = 30) of 0.8 mL of ice-cold washing medium in 1.5 mL Eppendorf tubes and store in metal tube-holders placed on ice.
2. Pellet RBCs by centrifugation (5 min 4000 × *g* at 10 °C) and remove buffy coat.
3. Two batches of RBC suspensions are formed: wash one of them (2/3 of all cells) with Ca^{2+}-containing incubation medium and wash the other (the remaining 1/3) in a Ca^{2+}-free EGTA-containing incubation medium.
4. Then resuspend Packed RBCs to a hematocrit of ~50%. Hematocrit of each blood suspension is measured in triplicate using hematocrit capillaries.
5. Prepare vials for scintillation counter for the detection of ^{86}Rb in RBCs and the incubation medium. For each sample prepare three vials for the detection of the amount of radioactive tracer in the incubation medium and five further vials for the detection of ^{86}Rb accumulation in RBCs at various time points. Altogether 48 vials for six samples are filled with water for ^{86}Rb detection using Cerenkov radiation (*see* **Note 11**).
6. Prepare the following six sets of incubation medium (volume: 2.4 mL) for six samples:
   - Ca^{2+}-containing control: 2.4 mL medium supplemented with 1.8 mM $CaCl_2$ and 3 μL DMSO (solvent for TRAM-34);
   - TRAM-34-containing medium supplemented with 1.8 mM $CaCl_2$ and 3 μL TRAM-34 stock solution on DMSO (final concentration 10 μM);

Fig. 5 (continued) finally MK-801 for eNMDAR block as indicated. The current traces in panel (**a**) correspond to the time points at 2 min, 10 min, and 18 min, respectively, in panel (**c**). Panel (**b**) summarizes the statistics of whole cell current measurements. Panel (**d**) shows the I–V curves of responding RBCs under the conditions indicated (data are mean of 10 measurements). The slope of the curves in the range between −30 and 0 mV was analyzed by linear regression and the results are presented in panel (**e**), revealing significant differences in the whole cell conductance at the physiologically relevant membrane potential for the different experimental conditions. This figure is a reprint from the American Journal of Physiology—Cell Physiology [2]

- Ca^{2+}-free incubation medium supplemented with 100 μM EGTA and 3 μL DMSO;
- Ca^{2+}-containing medium with 30 μL of NMDA-glycine stock solution and 3 μL DMSO;
- Ca^{2+}-containing medium supplemented with 30 μL NMDA-glycine stock solution and 3 μL of TRAM-34 stock solution;
- Ca^{2+}-free medium supplemented with NMDA-glycine solution 3 μL DMSO.

7. Warm these six sets of media up to 37 °C in a shaking water bath (*see* **Note 12**).
8. Add 0.4 mL aliquots of 50% Hct RBC suspensions washed in corresponding media to the pre-warmed sets of media and allow binding the inhibitors of Na,K-ATPase and Gardos channels (ouabain, TRAM-34) for 15 min under continuous shaking.
9. Initiate unidirectional $K^+(^{86}Rb^+)$ influx measurement by administering 15–30 μL of stock solution of ^{86}RbCl to each sample (*see* **Note 13**).
10. Incubate RBC suspension at 37 °C under continuous gentle shaking.
11. Collect 0.4 mL aliquots of the RBC suspension 10, 20, 30, 60, and 90 min after the ^{86}RbCl administration.
12. Stop ^{86}Rb uptake immediately by cooling and dilute 0.4 mL of RBC suspension with 0.8 mL of an ice-cold washing solution.
13. Precipitate RBCs by centrifugation in an Eppendorf centrifuge precooled to +4 °C (1.5 min 6000 × *g*).
14. Remove the supernatant and resuspend RBCs in 1 mL of an ice-cold washing medium and centrifuge again as in **step 13** just above.
15. Repeat **step 14** just above twice. Washed pellet of RBCs free from the extracellular tracer is finally obtained.
16. After the final aliquot of RBC suspension is collected 90 min after the onset of incubation with the tracer (*see* **Note 14**), collect the remaining suspension and pellet RBCs by centrifugation (1.5 min 6000 × *g*). Three 10 μL aliquots of the cell-free incubation medium (supernatant) are collected for the assessment of the ^{86}Rb content of the medium (A_{med}, Eq. 2). As the amount of ^{86}Rb taken in by RBCs in Cl^--free ouabain-containing medium is negligibly small compared to the extracellular pool of radioactive tracer, it does not change over the 90 min of the experiment. Thus, it is sufficient to estimate the extracellular ^{86}Rb levels once at the end of the experiment.

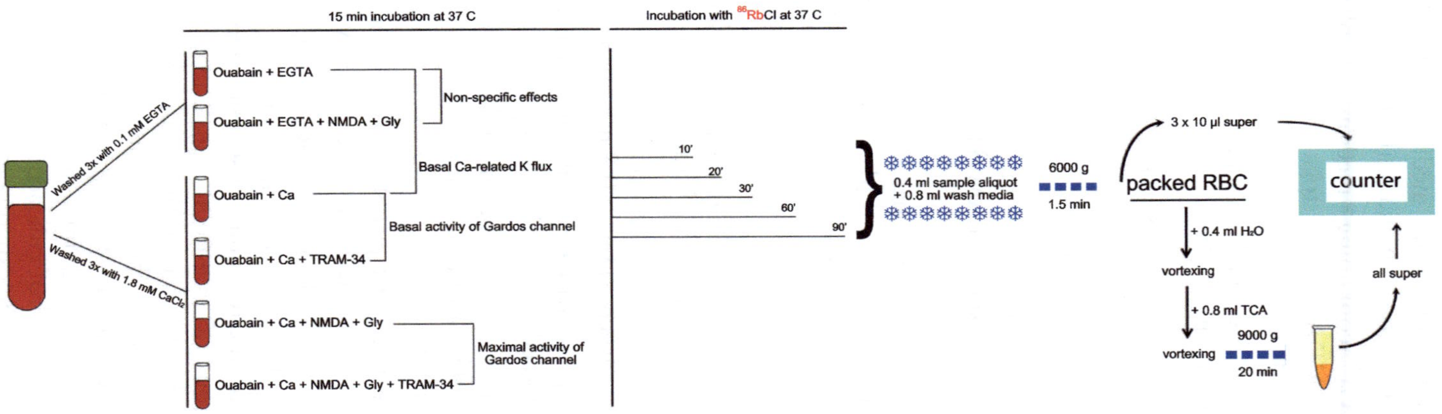

Fig. 6 Schematic representation of the handling procedure for blood samples used to evaluate the impact of eNMDARs on the activity of Gardos channel using radiolabeled tracer assay

17. Lyse washed RBC and deproteinize to ensure accurate detection of ^{86}Rb levels in a scintillation counter. Lysis of packed cells is achieved by mixing (under continuous vortexing) with 0.4 mL of H_2O to lyse the cells. Then add 0.8 mL TCA under constant vortexing to ensure effective denaturation. Then separate the denatured proteins by centrifugation (20 min 9000 × *g*) and transfer the colorless transparent supernatants to the vials filled with water for the detection of (A_{cells}, Eq. 2).
18. Calculate unidirectional K^+ influx as follows. Ratio of radioactivity of ^{86}Rb in 1 mL of RBCs (A_{cell}/V_{cell} of packed RBC in the aliquot in 1 L) to the radioactivity of 1 L medium (A_{med}). V_{cell} is calculated from hematocrit of stock RBC suspension. t in Eq. 2 stands for the duration of ^{86}Rb uptake was monitored. This ratio is plotted over time of incubation. Uptake rate of ^{86}Rb by 1 L RBCs per *h* is calculated from the linear part of the curve. Multiplication of the obtained uptake rate of the tracer by the extracellular K^+ concentration (4 mM) gives the value of unidirectional K^+ influx in Mole per L cells per hour.

$$J = \frac{A_{cell}[K^+]_e}{A_{med}\ V_{cell}\ t} \quad (2)$$

K^+ flux mediated by Ca^{2+}-sensitive K^+ (Gardos channel) is calculated as a difference of flux between that in control sample and the Ca^{2+}-free sample or the sample supplemented with the blocker of the channel TRAM-34 (Fig. 6). Comparison of the activity of Gardos channels in the absence and presence of the saturating concentrations of NMDA receptor agonists points to the amplitude of Ca^{2+} uptake via these receptors mediate.

4 Notes

1. We recommend using 0.1% bovine serum albumin (BSA) in all media for washing and maintaining human RBC. BSA supplementation for maintaining native RBC shape, and biochemical and hemorheological properties [9].
2. In contrast to all other measurements within this chapter that can be performed on standard laboratory equipment, independent of the manufacturer, automated patch-clamp requires device-specific chips that need to be compatible with RBC. The small size of the RBC and their ability to pass constrictions with small diameters (e.g., capillaries, spleen) require planar chips with wholes (equivalent to pipette-tip diameters) smaller than for most other cell types. They need to be in the range of 5–8 MΩ in solutions of physiological ionic strength. We have

good experience with the Patchliner (Nanion Technologies, Munich, Germany) (Fig. 5, [7]) and performed a proof of principle with the QPatch (Sophion, Copenhagen, Denmark). The Ionflux (Fluxion Biosciences, San Francisco, USA) failed in an initial test and for the machines from Molecular Devices (Silicon Valley, USA) we have no information available, just to mention the major suppliers of automated patch-clamp systems.

3. RBCs of patients may need lower acceleration and centrifugation speed to avoid hemolysis of most fragile cells.
4. Duration and speed for fractionation may be adjusted to achieve better fractionation for patients [4] and athletes [10].
5. 1 μL of blood per sample can be used for healthy donors with hematocrit of 40% and higher, for anemic patients' blood volume should be increased according to their hematocrit levels.
6. RBCs from patients with high cell density precipitate faster than healthy RBCs.
7. Time of observations may vary between 10 and 15 min.
8. Shear stress causes acute transient Ca^{2+} uptake events. Thus, even shaking of samples in carousel alters Ca^{2+} levels and position of samples within the carousel may affect the readout particularly in RBCs with abnormally high numbers of NMDARs. It is not excluded that some fragile cells get lost when flushed through the system before they are recorded [5]. Double staining may also affect the readouts [11].
9. Since automated patch-clamp does not allow visual inspection of the cell, it has to be ensured that a pure RBC suspension without "contamination" of other blood cells [5] is used. To ensure such a condition we advise the following filtering of the cell suspension [12]. 1.8 g of α-cellulose and microcrystalline-cellulose were added to 100 mL of 0.9% phosphate-buffered saline (PBS, Sigma Aldrich). 10 mL of this solution are filled into a syringe furnished with a filter paper at the bottom and pushed through the filter that the cellulose aggregates at the bottom form. A mixture of 9 mL PBS and 1 mL plasma was then filled into the syringe and let percolate through the filter as well. Finally, 1 mL of a 1:1 mixture of the RBC fraction and PBS were pushed through the filter and the filtrate was disposed.
10. Please keep in mind that the abundance of eNMDARs in mature RBCs of healthy people is very low and not all cells of the population contain eNMDARs, i.e., less than one third of all RBCs respond to NMDA stimulation (Fig. 5).

11. When using multiwell plate-based counters ^{86}Rb detection may pose problems due to the cross-well reactivity. Thus, if a vial-based counter is not available, wells adjacent to those containig high activity samples of the extracellular medium should remain empty. It is also true for the scintillation counters that use vials. Samples contain tracer from the incubation medium should be flanked with vials containig water alone. (Note ^{86}Rb ($T_{1/2}$ 18.6 days) may be detected using Cerenkov radiation [13] as its β-particle energies of 1.775 MeV are higher than 500 keV [14]. Of note, this Rb isotope also emits γ radiation with an energy of 1.018 MeV.
12. Handling of six samples in parallel requires some skills. One may run two samples at once, e.g., 1–4 followed by 2–5 and 3–6.
13. When working with $^{86}RbCl$ precautions have to be taken to avoid radioactive contamination. Tips, tubes, and all solid waste that came in contact with ^{86}Rb-containing liquids are collected and stored for decontamination. Liquid waste is collected separately and stored for decontamination. Precautions are taken to avoid spilling of liquid waste into the running water (e.g., vacuum pumps used for removing the supernatants by adding a capturing trap for liquid between the Büchner flask and the water pipe).
14. One has to check for the linearity of $^{86}Rb^+$ uptake over time to be able to assess the influx rate without interference of $^{86}Rb^+$ loss. If $^{86}Rb^+$ uptake is not a linear function of time at the later time points, the measurements have to be readjusted and limited to, e.g., 70 min.

Acknowledgments

AB and LK are supported by the European Union's Horizon 2020 research and innovation programme under grant agreement No 675115—RELEVANCE—H2020-MSCA-ITN-2015/H2020-MSCA-ITN-2015 and the European Union's Seventh Framework Programme for research, technological development and demonstration under grant agreement No 602121, CoMMiTMenT.

References

1. Makhro A, Wang J, Vogel J, Boldyrev AA, Gassmann M, Kaestner L, Bogdanova A (2010) Functional NMDA receptors in rat erythrocytes. Am J Physiol Cell Physiol 298(6): C1315–C1325. doi:10.1152/ajpcell.00407.2009
2. Makhro A, Hanggi P, Goede JS, Wang J, Bruggemann A, Gassmann M, Schmugge M, Kaestner L, Speer O, Bogdanova A (2013) N-methyl-D-aspartate receptors in human erythroid precursor cells and in circulating red blood cells contribute to the intracellular calcium regulation. Am J Physiol Cell Physiol 305

(11):C1123–C1138. doi:10.1152/ajpcell.00031.2013

3. Bogdanova A, Makhro A, Wang J, Lipp P, Kaestner L (2013) Calcium in red blood cells-a perilous balance. Int J Mol Sci 14 (5):9848–9872. doi:10.3390/ijms14059848
4. Hanggi P, Makhro A, Gassmann M, Schmugge M, Goede JS, Speer O, Bogdanova A (2014) Red blood cells of sickle cell disease patients exhibit abnormally high abundance of N-methyl D-aspartate receptors mediating excessive calcium uptake. Br J Haematol 167 (2):252–264. doi:10.1111/bjh.13028
5. Minetti G, Egee S, Morsdorf D, Steffen P, Makhro A, Achilli C, Ciana A, Wang J, Bouyer G, Bernhardt I, Wagner C, Thomas S, Bogdanova A, Kaestner L (2013) Red cell investigations: art and artefacts. Blood Rev 27 (2):91–101. doi:10.1016/j.blre.2013.02.002
6. Makhro A, Huisjes R, Verhagen LP, Mañú-Pereira MM, Llaudet-Planas E, Petkova-Kirova P, Wang J, Eichler H, Bogdanova A, van Wijk R, Vives Corrons J-L, Kaestner L (2016) Red cell properties after different modes of blood transportation. Front Physiol 7:228. doi:10.3389/fphys.2016.00288
7. Petkova-Kirova P, Lipp P, Kaestner L (2014) Novel automated technique to study ion channels in erythrocytes. Acta Physiol (Oxf) 210 (Suppl. 695):186
8. Kaestner L, Tabellion W, Weiss E, Bernhardt I, Lipp P (2006) Calcium imaging of individual erythrocytes: problems and approaches. Cell Calcium 39(1):13–19. doi:10.1016/j.ceca.2005.09.004
9. Reinhart WH, Piety NZ, Deuel JW, Makhro A, Schulzki T, Bogdanov N, Goede JS, Bogdanova A, Abidi R, Shevkoplyas SS (2015) Washing stored red blood cells in an albumin solution improves their morphologic and hemorheologic properties. Transfusion 55 (8):1872–1881. doi:10.1111/trf.13052
10. Makhro A, Haider T, Wang J, Bogdanov N, Steffen P, Wagner C, Meyer T, Gassmann M, Hecksteden A, Kaestner L, Bogdanova A (2016) Comparing the impact of an acute exercise bout on plasma amino acid composition, intraerythrocytic Ca2+ handling, and red cell function in athletes and untrained subjects. Cell Calcium 60:235–244. doi:10.1016/j.ceca.2016.05.005
11. Wesseling MC, Wagner-Britz L, Boukhdoud F, Asanidze S, Nguyen DB, Kaestner L, Bernhardt I (2016) Measurements of intracellular Ca2+ content and Phosphatidylserine exposure in human red blood cells: methodological issues. Cell Physiol Biochem 38 (6):2414–2425. doi:10.1159/000445593
12. Beutler E, West C, Blume KG (1976) The removal of leukocytes and platelets from whole blood. J Lab Clin Med 88(2):328–333
13. Francois B (1973) Detection of hard beta-emitting radionuclides in aqueous solutions using Cerenkov radiation: a review article. Int J Nucl Med Biol 1(1):1–14
14. Lide DR (ed) (2000) Handbook of chemistry and physics. CDC Press, Boca Raton, Anna Arbor, London, Tokyo

Chapter 16

NMDA Receptors as Voltage Sensors

Roustem Khazipov

Abstract

The membrane potential is an essential parameter of a living cell. However, measurements of the membrane potential using conventional techniques are associated with a number of artifacts. Cell-attached recordings of the currents through NMDA receptor channels enable noninvasive measurements of the neuronal membrane potential. This approach overcomes the problem of a leak conductance introduced during intracellular sharp electrode recordings and whole-cell patch-clamp recordings. Here, we describe the procedures of using cell-attached recordings of NMDA receptor channels to measure the true membrane potential.

Key words NMDA receptor, Patch-clamp, Cell-attached, Membrane potential

1 Introduction

Electrical potential across the cytoplasmic membrane, or a resting membrane potential (E_m), is a fundamental property of a living cell. In neurons and other excitable cells, the resting membrane potential plays a key role in electrogenesis by setting a default state of the voltage-gated and transmitter-activated ion channels of the plasma membrane and it acts as a driving forceDriving force for transmembrane co-transporters of ions and other molecules. Therefore, knowledge of the E_m values is critical for understanding the excitable cell functions. However, the measurement of the E_m values is not a trivial task, as each measure introduces some amount of error. Conventional approaches for the E_m measurements are the intracellular recordings using sharp electrodes and whole-cell patch-clamp recordings. Yet, these approaches may introduce strong errors in E_m measurements through the (1) alterations in the ionic composition of the intracellular milieu (dialysis problem) and (2) introduction of a leak conductance at the contact between the electrode and a cell membrane, which attains 500 MOhms during intracellular recordings and several GOhms during whole-cell recordings (leakage problem). Both the artifacts are more

Nail Burnashev and Pierre Szepetowski (eds.), *NMDA Receptors: Methods and Protocols*, Methods in Molecular Biology, vol. 1677, DOI 10.1007/978-1-4939-7321-7_16, © Springer Science+Business Media LLC 2017

pronounced in small cells, which undergo more efficient dialysis due to a smaller intracellular volume and which are more sensitive to the leakage error due to a higher membrane resistance [1]. Whereas these artifacts are general, in some extreme cases the size of the error can be dramatic. For example, in a newborn rat the hippocampal pyramidal cells with a membrane resistance in the gigaohm range, which is comparable to the gigaohm seal resistance, the E_m values measured using whole-cell or perforated patch recordings are only about a half compared to the E_m values measured noninvasively [2].

To avoid these problems, several noninvasive approaches to assess E_m have been developed based on the cell-attached recordings of the ionic channels or the integral currents. During the cell-attached recordings, a cell membrane remains intact and the problems of dialysis and a leak conductance are avoided. The rationale of this approach is based on the assumption that E_m equals the difference between the reversal potential of the currents through a given type of ionic channels and the equilibrium potential of the currents through these channels. To achieve this aim, voltage gated potassium channels and NMDA channels were used [2–5]. In this chapter, we describe the procedure of E_m estimation from cell-attached recordings of NMDA receptor channels which has been successfully used in various types of neurons from different species.

2 Materials

Cell-attached patch-clamp recordings of NMDA channels can be performed from neurons in any type of preparation including brain slices, isolated neurons, and neurons in cultures, and also in vivo using appropriate regular electrophysiological setups.

1. Patch electrodes: Any type of electrodes for patch-clamp recordings can be used. We pull electrodes from standard glass capillaries using two-stage puller to obtain the pipettes with a resistance of 7–10 MOhm when filled with the pipette solution. The tip of the pipette should be covered with silicone or wax to reduce the pipette capacitance and noise, particularly when long parts of the electrode contact with the bathing liquid. For the in vivo recordings from the surface structures with the electrodes placed into the brain to a depth of less than 1 mm, this step is not important. Fire polishing of the tip improves gigaseal formation but it is not mandatory. Longer pipette necks are required for recordings from the deep structures.

2. Pipette solution: 140 mM NaCl, 2.5 mM KCl, 1.8 mM $CaCl_2$, 10 mM HEPES, 0.01 mM NMDA, 0.01 mM glycine, 0.001 mM strychnine, pH adjusted to 7.4 by HEPES-Na (*see* **Note 1**). Ultra-pure water and powders should be used. After the preparation, the solution is filtered through 0.2 μm filters and stored in 1 mL aliquots at −20 °C. On the day of its use an aliquot should be thawed. No additional filtering is required.

3 Methods

3.1 Cell-Attached Recordings and Estimation of E_m

1. Conventional cell-attached recordings are performed under visual control or blindly. After gigaseal formation, currents through NMDA channels of about 4–5 pA are immediately visible at a holding potential of 0 mV and are easily distinguishable from other types of channels by their conductance and electrographic phenotype [2] (Fig. 1a). The recordings are performed with the parameters set for single channel recordings. We set an amplifier in whole-cell voltage clamp mode and use amplification 50 mV/pA, 2.5 kHz low pass filter. When using Axopatch200 series amplifiers, a capacitance mode is used to reduce the baseline noise. NMDA channels are recorded at the range of holding potentials from +20 mV to −90 mV in 10 or 20 mV increments for 10–30 s at each holding potential (Fig. 1b), depending on the ongoing frequency of NMDA channel openings to obtain about a hundred of single channel openings per each holding potential (*see* **Note 2**). The current recordings are acquired at 10 kHz together with the holding potential values.
2. After the recordings are completed, and if the attached cell did not spontaneously go into the whole cell, the same pipette can be used again for repatching another cell. The pipette 100–200 μm should be moved back from the cell, above the slice, and short pulse of positive pressure by 20–50 mL syringe should be applied. The pipette resistance is measured. If the pipette resistance did not increase compared to the values before patching, then another cell should be patched. If the gigaseal fails to form, the pipette should be changed. If not, we proceed in the same way to sample the needed number of cells. In our hands from 2 to 8 cells can be reliably recorded with the same pipette. We should keep in mind that not all patches contain NMDA channels.
3. During the data analysis, amplitudes of currents through NMDA receptors are calculated at different holding potentials, and the current-voltage relationships are fit with the linear function to estimate the reversal potential (Fig. 1c). Because

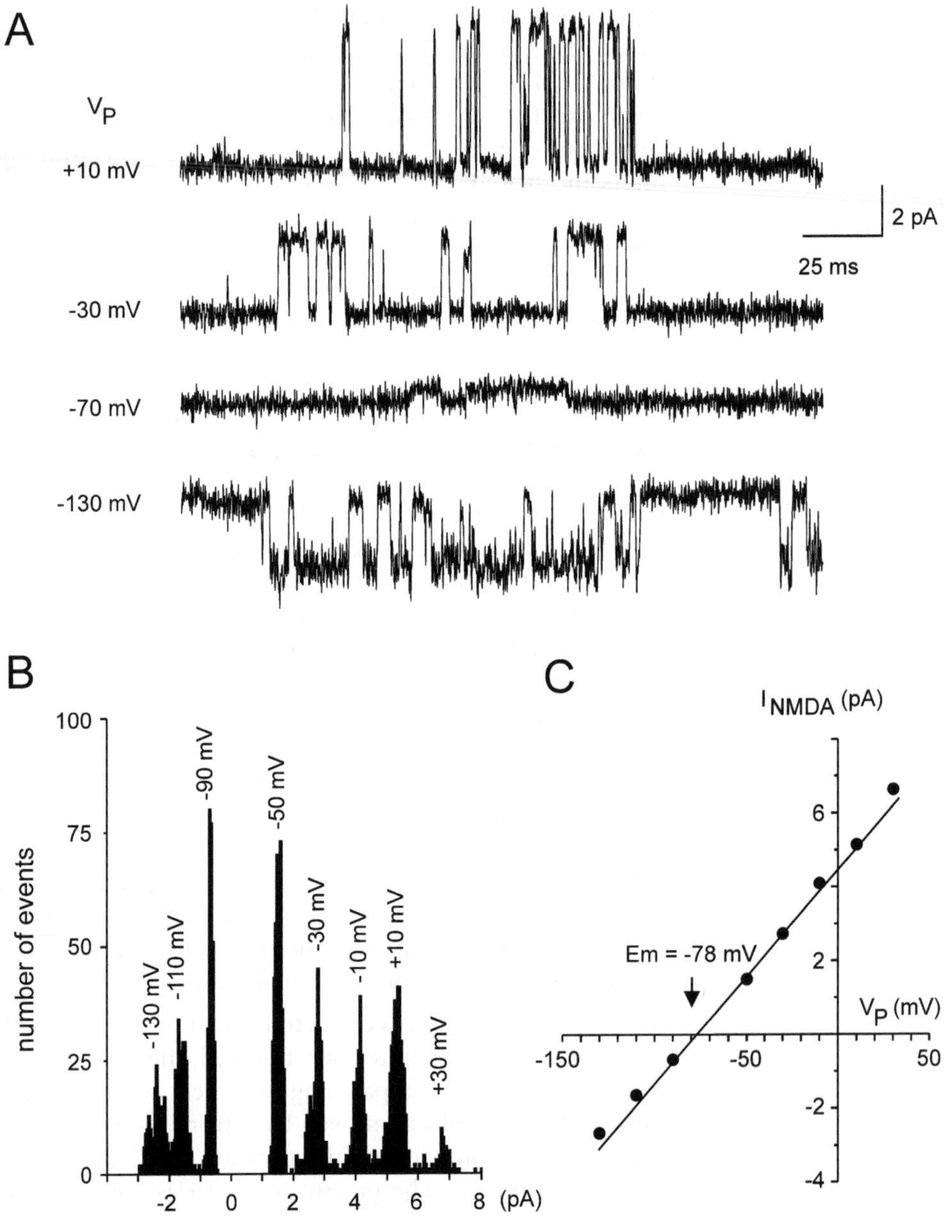

Fig. 1 The membrane potential determined by the cell-attached recordings of NMDA channels. (**a**) The cell-attached recordings of NMDA channels from P2 CA3 pyramidal cell at various pipette potentials (V_P); note that upward openings in the cell-attached mode correspond to inward currents. (**b**) The current amplitude histograms through single NMDA channels at different pipette potentials. (**c**) The current-voltage relationship of NMDA channels recorded in the cell-attached mode. Note that the currents through NMDA channels reverse at $V_P = -78$ mV; assuming that NMDA receptor-mediated currents reverse at a membrane potential of 0 mV, the resting membrane potential of this cell should be -78 mV. *Reproduced from ref.* 2

the currents through NMDA receptors reverse near 0 mV membrane potential [6], the holding pipette potential at which the currents through NMDA channels reverse in the cell-attached recordings corresponds to E_m (*see* **Notes 3** and **4**).

3.2 Dynamic Monitoring of E_m

1. Once the conductance of NMDA channels in a given patch is established, the dynamic changes in E_m can be monitored through the recording session. Using simultaneous recordings of E_m, using NMDA channels in the cell attached mode as voltage sensors and the recordings of population activity, using extracellular electrodes one can assess E_m dynamics in relation to the network activity or monitor the fluctuations of E_m in spontaneously active neurons [2].

3.3 Estimation of the Reversal Potentials of Currents through GABA Channels

In addition, this technique can be used to monitor the effects of various compounds on E_m, such as depolarizing actions of GABA [7, 8].

1. Combination of the cell-attached recordings of NMDA and GABA(A) channels from the same neuron enables estimation of the reversal potentials of currents through GABA(A) channels (E_{GABA}). In contrast to NMDA currents, in which the reversal potential is always near 0 mV on the membrane, E_{GABA} is a more variable parameter, and the cell-attached recordings of GABA(A) channels only provide an estimate of the driving force acting on the currents through GABA(A) channels (DF_{GABA}). Estimation of E_{GABA} from the cell-attached recordings requires measurements of both E_m and DF_{GABA}. In practice, these measurements can be achieved either by simultaneous cell-attached recordings of NMDA and GABA(A) channels during dual cell-attached recordings or through sequential patching of neurons for the estimation of E_m (using the cell-attached recordings of NMDA currents) and DF_{GABA} (equals to [minus] pipette potential at which GABA (A) currents reverse during the cell-attached recordings). E_{GABA} can be further calculated as $E_{GABA} = E_m + DF_{GABA}$. This noninvasive approach to measure E_m and E_{GABA} noninvasively is becoming popular for various applications including disease models [9–13].

4 Notes

1. The pipette solution mimics an extracellular medium but it should contain minimal amounts of magnesium to avoid magnesium block.
2. In our initial descriptions, we were recording also at more negative potentials (<-90 mV) to reverse NMDA currents.

However, we found that this part of the IV curve is not informative because of: (1) abundance of the other types of channels at the negative holding potentials that open during the patch depolarization and contaminate the trace and (2) blockade of NMDA channels by magnesium from inside the cell. Therefore, the linear part in the range of about 100 mV within the range of the holding potentials from +20 to −80 mV is sufficient to approximate the reversal potential of the currents through NMDA channels and to deduce E_m.

3. The cell-attached recordings are not entirely noninvasive. First, ambient opening of channels in the patch (most frequently potassiumK^+ channels) including NMDA channels provides electrical access to the cell. Therefore, cell-attached recordings can be also regarded as whole-cell recordings but with very high series resistance. Evidently, this is a major pitfall of the approach, and the estimation of the error resulting from this artifact is difficult to assess. In practice, this error can be monitored by comparing the frequency of action currents at different holding potentials. If the access resistance in the cell attached recordings is high, firing frequency does not depend on the holding potential and this can be enough to consider that cell-attached recordings do not influence E_m [9, 14].
4. The E_m values obtained using the cell-attached recordings are typically more negative than the values obtained using sharp electrodes, or whole-cell and gramicidin perforated patch recordings. The model studies indicate that this is primarily due to the leak conductance through the contact between the electrode and the membrane introduced during invasive recordings, and that the error is proportional to the ratio between the cell membrane and the contact resistance. The results of the model have been confirmed during simultaneous cell-attached recordings of NMDA channels and whole-cell recordings from the same neuron [15] (Fig. 2). The difference between the E_m values measured using the cell-attached recordings of NMDA channels and the whole-cell recordings attains nearly 30 mV in neonatal rat hippocampal pyramidal cells, which have a membrane resistance in the gigaohm range. The error introduced during the whole-cell recordings is not negligible (10–13 mV) either in adult neurons, that is also in agreement with similar comparisons performed using potassium channels as voltage sensors in cell-attached recordings [3].

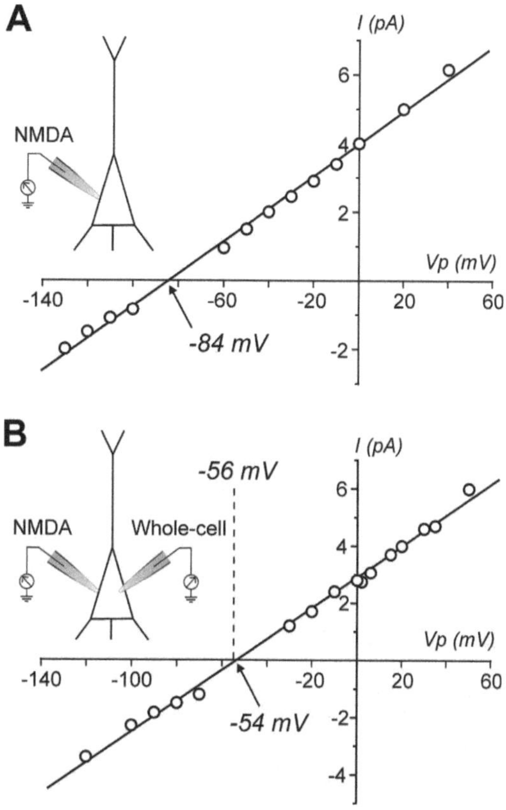

Fig. 2 Comparison of the membrane potentials measured using dual cell-attached recordings of NMDA channels and whole-cell recordings. (**a**) NMDA channels were recorded with a single patch pipette in the cell-attached mode. *I–V* relationships of NMDA currents give an E_m estimate of −84 mV (*arrow*). (**b**) Dual whole-cell and cell-attached recordings from the same cell estimated E_m at −56 mV (*dashed line*) and −54 mV (*arrow*), respectively. Note that the whole-cell recordings cause strong neuronal depolarization. V_p, pipette potential. *Reproduced from ref.* 15 *with permission*

Acknowledgments

Drs. Roman Tyzio and Anton Ivanov are acknowledged for their valuable contribution to the development of the technique and for their insightful comments and suggestions on the manuscript. This work was supported by INSERM (LIA to RK), the Program of Competitive Growth and the subsidy allocated to Kazan Federal University for the state assignment in the sphere of scientific activities (№ 6.2313.2017/4.6).

References

1. Barry PH, Lynch JW (1991) Liquid junction potentials and small cell effects in patch-clamp analysis. J Membr Biol 121(2):101–117
2. Tyzio R, Ivanov A, Bernard C, Holmes GL, Ben Ari Y, Khazipov R (2003) Membrane potential of CA3 hippocampal pyramidal cells

during postnatal development. J Neurophysiol 90(5):2964–2972

3. Fricker D, Verheugen JA, Miles R (1999) Cell-attached measurements of the firing threshold of rat hippocampal neurones. J Physiol 517(Pt 3):791–804
4. Verheugen JA, Fricker D, Miles R (1999) Non-invasive measurements of the membrane potential and GABAergic action in hippocampal interneurons. J Neurosci 19(7):2546–2555
5. Khazipov R, Ragozzino D, Bregestovski P (1995) Kinetics and Mg^{2+} block of N-methyl-D-aspartate receptor channels during postnatal development of hippocampal CA3 pyramidal neurons. Neuroscience 69:1057–1065
6. Nowak L, Bregestovski P, Ascher P, Herbet A (1984) Prochiantz a (1984): magnesium gates glutamate-activated channels in mouse central neurones. Nature 307:462–465
7. Khazipov R, Leinekugel X, Khalilov I, Gaïarsa J-L, Ben-Ari Y (1997) Synchronization of GABAergic interneuronal network in CA3 subfield of neonatal rat hippocampal slices. J Physiol 498(3):763–772
8. Leinekugel X, Medina I, Khalilov I, Ben-Ari Y, Khazipov R (1997) Ca^{2+} oscillations mediated by the synergistic excitatory actions of GABAA and NMDA receptors in the neonatal hippocampus. Neuron 18(2):243–255
9. Tyzio R, Khalilov I, Represa A, Crepel V, Zilberter Y, Rheims S et al (2009) Inhibitory actions of the gamma-aminobutyric acid in pediatric Sturge-weber syndrome. Ann Neurol 66(2):209–218
10. Tyzio R, Cossart R, Khalilov I, Minlebaev M, Hubner CA, Represa A et al (2006) Maternal oxytocin triggers a transient inhibitory switch in GABA signaling in the fetal brain during delivery. Science 314:1788–1792
11. Tyzio R, Nardou R, Ferrari DC, Tsintsadze T, Shahrokhi A, Eftekhari S et al (2014) Oxytocin-mediated GABA inhibition during delivery attenuates autism pathogenesis in rodent offspring. Science 343(6171):675–679
12. Khazipov R, Valeeva G, Khalilov I (2015) Depolarizing GABA and developmental epilepsies. CNS Neurosci Ther 21(2):83–91
13. Valeeva G, Valiullina F, Khazipov R (2013) Excitatory actions of GABA in the intact neonatal rodent hippocampus in vitro. Front Cell Neurosci 7:20
14. Khalilov I, Minlebaev M, Mukhtarov M, Khazipov R (2015) Dynamic changes from depolarizing to hyperpolarizing GABAergic actions during Giant depolarizing potentials in the neonatal rat hippocampus. J Neurosci 35 (37):12635–12642
15. Tyzio R, Minlebaev M, Rheims S, Ivanov A, Jorquera I, Holmes GL et al (2008) Postnatal changes in somatic gamma-aminobutyric acid signalling in the rat hippocampus. Eur J Neurosci 27(10):2515–2528

Chapter 17

Development of a Computational Approach/Model to Explore NMDA Receptors Functions

A. Florence Keller, Jean-Marie C. Bouteiller, and Theodore W. Berger

Abstract

Modern laboratory techniques allow studying NMDA receptors (NMDAR) either anatomically with specific antibodies coupled to sophisticated confocal microscopy, or physiologically by live imaging or electrophysiological techniques. However, NMDARs are not fixed in time and space and changes in their composition and/or distribution on the post-synaptic membrane may significantly impact the synaptic strength and overall function. The computational modeling approach therefore constitutes a complementary tool for investigating the properties of biological systems based on the knowledge provided by the lab experiments.

Here, we describe a general computational method aiming at developing kinetic Markov-chain based models of NMDARs subtypes capable of reproducing various experimental results. These models are then used to make predictions on additional (non-obvious) properties and on their role in synaptic function under various physiological and pharmacological conditions. For the purpose of this book chapter, we will focus on the method used to develop a NMDAR model that includes pharmacological site of action of different compounds. Notably, this elementary model can subsequently be included in a neuron model (not described in detail here) to explore the impact of their differential distribution on synaptic functions.

Key words Computational model; kinetic models, Ordinary differential equation, NMDA receptors model, Pharmacology, Synaptic function, Optimization algorithm

1 Introduction

1.1 Computational Neuroscience Short History

The term of computational neuroscience was introduced in 1985 by Eric L. Schwarz, although the early roots of the field are found in the work of many pioneers including Hodgkin & Huxley and Hubel & Wiesel in the first half of the twentieth century [1, 2]. Given the complexity of the nervous system and the elements it comprises, one of the many advantages of the computational approach is its ability to isolate the system under consideration and allow for its quantitative and systematic characterization. One approach consists in modeling neurons and neuronal networks by focusing on electrical properties as exemplified by the work of [3]. Another approach consists in using computational simulation to

Nail Burnashev and Pierre Szepetowski (eds.), *NMDA Receptors: Methods and Protocols*, Methods in Molecular Biology, vol. 1677, DOI 10.1007/978-1-4939-7321-7_17,

understand individual receptor behavior through the description of its different kinetic states (e.g., model of $GABA_A$ receptor [4]).

The NMDA receptor is a perfect example that illustrates the capabilities of the methodology. Indeed, the NMDAR is a relatively complex receptor characterized by orthosteric binding sites (glycine/D-serine and glutamate) and many allosteric binding sites [5, 6]. The voltage dependency of its associated channel is an additional constraint that further modulates its conductance (i.e., electrical functional response) [7].

1.2 How to Model a Receptor?

A simple way to model voltage-gated ion channels (ex., Ca^{2+}, Na^{+}, Cl^{-} channels) or ligand-gated channels such as the NMDAR is to use the Hodgkin–Huxley (HH) formalism. This formalism consists of a conductance-based model where voltage-gated ion channels are represented by electrical conductance that depends on both voltage and time [8]. Such models are useful to explore the impact of their conductance on neuronal activity [9].

Although HH models can yield predictive responses, they remain fairly descriptive, and the gating parameters do not represent specific kinetic states (e.g., unbound, open, deactivated, desensitized, etc.) and cannot describe various aspects of channel behavior. In the case of the NMDA receptor, the more general form of Markov models (MM) [10] provides more flexibility for the description of specific kinetic states of the receptor and its associated ionic channel. Importantly, MM methodology also allows modeling the interactions with additional molecular species, modulators, agonists, and antagonists, whether they act in a competitive manner or on an allosteric binding site. MM models therefore constitute useful means to explore the complexity of the receptor transitions states and to determine/imagine the configuration that would explain physiological properties observed using multiple recording methodologies.

1.3 Multiscale Approach

Besides their specific inherent properties, NMDA receptors have a profound impact on central nervous system (CNS) function for two main reasons: (1) their prevalence at the major excitatory synapses and (2) their role as co-incidence detector between pre- and postsynaptic activity which leads to calcium entry in the postsynaptic spine, thereby triggering a cascade of events that arguably shape long-term potentiation and the cellular basis of learning and memory [11, 12]. In order to go a step further in the exploration of NMDARs functions, the computational approach allows for the integration of the receptor models in models of synapse, neuron, and population of neurons to evaluate their role in this hierarchical and temporal hierarchy of scales (Fig. 1) [13–15].

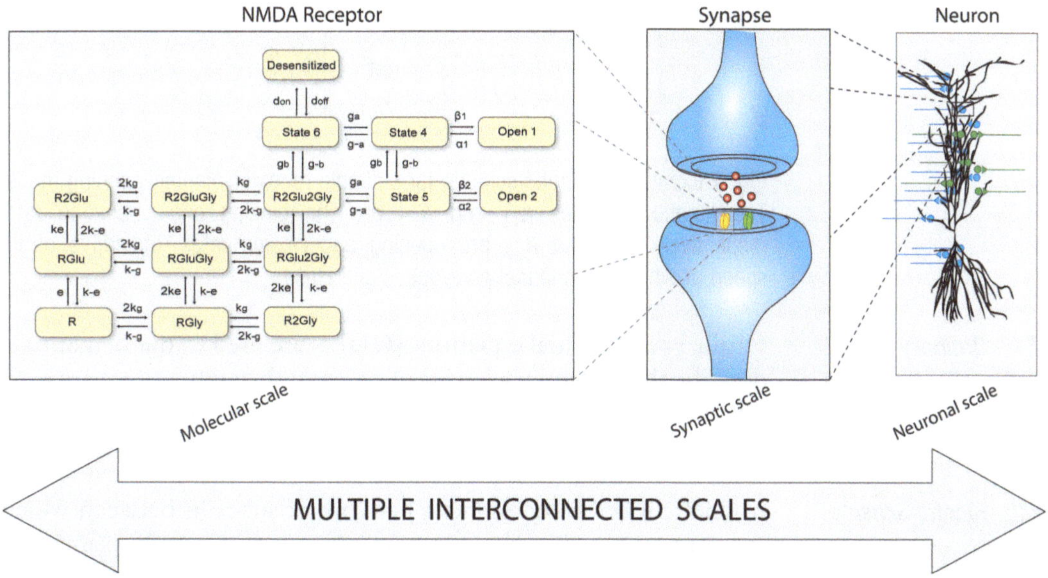

Fig. 1 Illustration of the multiscale approach. An elementary model such as the NMDAR is integrated in a model of synapse, itself connected to a model of neuron, exemplified here by a typical CA1 pyramidal neuron

1.4 The Procedure to Develop a Model Comprises Five Major Steps

First, we define the goal (purpose) of the model. In the case of the NMDAR, we wanted to explore its voltage- and Magnesium-dependence. In addition, we wanted to test the impact of Memantine, an allosteric antagonist, on the NMDARs functions. To achieve these goals we performed a thorough data mining and data analysis, including listing the existing NMDARs models. After careful data analysis, the NMDAR kinetic model structure was chosen, and the equations were written with their corresponding sets of kinetic parameters. Manual and automatic optimizations were then applied to adjust and sometimes find additional unknown parameters to improve the model's abilities to fit reported experimental results. Additional transitions states, equations, and parameters were then added to the model to implement the effect of memantine on its functions. The last step consists in integrating this new model into a multiscale modeling framework to emulate the biological substrate in which the model evolves; this could consist of a synapse, an entire neuron, or even a network of neurons. Performing this integration step provides insights into the effects of receptor-level changes on higher levels of hierarchical complexity.

2 Materials

Below is a list of mathematical constructs, tools, and techniques necessary to perform the simulations mentioned above:

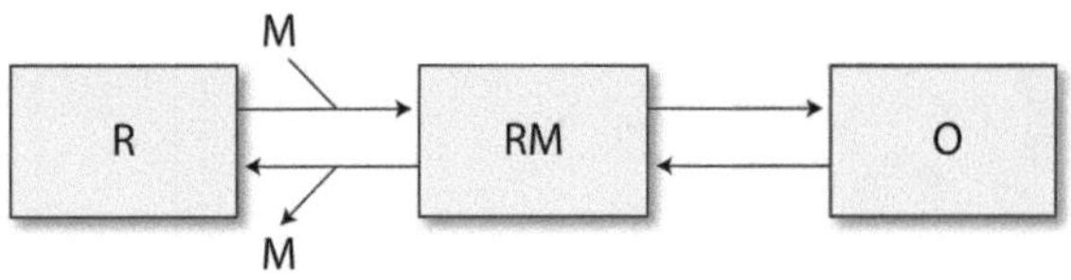

Fig. 2 Example of kinetic scheme for a simple receptor-binding-opening model. The kinetic variables k_{+1} and k_{-1} are the binding and unbinding rates respectively. k_{+2} and k_{-2} are opening and closing rates, indicative of the speed at which the channel pore opens and close

2.1 Ordinary Differential Equations (ODEs)

Ordinary differential equations (ODEs) are used to mathematically describe the dynamic behavior of a system through its derivative. In our case, ODEs constitute the fundamental building blocks of the kinetic scheme that describe the overall behavior of the receptor.

2.2 Kinetic Scheme

A kinetic scheme represents a network of states in between which the connections describe dynamical processes. A simple example of kinetic state could consist of the binding of a molecular species M to a receptor R which leads to the creation of a bound state RM, which results in the opening of an associated channel (O state) (Fig. 2).

The dynamics for the RM state are described by the differential equation:

$$\frac{d\mathrm{RM}}{dt} = k_{+1}.M.R - k_{-1}.\mathrm{RM} + k_{-2}.O - k_{+2}.\mathrm{RM}$$

The matrix representation for the dynamic behavior of the entire system presented above is

$$\begin{bmatrix} dR/dt \\ d\mathrm{RM}/dt \\ dO/dt \end{bmatrix} = \begin{bmatrix} -M.k_{+1} & k_{-1} & 0 \\ M.k_{+1} & -k_{-1} - k_{+2} & k_{-2} \\ 0 & k_{+2} & -k_{-2} \end{bmatrix} \begin{bmatrix} R \\ \mathrm{RM} \\ O \end{bmatrix}$$

More details on Markov kinetic state models may be found in [16] and *see* **Note 2** on the limitation of the design of kinetic states.

2.3 Solvers

The dynamic behaviors represented by kinetic schemes are defined by changes in derivatives that take place between two distinct moments in time. Determining the solutions to this kind of problem requires the use of numerical solvers, which provide an approximation for the system states at $t+1$ given its values at time t. A wide range of solvers exist (explicit, implicit, explicit-implicit, fixed time-step, or variable time-step, etc.) that yield different results and performances. For more details, we invite the reader to read [17–19].

2.4 Systems Biology Markup Language (SBML)

SBML constitutes an attempt at defining a common intermediate format—lingua franca—to facilitate communication of the most essential aspects of models. Adoption of the SBML format offers several benefits that include: (1) enabling the use of multiple tools without rewriting models, (2) enabling models to be shared and published in a form other researchers can use—even though they may use different software environments, and finally (3) extending the lifetime of models beyond that of the software used to create them. Notably, other formats exist and propose alternative attempts at providing a standardized format. They include NeuroML, LEMS [20], CellML [21], and Antimony [22], each yielding various benefits and drawbacks, and with varying levels of adoption.

2.5 SBML Simulation Engine

Many tools (e.g., Libroadrunner, COPASI, etc.) exist for the simulation of models specified in the SBML format. LibRoadrunner is the preferred open-source tool for its overall ease of use and outstanding performance. LibRoadrunner is an extensible, high-performance, cross-platform, open-source simulation engine for simulating SBML models. Under the hood, it uses a custom Just-In-Time (JIT) compiler to compile the SBML models into very fast native machine code, making it appropriate for solving very large models. Additional details on simulation engine features and performances may be found in [23].

2.6 NEURON

NEURON is a simulation environment designed to develop and emulate neurons and networks of neurons. It is particularly useful for building, managing, and using models designed to work closely to experimental data, and especially those that involve cells with complex anatomical and biophysical properties. One of NEURON's notable benefits is the clear separation between biological and purely computational considerations, enabling the user to focus on addressing the neuroscience research-level question. Importantly, NEURON is a mature platform that has benefited from the community's feedback for over 30 years. Exhaustive description of this simulation platform may be found in [24, 25].

2.7 Programming Language

Several programming languages or environments exist to support in-silico exploration of biological systems. Among the preferred contenders are Matlab and Python, which has grown over the years to provide a large number of useful libraries to the scientific community. Of importance, given Python's large user base and active community, number of tools provides application program interfaces (API) that support it. Notably, Python directly interacts with Neuron, the RHENOMS® platform (Rhenovia Pharma, Mulhouse, France) and libRoadrunner.

2.8 Optimization

Many useful automatic optimization methods exist, among which are the well-established gradient descent and genetic optimization. Of note, the multi-objective optimization method called EMOO [26] was successfully applied to the optimization of parameters for the NMDA-R model to extend its prediction abilities to novel sets of reported experimental results [27].

2.9 Computational Environment and Calculations

Models of receptors and synapses were implemented and validated using RHENOMS®, and LibRoadrunner [23]. Models of cells and networks were coded in the NEURON simulation environment [24], using parallel implementation, and MPI. Simulations were performed using the variable-step algorithm cvode, using Rhenovia's computing facility (eight nodes consisting of dual processor octocore Intel® Xeon® CPUs @ 2.00 GHz, and 32 Gb).

3 Methods

1. We aimed at developing a generic model of NMDAR that would fit most of the functional characteristics described by experimental results. This includes its affinity for glutamate and co-agonist glycine, the activation and deactivation rates, the desensitization to repeated stimulations, as well as the specific voltage and magnesium dependency (*see* **Note 3** for limitations of the capabilities of one model to integrate all/most of described characteristics). This model can be simulated alone to study its intrinsic behavior, for example to decipher how a compound interacts with the different transition states. Figure 3 provides an example of receptor-level study, by measuring the evolution of receptor-channel current as a function of the Magnesium concentration.

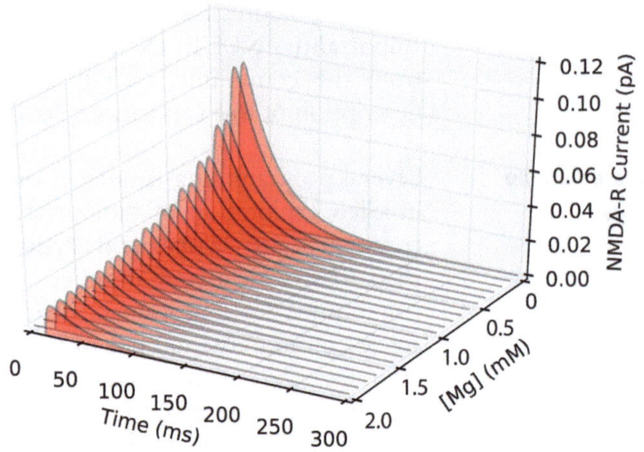

Fig. 3 Simulations results showing NMDA-R current as a function of extracellular [Mg^{2+}] in response to the application of a pulse of glutamate at time equal 20 ms. Glutamate concentration then rises to 8 microM and returns to zero after 1 ms. Results obtained in simulated voltage clamp protocol with postsynaptic voltage held at $V = -70$ mV. These 20 simulations were run with LibRoadrunner. Each simulation took in average 25 ms on a PC workstation

2. The model can also be integrated in a synapse/neuron model to generate higher levels predictions. Notably, this is not a trivial step and it may lead to challenges associated with the integration of an elementary model into higher hierarchical entities such as a neuron model (*see* **Note 4**). Figure 4 shows the results obtained at the level of a CA1 pyramidal neuron comprising synapses in which the NMDA receptor is antagonized at 50% inhibition levels with the competitive antagonist APV compared to control (i.e., no antagonist) levels of excitability [26].

   Furthermore, the generic model may also serve as the starting point to further develop subtypes NMDARs, in particular NR2A- and NR2B-containing ones to study the role of each subtype receptor under physiological, pathological, and pharmacological conditions (*see* the study in [14]).
   In this chapter, we will illustrate this method by showing the development of the generic NDMAR.
3. An important initial step is to perform a careful and thorough data mining and data analysis of both the experimental and computational studies to collect information (1) on the receptor functional properties, revealed by imaging, single-channel, and whole-cell electrophysiological recordings from either cell line, cell culture or brain slices, (2) on anatomical composition and ultrastructural distribution, (3) on the mode of action of the exogenous compound of interest (orthosteric or allosteric modulators), (4) on the different existing computational models. This work is essential to start the process of modeling, but will also be run throughout the whole process. Experimental results to be used for the validation of the numerical model are chosen. Important parameter values are gathered, such as binding and unbinding constants for the agonist Glutamate and the co-agonist Glycine, mode of action of the magnesium, specific effect of the Memantine, an allosteric antagonist (Fig. 5). Notably, many parameters are not available and the experimental results obtained for a same experiment are not always similar (*see* **Note 1**). With the perspective of including this receptor into a synapse/neuron model, it is also important to know its anatomical distribution and to estimate their number, extrapolated from the current recorded in patch-clamp for example.
4. After a careful analysis of the different models, we chose to use that of Schorge et al. [28]. We implemented the kinetic scheme of the NMDA receptor previously published by the authors using GUI-based design tools for visual model development such as Narrator (http://www.narrator-tool.org/), JDesigner (http://sbw.kgi.edu/) and CellDesigner (http://www.celldesigner.org/) (Fig. 6).

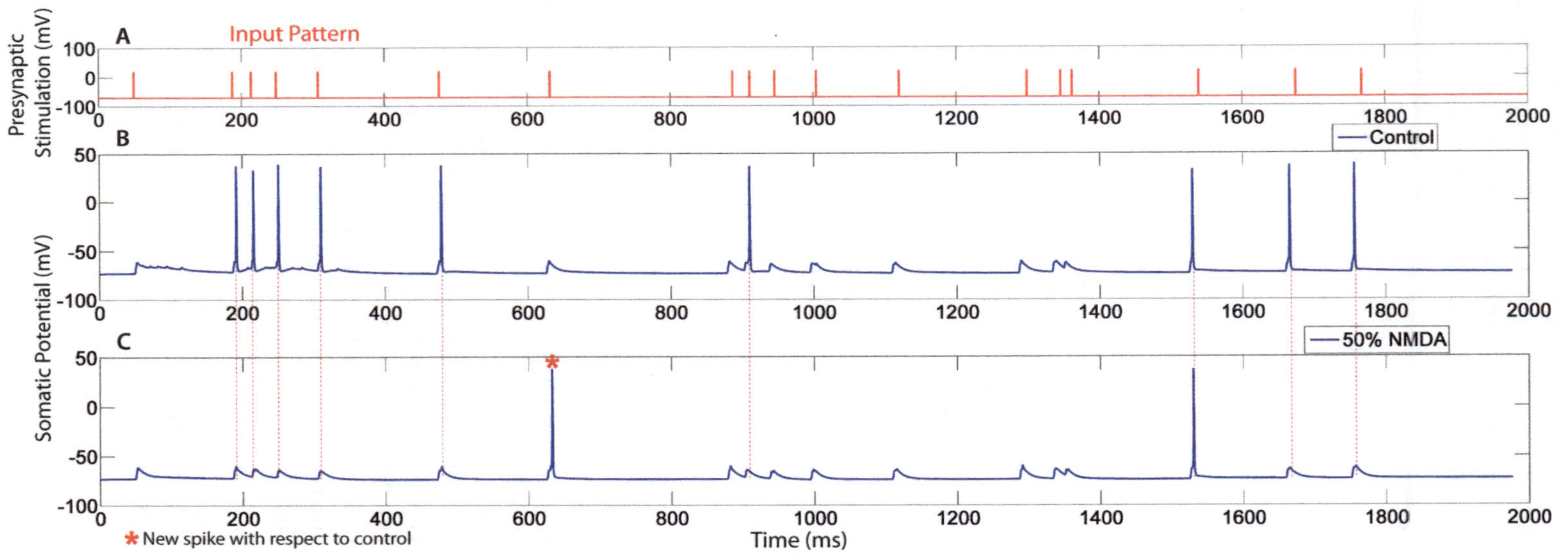

Fig. 4 Numerical recording of the electrical/spiking activity of a CA1 pyramidal neuron comprising synapses in which the NMDA receptor is antagonized at 50% inhibition levels with the competitive antagonist APV, as compared to control (i.e., no antagonist) levels of excitability [from Bouteiller et al. (2012) [33].

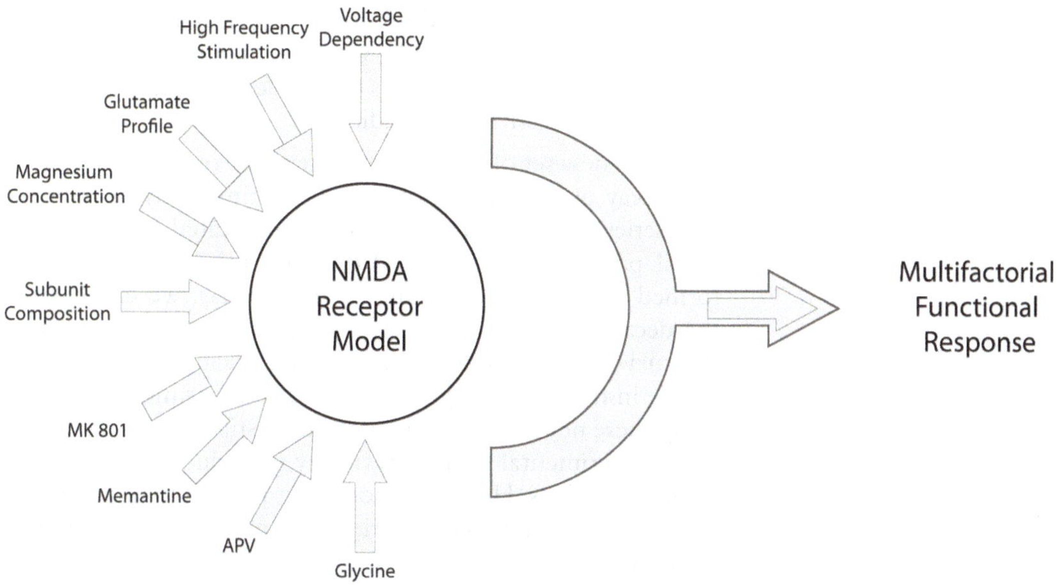

Fig. 5 A comprehensive model ideally provides a complete and quantitative view of the multidimensional space of parameters that influence the function of the biological system under consideration

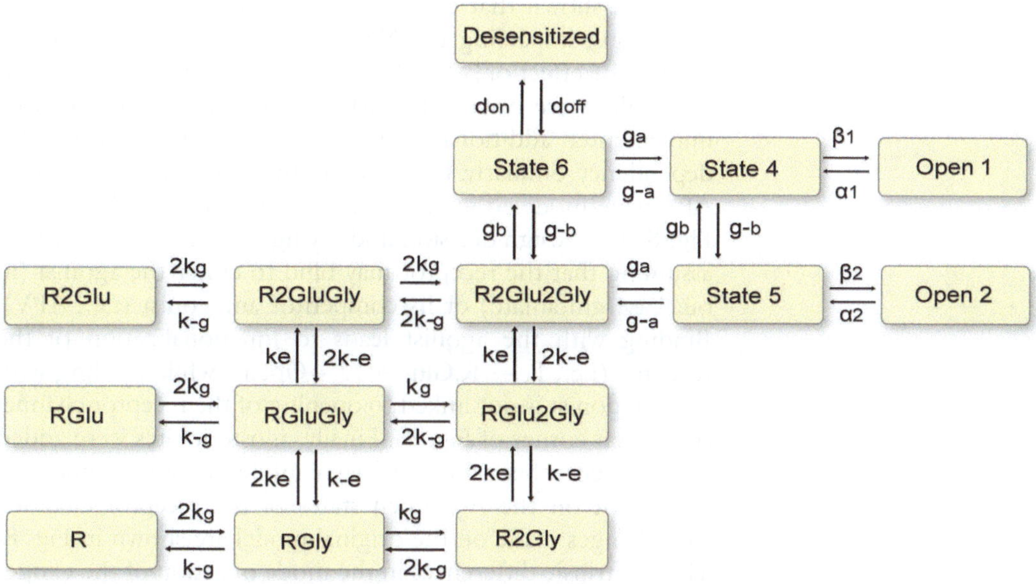

Fig. 6 Kinetic scheme of the NMDAR model (adapted from Ambert et al. (2010) [29])

Although the model correctly fit number of experimental results obtained with a single application of agonist, we found that it was unable to replicate other results obtained by long and repetitive applications of glutamate. Indeed, the model did not exhibit desensitization as observed experimentally. To address this issue, we modified several kinetic parameters of the model. In Ambert et al. [29], we did it manually as there were few parameters. However, this can be done using an optimization algorithm (*see* in Subheading 2).

With this new set of kinetic parameters, the overall rising phase and decay of the current observed experimentally in response to a series of pulses are more closely replicated than with the initial parameters (Fig. 7). Although optimization was performed on multiple objectives (i.e., criteria), we emphasized the decay component as it strongly influences the area under the curve—and therefore the resulting calcium concentration profile in the spine. Importantly, in this case, simulated results using these new sets of parameter values still reproduced all the other experimental results previously reproduced by the initial parameter values. However, modifying a model to fit additional experimental results is often tricky and nontrivial (*see* **Notes 1, 2** and **3**).

The NMDAR is highly sensitive to external magnesium (Mg^{2+}) concentration, being blocked in its presence, despite the binding of agonists. The depolarization of the cell is required to relieve this block to allow current through the channel pore. It has been shown that external [Mg^{2+}] can vary under specific conditions, impacting the NMDAR function. In the original model, the effect of [Mg^{2+}] was not included since experimental results were obtained in Mg-free solutions. We therefore implemented additional transition states to functionalize this dependency. Similarly, the effects of the competitive antagonist APV were modeled with the addition of states, as depicted in Fig. 8. The design decision underlying this addition consists in assuming that the receptor may bind to either the agonist (in our case glutamate) or its competitor antagonist (i.e., APV). Binding with the agonist leads to functionalization of the receptor (i.e., R → R.Glu →... →Open), while binding with the antagonist is not linked to opening of the receptor channel (lower left corner of Fig. 8). Finally, allosteric sites were added to represent the action of Memantine, a non-competitive antagonist on the two open states of the receptor channel. The changes made on the original model are shown in Fig. 8. These intrinsic differences in the mode of action of the exogenous modulators lead to very distinct functional outcomes. We redirect the reader to [29] for additional details on the insights provided by this model on receptor function following administration of either compound. Of importance, modifying the model and its state transitions and parameter values is not trivial (*see* **Note 2**).

To go further in the study of the role of NMDA receptors on synaptic and neuronal functions, one can connect the model to

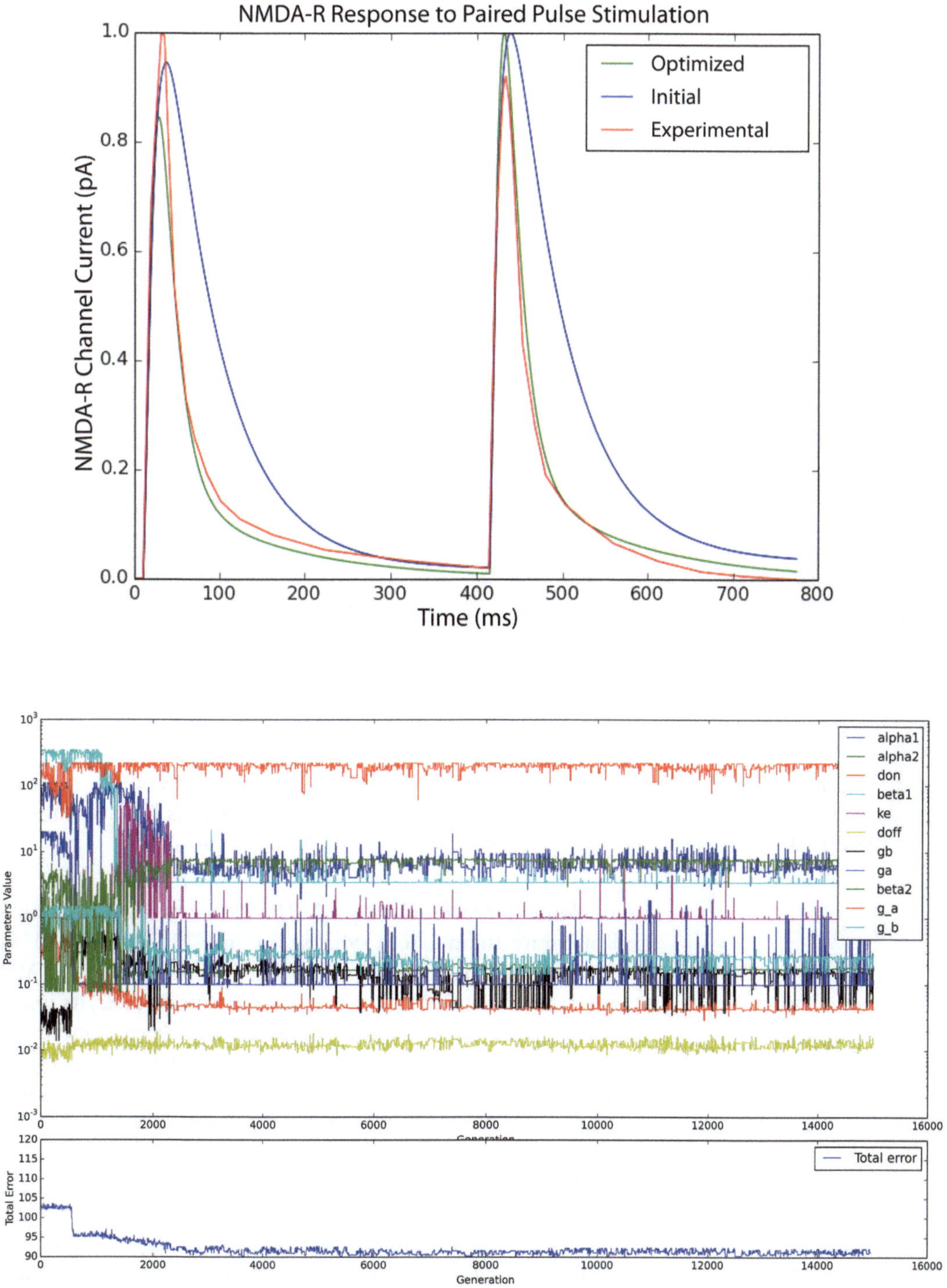

Fig. 7 Result obtained using the EMOO optimization procedure. *Top*: response of the NMDAR model to paired pulse stimulation. Comparison between experimental results, results obtained with the model comprising its original parameter values, and the model with modified (optimized) values. The optimized response remains closer to the experimental output. *Bottom*: Variations of the parameter values as a function of the number of optimization generation. The bottom portion outlines the overall decay in error between optimized output and digitized experimental results

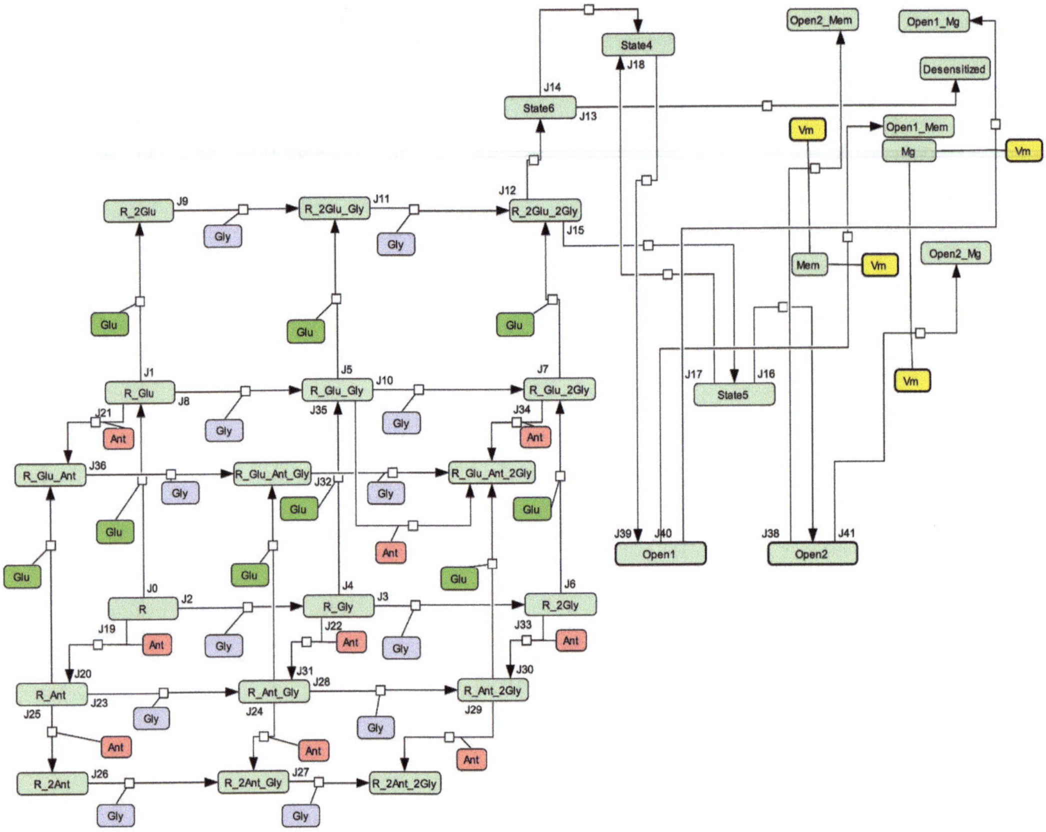

Fig. 8 NMDAR 15 with additional states to account for competitive antagonist (ANT in the scheme), magnesium (noted Mg) and memantine (MEM) as channel blockers. The initial model is the one contained in the *yellow* background. Open states for the receptor channel are labeled *Open1* and *Open2*

other elementary models (e.g., AMPARs, intracellular calcium) and integrate them into a synapse/neuron [13–15, 30]. This step requires the identification of what is defined in the SBML standard as the *boundary conditions* of the model, i.e., its input variables, as well as its output variables. Boundary conditions for the NMDA receptor model presented above are: glutamate concentration profile, glycine concentration, memantine concentration, APV concentration, and voltage. Voltage may be provided by a model of excitatory postsynaptic potential which takes into account changes in synaptic conductances (e.g., through AMPA receptor channel, voltage-dependent channels, etc.) and the characteristics of the neuron in which it is embedded (e.g., via back-propagated action potential, dendritic spikes, etc.). The output variables of our NMDAR model will constitute boundary conditions (i.e., input variables) for other models. For example, calcium current will constitute one of the inputs to a model that calculates the calcium concentration inside the postsynaptic spine, which may subsequently be used as an input to a model that determines changes in long-term potentiation/depression [11, 12]. Concretely, this integration step may be done using different tools and modeling environments such as the NEURON or RHENOMS®/EONS, or through direct scripting. Of importance, this integration step leads to important computational considerations outlined in **Notes 5** and **6**.

4 Notes

1. Availability and variability of experimental data.
   - All parameters required to fit the kinetic models are not experimentally measurable. If it is possible to measure/estimate the binding and unbinding constants for ligands, a great variability in the results reported is often observed in the literature. In addition, the access to other transitions states is restricted and it is very difficult to predict their exact number (*see also* **Note 2**).
   - Furthermore, biological results are often obtained in specific preparations—expression system, cell line, cell culture (e.g., ex vitro brain slices, in vivo recordings, etc.)—and under distinct experimental conditions, with in each case specific composition of intra- and extra-cellular medium (for example, the concentration of Mg^{2+} is not always the same), as well as different stimulation protocols (e.g., electrical versus exogenous application of saturated concentration of agonists).
2. Pertinence/reality of transition states and kinetic constants. The initial structuring of kinetic scheme of the NMDARs is based on the existing biological information and aimed at reproducing one (or a small set of) specific characteristics. Consequently, simple models rapidly exhibit limitations in their ability to emulate complex features of the receptor, and additional states are needed to integrate additional characteristics. These additional transition states are introduced to improve prediction of a larger number of biological results. Yet the intermediate transition states are abstract states for which it is virtually impossible to obtain a quantifiable experimental measure.

The same limitations apply to finding the kinetic constants between transitions states. Besides the binding/unbinding constants that can be experimentally measured, the other ones are indirectly estimated and in complex models, often represent an interpreted view of the mechanisms thought to take place in the biological system modeled.

3. Limitations of the model

   In the computational world, it is commonly said that all the models are wrong, but some of them can help further understand a complex problematic. It is indeed highly complex to develop a model which will take into consideration a large number of characteristics and which can reproduce a large number of experimental results. For example, a more detailed NMDA receptor model may also consider zinc concentration, the gradient concentrations that exist across the cell membrane for sodium, potassium, and calcium the main ions for which the channel is permeable. It may also include the effects of many psychoactive drugs such as phencyclidine (PCP), alcohol (ethanol) and dextromethorphan (DXM), as well as the effects of the anesthetic drugs ketamine and nitrous oxide. Due to the multifactorial nature of receptor functional outcome, a given model often exists under relatively fixed conditions.

4. Large number of models

   A consequence of the previous note is that numbers of models may be built, each reproducing one or several experimental results. For the NMDARs, Schorge et al. [28] and Erreger et al. [31] tested several models to explain the differences in channel opening observed between subtypes of NMDARs, supported by different experimental observations.

   In an attempt to extend prediction levels to conditions involving paired-pulse stimulations, we made modifications to the original model [29]. Such optimization methodologies are unfortunately rarely used, and oftentimes, there are as many models as modeling studies as each study and model focuses on a specific behavior or property of the NMDARs and verifies a specific and unique hypothesis.

5. Scaling issue: prediction errors

   Multiple connections between different models can directly lead to computational challenges of at least two natures. The first one is the propagation and amplification of prediction errors: the biological models of biomolecular, synaptic, and neuronal nature are highly nonlinear, and small prediction errors at the receptor level may lead to large errors in neuronal spiking properties for example. The second challenge is of numerical nature: as the number of calculations increases, so does the instability of the system of differential equations. This may consequently lead to stiffness and potential numerical errors [17].

6. Scaling issue: computational resources and calculation time

   The multiscale approach that consists in placing elementary models such as NMDAR in synaptic constructs, and then these synaptic constructs (which can be of inhibitory or excitatory nature) onto neuronal cells can quickly result in a sizable number of equations. This increased complexity can rapidly strain even the most advanced computational resources, and lead to excruciatingly slow computations making (and/or memory-related errors). Of importance, utilization of input-output modeling methodology may help decrease the computational complexity while maintaining prediction levels [32]. This methodology consists in the functional characterization of the model, and focuses the numerical expense on producing the desired functional output, rather than on the mechanistic components that comprise the model, thereby reducing simulation times—thereby improving scalability.

References

1. Churchland PS, Koch C, Sejnowski TJ (1993) What is computational neuroscience? In: Schwartz EL (ed) Computational neuroscience. MIT Press, pp 46–55
2. Schwartz EL (1990) Computational neuroscience. MIT Press, Cambridge, MA. ISBN 0-262-19291-8
3. Cutsuridis V, Wennekers T (2009) Hippocampus, microcircuits and associative memory. Neural Netw 22(8):1120–1128. doi:10.1016/j.neunet.2009.07.009
4. Destexhe A (1999) Can GABAA conductances explain the fast oscillation frequency of absence seizures in rodents? Eur J Neurosci 11 (6):2175–2181
5. Ogden KK, Traynelis SF (2011) New advances in NMDA receptor pharmacology. Trends Pharmacol Sci 32(12):726–733. doi:10.1016/j.tips.2011.08.003
6. Santangelo RM, Acker TM, Zimmerman SS, Katzman BM, Strong KL, Traynelis SF, Liotta DC (2012) Novel NMDA receptor modulators: an update. Expert Opin Ther Pat 22 (11):1337–1352. doi:10.1517/13543776.2012.728587
7. Zito K, Scheuss V (2009) NMDA receptor function and physiological modulation. Encycl Neurosci 2009:1157–1164. doi:10.1016/B978-008045046-9.01225-0
8. Hodgkin AL, Huxley AF (1952) A quantitative description of membrane current and its application to conduction and excitation in nerve. J Physiol 117(4):500–544
9. Kirli KK, Ermentrout GB, Cho RY (2014) Computational study of NMDA conductance and cortical oscillations in schizophrenia. Front Comput Neurosci 8:133. doi:10.3389/fncom.2014.00133
10. Colquhoun D (1973) The relation between classical and cooperative models for drug action. In: Rang HP (ed) Drug Receptors. University Park Press, Baltimore, MD, pp 149–182
11. Lynch G, Baudry M (1984) The biochemistry of memory: a new and specific hypothesis. Science 224(4653):1057–1063
12. Lynch G, Baudry M (2015) Brain and memory: old arguments and new perspectives. Brain Res 1621:1–4. doi:10.1016/j.brainres.2014.12.052
13. Bouteiller JM, Baudry M, Allam SL, Greget RJ, Bischoff S, Berger TW (2008) Modeling glutamatergic synapses: insights into mechanisms regulating synaptic efficacy. J Integr Neurosci 7 (2):185–197
14. Keller A, Ambert N, Legendre A, Bedez M, Bouteiller JM, Bischoff S, Baudry M, Moussaoui S (2016) Impact of synaptic localization and subunit composition of ionotropic glutamate receptors on synaptic function: modeling and simulation studies. IEEE/ACM Trans Comput Biol Bioinform. doi:10.1109/TCBB.2016.2561932
15. Stamboulian-Platel S, Legendre A, Chabrol T, Platel JC, Pernot F, Duveau V, Roucard C, Baudry M, Depaulis A (2016) Activation of GABAA receptors controls mesiotemporal lobe epilepsy despite changes in chloride transporters expression: in vivo and in silico approach. Exp Neurol 284(Pt A):11–28. doi:10.1016/j.expneurol.2016.07.009
16. Bowman GR, Pande VS, Noé F (eds) (2014) An introduction to Markov state models and their application to long timescale molecular simulation, Advances in experimental medicine and biology, vol 797. Springer, Dordrecht

17. Sarmis MBJ, Ambert N, Legendre A, Bischoff S, Haeberlé O, Baudry M (2013) Assessing numerical resolution methods performance for kinetic models of receptors and channels. J Comput Sci Syst Biol 6:150–164. doi:10.4172/jcsb.1000112
18. Dahlquist G, Björck Å (2012) Numerical methods, Dover books on mathematics. Courier corporation, 2012. ISBN: 0486139468, 9780486139463
19. Hindmarsh ACBPN, Grant KE, Lee SL, Serban R, Shumaker DE, Woodward CS (2005) SUNDIALS: suite of nonlinear and differential/algebraic equation solvers. ACM Trans Math Softw 31(3):363–396
20. Gleeson P, Crook S, Cannon RC, Hines ML, Billings GO, Farinella M, Morse TM, Davison AP, Ray S, Bhalla US, Barnes SR, Dimitrova YD, Silver RA (2010) NeuroML: a language for describing data driven models of neurons and networks with a high degree of biological detail. PLoS Comput Biol 6(6):e1000815. doi:10.1371/journal.pcbi.1000815
21. Miller AK, Marsh J, Reeve A, Garny A, Britten R, Halstead M, Cooper J, Nickerson DP, Nielsen PF (2010) An overview of the CellML API and its implementation. BMC Bioinformatics 11:178. doi:10.1186/1471-2105-11-178
22. Smith LP, Bergmann FT, Chandran D, Sauro HM (2009) Antimony: a modular model definition language. Bioinformatics 25 (18):2452–2454. doi:10.1093/bioinformatics/btp401
23. Somogyi ET, Bouteiller JM, Glazier JA, Konig M, Medley JK, Swat MH, Sauro HM (2015) libRoadRunner: a high performance SBML simulation and analysis library. Bioinformatics 31(20):3315–3321. doi:10.1093/bioinformatics/btv363
24. Hines ML, Carnevale NT (1997) The NEURON simulation environment. Neural Comput 9(6):1179–1209
25. Carnevale NT, Hines ML (2006) The NEURON book. Cambridge University Press. ISBN: 1139447831, 9781139447836
26. Bahl A, Stemmler MB, Herz AV, Roth A (2012) Automated optimization of a reduced layer 5 pyramidal cell model based on experimental data. J Neurosci Methods 210 (1):22–34. doi:10.1016/j.jneumeth.2012.04.006
27. Bouteiller JM, Feng Z., Onopa A, Berger T.W. (2015) Maximizing predictability of a bottom-up complex multi-scale model through systematic validation and multi-objective multi-level optimization. doi:10.1109/NER.2015.7146619
28. Schorge S, Elenes S, Colquhoun D (2005) Maximum likelihood fitting of single channel NMDA activity with a mechanism composed of independent dimers of subunits. J Physiol 569 (Pt 2):395–418. doi:10.1113/jphysiol.2005.095349
29. Ambert N, Greget R, Haeberle O, Bischoff S, Berger TW, Bouteiller JM, Baudry M (2010) Computational studies of NMDA receptors: differential effects of neuronal activity on efficacy of competitive and non-competitive antagonists. Open Access Bioinformatics 2:113–125. doi:10.2147/OAB.S7246
30. Greget R, Dadak S, Barbier L, Lauga F, Linossier-Pierre S, Pernot F, Legendre A, Ambert N, Bouteiller JM, Dorandeu F, Bischoff S, Baudry M, Fagni L, Moussaoui S (2016) Modeling and simulation of organophosphate-induced neurotoxicity: prediction and validation by experimental studies. Neurotoxicology 54:140–152. doi:10.1016/j.neuro.2016.04.013
31. Erreger K, Dravid SM, Banke TG, Wyllie DJ, Traynelis SF (2005) Subunit-specific gating controls rat NR1/NR2A and NR1/NR2B NMDA channel kinetics and synaptic signalling profiles. J Physiol 563(Pt 2):345–358. doi:10.1113/jphysiol.2004.080028
32. Hu EY, Bouteiller JM, Song D, Baudry M, Berger TW (2015) Volterra representation enables modeling of complex synaptic nonlinear dynamics in large-scale simulations. Front Comput Neurosci 9:112. doi:10.3389/fncom.2015.00112
33. Bouteiller JM, Legendre A, Allam SL, Ambert N, Hu EY, Greget R, Keller AF, Pernot F, Bischoff S, Baudry M, Berger TW (2012) Modeling of the nervous system: from modulation of glutamatergic and gabaergic molecular dynamics to neuron spiking activity. Conf Proc IEEE Eng Med Biol Soc 2012:6612–6615. doi:10.1109/EMBC.2012.6347510

Index

Nail Burnashev and Pierre Szepetowski (eds.), *NMDA Receptors: Methods and Protocols*, Methods in Molecular Biology, vol. 1677, DOI 10.1007/978-1-4939-7321-7,

D

MIX
Papier aus verantwortungsvollen Quellen
Paper from responsible sources
FSC® C105338

If you have any concerns about our products,
you can contact us on
ProductSafety@springernature.com

In case Publisher is established outside the EU,
the EU authorized representative is:
Springer Nature Customer Service Center GmbH
Europaplatz 3, 69115 Heidelberg, Germany

Printed by Libri Plureos GmbH
in Hamburg, Germany